Lecture Notes in Computer Science 3008

Springer
Berlin
Heidelberg
New York
Hong Kong
London
Milan
Paris
Tokyo

Stephan Heuel

Uncertain Projective Geometry

Statistical Reasoning for Polyhedral Object Reconstruction

Springer

Author

Stephan Heuel
University of Bonn
Institute for Photogrammetry
Nussallee 15, 53115 Bonn, Germany
E-mail: stephan@heuel.org

Inaugural-Dissertation to obtain the degree Dr.-Ing.
Agricultural Faculty
Rheinische Friedrich-Wilhelms-Universität, Bonn, Germany

Referent: Prof. Dr.-Ing. W. Förstner
Coreferent: Prof. Dr. rer.nat. H.-P. Helfrich
Date of Exam: Friday, 13. Dezember 2002

Library of Congress Control Number: 2004104982

CR Subject Classification (1998): I.5, I.4, G.3, I.3.5, I.2.10

ISSN 0302-9743
ISBN 3-540-22029-1 Springer-Verlag Berlin Heidelberg New York

Springer-Verlag is a part of Springer Science+Business Media

springeronline.com

Typesetting: Camera-ready by author, data conversion by Boller Mediendesign
Printed on acid-free paper SPIN: 10999820 06/3142 5 4 3 2 1 0

Foreword

The last decade of computer vision can be characterized by the development of tools for geometric reasoning. Algebraic projective geometry, with its multilinear relations and its embedding into Grassmann-Cayley algebra, became the basic representation of multiple view geometry, on the one hand leading to a deep insight into the algebraic structure of the geometric relations and on the other hand leading to efficient and versatile algorithms for the calibration and orientation of cameras, for image sequence analysis as well as for 3D-object reconstruction. Prior to the use of projective geometry in computer vision, some kinds of features or information had to be extracted from the images. The inherent uncertainty of these automatically extracted image features, in a first instance image points and image line segments, required tracing the effects up to the inherently uncertain result. However, in spite of the awareness of the problem and the attempts to integrate uncertainty into geometric computations, especially joint final maximum-likelihood estimation of camera and object parameters, geometric reasoning under uncertainty rarely was seen as a topic on its own.

Reasons for the lack of integration of uncertainty may have been the redundancy in the representation of Euclidean entities using homogeneous coordinates leading to singularities in the statistical distributions, the non-linearity of the relations requiring approximations, which give biased results, and the dependency of both the numerical accuracy and the statistical properties on the coordinate systems chosen; another reason may have been the lower reward for performing good engineering.

The author provides a transparent integration of algebraic projective geometry and spatial reasoning under uncertainty with applications in computer vision. Examples of those applications are: grouping of spatial features, reconstruction of polyhedral objects, or calibration of projective cameras. He clearly cuts the boundary between rigorous but infeasible and approximate but practicable statistical tools for three-dimensional computations. In addition to the theoretical foundations of the proposed integration and the clear rules for performing geometric reasoning under uncertainty, he provides a collection of detailed algorithms that are implemented using now publicly available software.

The book will be of value to all scientists interested in algebraic projective geometry for image analysis, in statistical representations of two- and three-dimensional objects and transformations, and in generic tools for testing and estimating within the context of geometric multiple-view analysis.

Bonn, February 2004 *Wolfgang Förstner*

Preface

I wish to acknowledge the following people who directly or indirectly contributed to this work. First, I would like to thank my family, especially my parents, who have given me sufficient freedom to pursue my own interests and at the same time have shown me unconditional love.

I was very fortunate to have met wonderful people during my time at the University of Bonn. I refer in particular to Oliver who contributed to my life in numerous ways.

I also thank all the people at the University and the Institute of Photogrammetry in Bonn for a wonderful time I had there both as a student assistant and as a research assistant. In particular I wish to mention Kirsten Wolff, Andre Braunmandl, Heidi Hollander, Felicitas Lang, Wolfgang Schickler, Hardo Müller, Boudewijn Moonen, Thomas H. Kolbe, Petko and Annett Faber, Mirko Appel and many others. I also want to thank Thomas Läbe for coordinating the Java implementation of the algorithms described in this work.

I am deeply grateful that I had the opportunity to work with my advisor, Prof. Dr. Wolfgang Förstner. He was not only the initiator of this work and provided continuous support throughout the duration of the research, but more importantly he was a *Doktorvater* in the truest sense of the word. I extend my gratitude to Prof. Dr. Helfrich for his willingness to serve as a second reviewer for this work.

Lastly, I wish to express my thanks to Anette for her support during the last 15 months. I am still wondering why we did not meet earlier in life.

Bonn, October 2002 *Stephan Heuel*

Contents

List of Tables

List of Figures

List of Algorithms

1 Introduction

> *A better conceptual framework for representing and manipulating geometric knowledge can lead to implementation designs that provide more adequate separation between generic geometric principles and specific constraints of the application.*
>
> KAPUR AND MUNDY 1989B

1.1 Motivation

This thesis is about uncertain geometric reasoning by integrating projective geometry and statistics. It is motivated by geometric problems in image analysis, in particular by object reconstruction from single or multiple images.

Extracting information from images is a simple task for humans, that can already be accomplished at a very early stage: one year old children are able to recognize objects on pictures (PIAGET 1954). The visual sense is one of the dominant ways of perceiving the surrounding world. At first glance it is quite surprising that it is an extremely difficult task for computers to achieve a similar ability with respect to image interpretation. Forty years ago researchers at the MIT thought that it was at a complexity level of a PhD thesis to teach computers to see (ROBERTS 1963). Within the last decades it was realized that there are immense problems that need to be solved in order to accomplish this task and the field of Computer Vision emerged from this problem.

Among many reasons, the problems are due to the fact that we are only beginning to understand how the vision system of human beings work. We may explain the process on a biological level: it is known how light is mapped on our retina and stimulates perceptive neurons, but it is yet unclear how this mapping finally causes our mind to recognize objects. Opposed to low-level biological resp. neuronal research, the field of psychology tries to provide high-level explanations of the visual system, for example BIEDERMAN 1987 describes a model where the human recognizes objects by breaking it down into components, however no biological evidence is known for this model.

S. Heuel: Uncertain Projective Geometry, LNCS 3008, pp. 1-17, 2004.

Another early attempt to describe cognitive aspects are the Gestalt psychologists in the 1930s, who described objects and their relations by their saliency (*Prägnanz*). But again, the psychological models do not provide an instruction of how to build an artificial system with comparable visual abilities to human beings. For example, it is not clear how to describe objects like a chair to computers, such that it can somehow match the description with a digital image; there just exist too many different forms of chairs to adequately describe it by shape, texture, function etc. To summarize, one may say that we are living exemplars of a working visual system, but don't yet have a detailed explanation, why this system works. Though the high expectation of building artificial visual systems has not been fulfilled, the Computer Vision community still has made significant progress in restricted applications, e.g. the guidance of robots or vehicles to name only one of them.

Despite the previous observations, computers do have advantages over human beings when they are used as *precise measuring devices* instead of interpreters of images. Extracting precise measurements from images is one of the central traditional tasks in Photogrammetry and has been performed since the invention of photography. While it is possible to extract metric information from one image, usually two or more images are used in photogrammetric tasks. Multiple images allow to apply stereoscopic measuring techniques, which enables the retrieval of real 3D information about the scene. In the beginning of the last century, the first devices for stereoscopic measurements, so-called stereo-comparators, were invented and extended to analog plotting devices, which are still used for practical applications. In 1957 the concept of an analytical plotter was patented by Helava: an analytical plotter uses a small computer to establish the geometry between the analog images. About 1990 the first digital photogrammetric workstations have been introduced which store the images digitally so that the complete process of measuring can now be performed on a computer.

With the introduction of digital photogrammetric stations, the automation of the measurement process becomes more and more important. At this point it becomes clear that the future goals and problems of Computer Vision and Photogrammetry are largely overlapping and therefore both fields share more and more work in recent years. For photogrammetrists, there are two tasks that are the main goal of automation: first, the determination of the orientation parameters of the cameras at the time when the images were taken and second the automation of measuring the desired objects in the images. Both tasks are central for Photogrammetry as well as for Computer Vision. In this work we cover the measuring or *reconstruction* of buildings from multiple images and the determination of the *orientation* of single images. We focus on the reconstruction task and will see that the algebraic relations that we derive simplify the orientation task as a special case of reconstruction. We further assume that the buildings can be expressed by a polyhedral model.

The problem of image interpretation is now restricted to the reconstruction of a specific class of objects, its automation is still a difficult problem: the buildings that are observed can be arbitrarily complex, they are surrounded by unknown scenery and the images may vary in quality, scale and lighting conditions. Due to these and other problems, the automated extraction of buildings from images is still an on-going research task, cf. MAYER 1999. On the other hand, because of the high demand of up-to-date spatial data, there exist a number of commercial user-assisted systems for the reconstruction of man-made objects (BALTSAVIAS *et al.* 2001). Recently, user assisted systems adopt more and more automation techniques that have been proven to be successful and stable enough to be included in a working environment.

1.2 Objective of the Thesis

The objective of the thesis is the derivation of theoretical foundations, design and empirical verification of a *tool-box* for statistical geometric algorithms. The practical value of the tool-box is demonstrated by its application to the task of object reconstruction from multiple images.

The main idea of this work is the integration of a consistent geometric representation into a statistical framework: some of the geometric problems that are of interest for the reconstruction task are for example (i) the optimal estimation of a building corner or edge from a set of image observations and (ii) testing geometric relations between independently measured corners or edges, such as identity, incidence, parallelity or orthogonality. The need for a statistical framework arises from the fact that a solution to these problems have to be clearly defined in terms of performance and quality of the results (FÖRSTNER 1996, COURTNEY *et al.* 1997). Statistical concepts have the advantages that they are comparable across different systems, can be tested against empirical data and help in the evaluation of hypotheses. To simplify the statistical description, we use linearized models for estimation and testing (KOCH 1999). Although linear models may only appear sub-optimal at first sight, they can be used as long as it can be shown that they are sufficient with respect to the data and the task that needs to be solved. For example for a large number of geodesic tasks the use of linearized statistical models is beyond dispute. The advantage here is the fact that the full power of methods from linear algebra are at hand.

Obviously linear statistical models require a linear representation and manipulation of geometric data. The techniques known from projective geometry – a popular mathematical field in the 19th century – have been proven to be very convenient for a lot of geometric tasks in Photogrammetry and Computer Vision. In this work we first express the operations and formulas from projective geometry in terms of linear algebra and secondly include the derived linear formulas in statistical models to establish practical and working algorithms.

In the remainder of this chapter we describe the background of this work, then briefly review related work and finally give an overview of the rest of the thesis.

1.3 Reconstructions of Polyhedral Objects

The statistical geometric algorithms that are introduced in this thesis are motivated by the task of object reconstruction from multiple images: it will be shown that the algorithms enable automatic and user-assisted systems to generate the most plausible reconstruction from a given set of images. Though this thesis concentrates on the reconstruction task, it is important to note that other fields may also make use of this work, such as robotics or geometric drawing systems.

The reconstruction of objects from multiple images is a central task both in Computer Vision as well as in Photogrammetry. As reconstruction is embedded in a complete vision process, we distinguish between the terms extraction, detection, reconstruction and recognition of objects: the *detection* or localization of objects refer to methods that locate an object within a specific image area. Hence, the object itself or the object class is already known. In contrast to object detection, object *recognition* does not necessarily have one particular object or class of objects which is about to be detected. The term recognition is usually used to describe methods to identify objects. It is a problem of finding the right categories. The term *reconstruction* refers to methods for obtaining the shape of an object. In Computer Vision terms, this task is called "shape from X", where X may stand for motion, shading, stereo, etc. Finally, *extraction* refers to the whole process of detection, possibly recognition and a subsequent reconstruction of the object.

1.3.1 Object, Sensor, and Image Models

Since the early 70's there has been an on-going discussion on the general approach of attacking the problem of object extraction from images, which can be summarized as "models versus exemplars" (PONCE *et al.* 1996, BLAKE *et al.* 2001). The underlying question is how to represent the objects that are viewed in the images. On the one hand there are physically-based models, that serve as prototypes for a class of objects. These models are usually described with a comparable small set of parameters with known semantics. On the other hand the objects may be described only by a set of representative examples without using any semantics. This discussion is also often characterized by the pairs "object-based vs. view-based" or "parametric vs. non-parametric".

In this work we mainly follow the track of a model-based approach as we are specifically interested in exactly measuring the parameters of the unknown object. We distinguish between three kinds of models: an object model, a sensor model and an image model (BRAUN *et al.* 1995 and FUCHS 1998).

Object Model A quite generic model of a building is a *polyhedron* which consists of straight line segments between two points, and planar faces surrounded by line segments. This description fits to a large number of actual buildings. It has to be noted that a polyhedron does not contain any building-specific knowledge. An example for a more specific model is a parameterized saddle-roof building which has four parameters, namely length, width, height and roof-height. This parameter model is quite specific and can be uniquely related to a building. We deliberately choose a polyhedral model though; this way the algorithms in this thesis do not contain any building-specific knowledge, such that they can be applied to other applications. In a separate step, outside the scope of this work, the reconstructed polyhedra respectively its parts have to be labeled with building-specific meanings such as roof, vertical walls, etc., see for example LANG 1999 or TELEKI 1997. Besides the geometric constraints of the object, we assume that the surfaces of a polyhedron are opaque and have a diffuse reflectance property. This assumption ensures the actual observability of the polyhedral edges and corners.

Note that we may also define a scene model, that consists of a number of objects and possibly light sources, that influence the appearance of the scene.

Sensor Model A sensor model describes the process of capturing the image and thus characterizes the properties of the sensor important to the application. In Photogrammetry, one usually uses a geometric model of a pin-hole camera, resulting in a mapping from 3D to 2D that preserves straight lines. As real cameras introduce a small distortion to this model, one may correct these errors using a non-linear model such as a radial distortion model, cf. FÖRSTNER 2001B. One may extend the definition of a sensor model to any transformation from one domain to another, including algorithms, cf.WEIDNER 1997.

Image Model Finally the image model defines the geometric, radiometric and topological properties of the image. The image model is mainly defined by the object and sensor model: since we assume a straight-line preserving mapping, the image of a polyhedron is ideally a set of polygons that are connected to each other, obtaining a Mondrian-like image. Surfaces of polyhedra are mapped to blobs, that are surrounded by polygon edges but not necessarily simply connected due to occlusions. For the same reason, the mapping does not strictly preserve the neighborhood relationships.

Errors Introducing models of reality means to obtain an abstraction or generalization of the world. The abstraction introduces an error when measuring physical properties, the so-called *generalization error* or *bias*. For example, the introduction of polyhedral models to describe buildings naturally introduces errors since at a certain level of detail, no existing building fulfills a strict polyhedral model. A second source of errors are the actual errors in the measurement or *observation error*.

Both kinds of errors have to be taken into account, see for example WINTER 1996 and DAVID *et al.* 1996.

1.3.2 Representation of Polygons and Polyhedra

There exist a number of possible representations for a polyhedron, depending on the context in which the term is used. In mathematics, or more precisely in geometry, a polyhedron is a volume which is enclosed by plane patches, cf. ZEIDLER 1996. Another valid definition refers to polygons stating that a polyhedron consists of a set of polygons that are joined at their edges.

It is easy to see that a polygon in 2D may be defined similarly exchanging the word *plane* by *line*, *edges* by *point*, etc. This notion can be generalized by the term **polytope**, which COXETER 1989 defines as the general term of the sequence "*point, line segment, polygon, polyhedron, ...*" or more specifically:

Definition 1 (Polytope) *A polytope in an n–dimensional space is a finite region enclosed by a finite number of hyperplanes. A polygon is a polytope in 2 dimensions, a polyhedron a polytope in 3 dimensions.*

The advantage of this definition besides its generality is the fact, that it uses projective linear subspaces, namely hyperplanes. This will be conceptually useful when we exploit the projective geometry for the task of object reconstruction.

In Computer Graphics or Computational Geometry, the definition of a polyhedra is based on the underlying data-structure that represents the polyhedra: a 3D solid especially a polyhedron may be represented by a wire-frame model, a surface model or by a volume model, cf. ENCARNACAO *et al.* 1997.

- *wire-frame model*: A wire-frame model simply consists of a collection of 3D edges. There is no connection between the edges defined and no relationships to existing surfaces.
- *surface model*: A surface model is a collection of surfaces, where there is no relationship between the individual surfaces. The decision whether a point is inside or outside the 3D solid, in our case the polyhedron, cannot be made.
- *volume model*: A volume model is a complete description of a three-dimensional object, the decision whether a point is inside or outside can always be made.

In this work we are mainly concerned with wire-frame models and surface models.

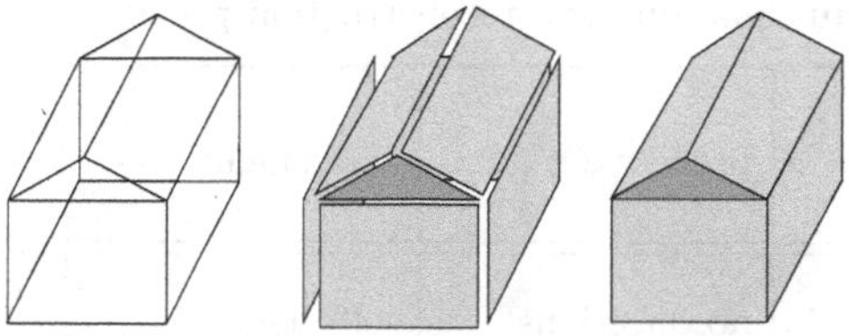

Fig. 1.1. Wireframe, surface and volume models. In this work we are mainly concerned with wireframe models and surface models.

1.3.3 Automated Reconstruction

What is the meaning of the attribute "automatic" when doing automatic reconstruction? As there are always human beings who will start the process and evaluate the final result, we ask this question with the perspective of a potential user of the system: what does a user expect from an automatic system? It usually depends on the level of her experience. Somebody with no or minimal experience in Computer Vision or Photogrammetry does not want to be disturbed by technical details such as thresholds or maximal precision but only wants to see the final result. On the other hand input parameters may be important for an expert: they can be used to control the reconstruction process. An expert might also be interested in intermediate results to change the behavior of the system.

A possible definition for the *scale of automation* is the amount of knowledge that is required to use the software. Opposed to the user-side knowledge is the "knowledge" that is encoded in the software, which lowers the amount of user interaction. We may distinguish between four degrees of scale: (i) *fully automatic*: no interaction needs to be done; (ii) *automatic*: simple yes/no decisions have to be made or a selection of two or more alternatives; (iii) *semi-automatic*: some quantitative parameters have be set by mouse or keyboard; (iv) *interactive*: the user continuously steers the acquisition, as with classical photogrammetric stations.

The acquisition process itself may be subdivided into three parts, where all of these three parts are ideally being automated: (a) initialization: specifying the input parameters, (b) process: the actual acquisition and (c) assessment: characterization of the results obtained by the system. Table 1.1 summarizes the different scales of automation with respect to the acquisition process.

Of course the ultimate goal is a fully automatic reconstruction started by pushing a button and finished by a machine-generate diagnosis of the results. This is hard to achieve at the moment and for practical purposes it may be advisable to include some minimal interaction to an automated system, see HEUEL AND NEVATIA 1995, FUA AND BRECHBÜHLER 1996. Vice versa, one may use the opposite approach and enhance interactive systems such that they reach a higher level of automation.

The discussion of automation does not only hold for complete systems, but also for modules that are used within the system: the modules must be capa-

Table 1.1. Different degrees of automation for a reconstruction process

process step → ↓ scale of automation	initialization	process	assessment
fully automatic	pushing the start button	no questions	self-diagnosis
automatic	selection of some qualitative initialization parameters	very few qualitative questions, if at all	confirmation resp. rejection of results
semi-automatic	simple qualitative and quantitative input parameters	some qualitative and quantitative input and adjustment steps	checking and possibly editing results
interactive	qualitative and quantitative parameters, depending on the complexity of the task	qualitative and quantitative input and adjustment steps, depending on the complexity of the task	permanent steering throughout the process

ble to assess their own results and give measures of its accuracy. The input parameters must be either simple and/or the module can be steered by an earlier module in a consistent manner.

1.4 Geometric Reasoning

In the previous section we have seen that the models we use in this work all have strong geometric properties: the image capturing process is usually modeled by a geometric pin-hole camera, in the images we observe polygons and finally we are mainly interested in the geometric parameters of the polyhedral object. Although there are other cues in the images that can be used for object reconstruction, such as radiometry and topology, we will largely concentrate on the geometric properties.

In this section we formulate the object reconstruction as a problem of inferring the most plausible object from the geometric properties of the image and the cameras, hence doing *geometric reasoning*.

1.4.1 Reasoning and Object Reconstruction

According to Russell and Norvig 1995, p. 163, *reasoning* or *inference* are terms that *"[...] are generally used to cover any process by which conclusions are reached"*. Usually *logical inference* or deduction is used as the primary method of reasoning. In the late 19th century, the geodesist, mathematician and philosopher Charles S. Peirce (1839-1914) distinguished between three different methods of reasoning: deduction, induction and abduction, cf. 1.2. Deduction is a method of reasoning by which one strictly infers a conclu-

Table 1.2. Examples for three different methods of reasoning, see Peirce 1931.

abduction (generation of hypotheses)	All beans from this bag are white	These beans are white	Therefore, these beans are from this bag.
induction (generalization)	This bean is from this bag, and this bean, and this, ...	These beans are all white.	Therefore, all beans from this bag are white
deduction (logical inference)	All beans from this bag are white.	These beans are from this bag.	Therefore, these beans are white.

sion from a set of given (and true) facts by employing the axioms and rules of inference for a given logical system. Mathematical theorems are usually proven using deductive argumentation, because the conclusion is analytical and uniquely follows from the given facts.

On the other hand, induction is a method of reasoning by which one infers a generalization from a series of instances. This inference is only strict if the series of instances cover the whole space of possible instances, therefore being a complete induction.

Finally, abduction is a method of reasoning by which one infers an explanation of observations ("the beans are from this bag") given an hypothesis ("all beans from this bag are white"). Obviously, this method of inference is not strictly truth-conserving. Peirce argued that in contrast to deduction, abduction reveals new insights into reality. Usually the interpretation of abduction is now generalized to give not only one but the possible best explanation of observations given an hypothesis.

The problem of object recognition from images can not be classified in one of the strict inference methods deduction or complete induction. It starts with the fact that a detection algorithm is only able to state a hypothesis that a building has to be reconstructed at a particular location, it can not logically deduce that there must be an object without much more constraints such as a close-world assumption. The same argument applies for the reconstruction of objects, as there is no stringent sequence of inferences that an object has only one fixed set of parameters.

It can be argued that the task of object reconstruction from a set of images is essentially a mixture of weak induction and abduction: at first, we have to deal with observations from the real world, thus making a generalization of a series of instances, which can be classified as inductive inference. Then in a second step we have to build hypotheses and generate the most plausible explanation from the observations given the set of hypotheses.

1.4.2 Reasoning within Geometry

In traditional Artificial Intelligence (AI), one mostly deals with symbolic representation of knowledge, i.e. some data representation. *Geometric reasoning* deals with specific kind of knowledge, namely geometric information. KAPUR AND MUNDY 1989B, p. 3 define the domain of geometric reasoning as "[...] an approach for representing geometric and topological properties and [...] geometric procedures that manipulate these representations.". As mentioned at the beginning of this chapter they argue that "[...] a better conceptual framework for representing and manipulating geometric knowledge can lead to implementation designs that provide more adequate separation between generic geometric principles and specific constraints of the application.".

Traditionally, geometric reasoning was identified with mechanically proving theorems in geometry, see GELERNTER 1963 and numerous researchers have been working on this field, cf. RICHTER-GEBERT AND WANG 2000. On the other hand, reasoning methods dealing with geometry have been extensively used in applications such as robotics and motion planning, computer vision and solid modeling DURRANT-WHYTE 1989. Considering the above definition, geometric reasoning in terms of object reconstruction means to extend the given geometric knowledge, i.e. the given images, to a 3D model of the unknown object.

In order to accomplish geometric reasoning for polyhedral object reconstruction, three important tasks have to be solved:

- find a way to *represent the geometric knowledge* in 2D and 3D in a consistent manner. Specifically we are interested in the representation of points, lines and planes. No special conditions or singularities should arise from the representation; for example the horizon in an image is essentially a mapping of a 3D line at infinity and should be represented in the same way as the mapping of any other 3D line.
- *manipulate* geometric entities to obtain new entities. This can be done either by unique operations such as the intersection of lines or by finding an estimate for an unknown object given a set of observations.

- deriving *hypotheses of relations* between geometric entities that are relevant for the application, for example the incidence relation of a line and a plane.

To restate the relation of reasoning and object extraction in the context of geometry, there are two inference methods that can be applied: first use induction, e.g. by testing relations between observed instances of points and lines and therefore establishing hypotheses of the underlying structure. In a second step we may use abduction, for example estimating a line given some previously hypothesized configuration of points and lines. The estimation of a geometric object can be performed by a least-squares method, which can be interpreted as finding the best explanation for the given observations.

1.4.3 Uncertain Reasoning

As explicated above, measurements of the real world are inherently uncertain: the observed error consists of an abstraction error due to the underlying model and an error in precision due to the imperfectness of measuring devices. A system that aims to infer new information from given measurements has to be able to represent and account for the effects of uncertainty.

An example might be given as follows: a robot has a mounted CCD camera, from which it infers information about its environment. The robot must be able to deal with uncertainty since it knows its position and the position of other objects only up to some error, so it is not certain about its position. Here, dealing with uncertainty means to (i) represent the errors of its position in an efficient manner, (ii) propagate these errors when moving the camera or changing the position and (iii) taking them into account when making decisions, such as *"Can I go through this door? Will I miss the door when I continue moving in the current direction?"*.

Recently, more and more researchers acknowledge that precise logic alone is not capable of solving these and many other tasks in a robust way, cf. MUMFORD 2000. In our application, we require that uncertainty should be an intrinsic part of geometric descriptions, cf. DURRANT-WHYTE 1989.

1.5 Previous Work

We categorize previous work that relates to this thesis in three categories: (i) projective geometry, (ii) geometric reasoning and statistics and (iii) polyhedral object reconstruction.

1.5.1 Projective Geometry in Photogrammetry and Computer Vision

Early references in Photogrammetry for the implicit techniques of Projective Geometry can be found in FINSTERWALDER 1899 and DAS 1949. In his short paper, Das describes a form of the collinearity equations with the abstract parameters of a projective camera matrix and thus being linear in the image and object coordinates. He applied the abstract collinearity equations to classical problems of Photogrammetry such as absolute and relative orientation of images. The approach of linearization was called DLT by ABDEL-AZIZ AND KARARA 1971. The relative orientation between two images in the context of projective geometry was described by THOMPSON 1968. Another approach for relative orientation using algebraic geometry includes VAN DEN HOUT AND STEFANOVIC 1976, for an in-depth review of work on orientation tasks including historical references, see WROBEL 2001.

The application of projective geometry to the problems of reconstruction of a real world scene from a set of uncalibrated images was introduced by the Computer Vision Community, a first tutorial introduction within this community was written by MUNDY AND ZISSERMAN 1992. One of the key advances was the realization, that a scene can be reconstructed from a set of uncalibrated images as all possible reconstructions are projectively equivalent (FAUGERAS 1992, HARTLEY *et al.* 1992). The remaining task to obtain a unique Euclidean reconstruction is to find a one-to-one transformation, that transforms ("stratifies") a projective reconstruction to the desired Euclidean representations. There exist a number of strategies for stratification, see e.g. FAUGERAS AND LUONG 2001 for a summary.

The results using projective geometry beginning in the early '90s have been successfully applied to a number of different applications, such as representation of different camera models (ALOIMONOS 1990, FAUGERAS 1993), computation of relative orientation of two cameras (TORR AND MURRAY 1997, HARTLEY 1995B, ZHANG 1998 where the basic work dates back to LONGUET-HIGGINS 1981), relative orientation of three views (SPETSAKIS AND ALOIMONOS 1990, SHASHUA AND WERMAN 1995, HARTLEY 1997), self-calibration (FAUGERAS *et al.* 1992, ZISSERMAN *et al.* 1995) and structure computation (POLLEFEYS *et al.* 2000). Probably the best references for an overview of the use of projective geometry for vision tasks can be found in the recent textbooks by HARTLEY AND ZISSERMAN 2000 and FAUGERAS AND LUONG 2001, the latter providing a more theoretical insight into the field.

A large subset of the work on projective geometry mostly centers on transformations, including camera transformations, relative orientations, especially for the uncalibrated case. In this work we mainly concentrate on the representation and manipulation of geometric primitives.

One particular interesting part of projective geometry when dealing with constructions of geometric primitives is the representation of projective points, lines and planes within the Grassmann-Cayley algebra, cf. CARLSSON 1994: the geometric primitives are represented by projective subspaces and it can be shown that the union and the intersection of these subspaces can be treated algebraically. The possibility of using algebraic expressions for geometric constructions is especially useful for formulating geometric problems involving multiple images since we can apply the geometric intuition for join and intersection to seemingly complicated algebraic expressions. Some of the applications of the Grassmann-Cayley algebra is the formulation of the trifocal tensor (TRIGGS 1995) or the modeling of cameras FAUGERAS AND PAPADOPOULO 1998.

1.5.2 Statistical Geometry and Reasoning

The work in Computer Vision which attacks problems with a reasoning approach is usually classified in the domain of knowledge-based image understanding – a comprehensive survey of this topic is written by CREVIER AND LEPAGE 1997. One particular advantage of reasoning systems is that using a set of algorithms with specified in- and output, a system can be tailored specific to its task. One problem when designing such a system is the representation of the uncertainty of the given knowledge. Generally, there exists quite a few different mathematical theories of evidence that can help to provide reasoning and control to an image understanding system, among them are fuzzy approaches (ZADEH 1979) and probabilistic approaches. The theory of fuzzy sets contains general fuzzy measures, from which the concept of probability measures is a subclass (KLIR AND FOLGER 1988).

This work follows a probabilistic approach of geometric reasoning. The actual reasoning process could be steered using a Bayesian network PEARL 1988, which SARKAR AND BOYER 1995 used for geometric knowledge base representation and grouping of geometric primitives of different types. In order tp apply these kind of networks, consistent representation of the primitives and an exact characterization of the performance of the algorithms is needed.

Since we restrict our application to polyhedral objects, we are interested in linear entities, although there exists work on a probabilistic modeling of general curves, see for example BLAKE *et al.* 1998. As opposed to representing geometric primitives, an example for the uncertain representation of projective invariants (MUNDY 1992) is the work by MAYBANK 1995. He develops a probability density function (pdf) for cross-ratios in images and defines a decision rule based on the pdf. This may be extended to more complicated invariants.

PORRILL 1988 developed a system called GEOMSTAT that uses a formalism for a generic statistical estimation using observed points and line segments by using a the most convenient representation for the parameter vectors and an unconstrained or minimal representation for uncertainty of the parameters. The uncertainty is modeled by a Gaussian distribution. To estimate unknown objects, a Gauß-Markov estimation model is used. The application of the system is the automatic reconstruction of wire-frame models from images.

Durrant-Whyte 1989 describes the uncertainty of entities by a point in parameters space with its pdf. Manipulating an entity means manipulating the underlying probability density function. He suggests a Gaussian approximation of this pdf. He also requires that the parameter representation of the entities should be statistically well-conditioned, meaning that there should be a one-to-one relationship between the parameters and the entity and that the parameters should change smoothly with smooth motions of the geometric entity.

Collins 1993 investigates the probability of projective objects and proposed to use a spherical distribution, the Bingham pdf. It is a second-order approximation to any antipodally symmetric density functions and thus serves a similar role on a sphere than a Gaussian on a plane. Errors of entities can be propagated, and it is shown how to perform statistical inference on incidence and equality relations with a Bayesian approach. These theoretical results are applied to grouping of 3D line- and plane-orientations from observed image features.

Seitz and Anandan 1999 introduce a technique for representing points, lines and planes as probability functions. In particular, they use the Gaussian density function and develop a consistent implicit matrix representation containing mean and second moments. Then intersections and transformations of the entities can be formulated as unique operations on the implicit representation and therefore directly transforms the uncertainty representation. This can be applied in a consistent way for all types of entities. However there is no apparent way to extend this work for join operations.

Clarke 1998 describes homogeneous covariance matrices for homogeneous vectors, which in turn are representatives of projective entities. He shows that it is feasible to apply first-order error propagation using these covariance matrices to obtain uncertainty information for constructed entities such as a 2D point from two 2D lines. This has been shown using Monte Carlo simulations to compare statistical and predicted covariance matrices.

Criminisi 2001 presents an approach of using one or more images as a measuring device based on projective geometry. From simple situations such as measuring planar surfaces from one image, he continues with more complicated three-dimensional measures and using multiple images instead of only one. As a measurement without a statement about its precision is practically useless, the uncertainty of the final result was computed from the uncertainty of the input data by first-order error-propagation. It was demonstrated that the assumption of a Gaussian density function is valid for a sufficiently precise prediction of the errors of the result.

To our knowledge, the work by Kanatani 1996 is probably the most general approach to the statistical estimation and testing of three-dimensional structures based on image data and geometric constraints. It includes points, lines in 2D and 3D and planes in 3D and demonstrates (i) how to correct these primitives to satisfy geometric constraints (including testing these constraints), (ii) optimal fitting of unknown primitives given some observations with applications to vision tasks and (iii) derives a criterion for selection of possible statistical models for the geometric data. Our work is similar to the first two of the three mentioned topics, but motivated by the Grassmann-Cayley algebra: the formulations of the testing and estimation problems become much simpler.

Finally it is worthwhile noting that the classical technique of bundle adjustment, extensively used in Photogrammetry, is acknowledged by the Computer Vision community to be the optimal general method for fitting cameras and structure, cf. TRIGGS *et al.* 2000, though the use of robust estimators is preferred in the vision context.

1.5.3 Polyhedral Object Reconstruction

One of the first works in polyhedral object reconstruction is the one by ROBERTS 1965. He introduced a data driven approach to identify and reconstruct the pose of polyhedral objects from digital images. At first, image features are detected based on the image intensities and then the best model in his database is selected that fits to the image features. Motivated by this example, researchers investigated the nature of line drawings of polyhedra with the hope to obtain useful results for recognition and reconstruction. Most notably is the work by CLOWES 1971, HUFFMAN 1971 and later SUGIHARA 1986 who developed methods for consistent labeling of polygonal edges that referred to the 3D topology. While this work is useful for line drawings, cf. BRAUN 1994, unfortunately it seems that it is quite hard to apply the reasoning to real imagery. This is mainly due to the fact that it is almost impossible to obtain accurate topological information from segmentation results of complex images.

Later on, the approach of fitting known polyhedral models to one or more uncalibrated images was refined for example by LOWE 1991. The main problems with such an approach are: (i) a database of possible models has to exist and (ii) the search for selecting the best model can be quite involving. Thus one tries to first find parts or invariants of polyhedra such as in ROTHWELL AND STERN 1996 or MOHR *et al.* 1992.

An important application of polyhedral object reconstruction is the reconstruction of man-made structures such as buildings from a set of aerial images. Here the advantage is that usually one has quite precise camera parameters, which can be used in the system. For an overview of recent work in this area, see MAYER 1999. As mentioned above, a polyhedron is not a building model per se, the semantics of its parts are missing: for example surfaces have to be labeled as roof surfaces. Still a lot of researchers use polyhedra as the underlying geometric model and build their semantic models upon it.

Reconstruction of buildings may not only rely on gray scale images, one can also use laser-data as in WEIDNER 1997, BRUNN AND WEIDNER 1998 or radar data cf. STILLA *et al.* 2001. But this kind of range-data was found to be especially suited for the detection task of buildings by discriminating possible areas with the underlying height information, see WEIDNER AND FÖRSTNER 1995 or JAYNES *et al.* 1997.

An automated building detection and reconstruction system based on rectangular blocks was introduced by LIN *et al.* 1994, where only one image is used and extracted features are grouped to higher aggregates until a building can be hypothesized. This work has been extended to multiple views, see NORONHA AND NEVATIA 2001, where the matching between the images is done on the feature level (line segments, junctions) and aggregate level (Parallels, Corners, U's).

More general classes of polyhedra than only rectangular shapes are treated for example by BIGNONE *et al.* 1996, who match 2D line-segments with color attributes to compute 3D line segments, that in turn are grouped to 3D surfaces, for which boundaries are sought in the images. In a last step, the surfaces are aggregated to polyhedral parts. MOONS *et al.* 1998 use a similar approach, but build the 3D surfaces using a triangulation of image edges, that are grouped and matched to complete surfaces using a geometrically and radiometrically steered region growing technique.

BAILLARD *et al.* 1999 first compute matching line segments in multiple images based on image correlation and geometric constraints, see also SCHMID AND ZISSERMAN 1997. Then they determine planar patches for each line by finding a projective transformation of image neighborhoods. Then they group 3D lines within these half-planes and close them by plane intersections. A different approach for 3D grouping of lines and corners solely based on geometrical and topological features is presented in HEUEL *et al.* 2000.

An important last step of a reconstruction system is the validation, that the chosen model actually fits to the image data. Some verification methods are using shadows HUERTAS *et al.* 1993 or walls NORONHA AND NEVATIA 2001. The verification of building hypotheses based on a set of models and the selection of the best model using a minimum description length (MDL) criterion can be found in KOLBE 1999.

Finally, we want to mention some of the user-assisted or semi-automatic approaches for polyhedra reconstruction, which do not aim at a full automation level. Within the photogrammetric area, there exists e.g. CYBERCITY MODELER (GRÜN AND WANG 1999) and the INJECT system (GÜLCH *et al.* 2000), which will be introduced in chapter 5. A system that was presented to the Computer Vision and Graphics community is FACADE cf. DEBEVEC *et al.* 1996, which especially aims on photo-realistic reconstruction. It uses parametrized models, that are interactively specified by the user with the aid of a precomputed image segmentation result. The final model is then optimized by a minimization function. There are two commercial systems with a similar purpose, IMAGEMODELER by Realviz (REALVIZ 2002) and PHOTOMODELER (PHOTOMODELER 2002).

1.6 Overview of Thesis

This work presents a unified approach for *representing*, *combining* and *estimating* points, lines and planes in 2D and 3D taking uncertainty information into account. To demonstrate the feasibility of the approach the derived methods are applied to the task of reconstructing polyhedra from multiple views. We use recent advances in the application of projective geometry to derive a simple algebraic framework for the representation, construction and expression of relations of simple geometric entities such as points, lines or planes. An important aspect is the explicit incorporation of uncertainty information to this framework.

Chapter 2 describes important aspects of projective geometry that are necessary for our task and introduce a representation for points and (infinite) lines in 2D and 3D and planes in 3D. These geometric entities will later be used to describe polygons and polyhedra. Furthermore basic geometric transformations between the entities are described.

Chapter 3 describes how to manipulate the introduced entities and how to establish relations between them: we start with the two basic operators of the Grassmann-Cayley algebra, join and intersection, and give simple bilinear algebraic expressions for them. We analyze the nature of these expressions and from there derive simple formulas for testing on incidence, identity, parallelity and orthogonality. Finally, an estimation scheme for unknown entities is presented based on a set of observations related to the unknown. This scheme can also be used for estimation of unknown transformations, yielding a general formulation for direct linear transforms, DLT.

Chapter 4 deals with the incorporation of uncertainty information in the algebraic framework of the last chapter. We propose an approximate uncertainty representation using a Gaussian pdf and analyze the error that is caused by the Gaussian assumption. It turns out that the approximation is valid under certain well-defined circumstances and enables us to formulate statistical versions of the methods from chapter 3, which are bundled in a system of procedures for statistically uncertain geometric reasoning, called `SUGR`, see section 4.7 on page 147.

Chapter 5 validates the usefulness of the `SUGR`-system by applying it to the task of polyhedral object reconstruction from multiple images. We demonstrate how to include the techniques both for user-assisted reconstruction systems or automated reconstruction systems and show some results. In particular it is demonstrated how `SUGR` can bridge the gap between interactive and automated approaches.

Finally we conclude with a short summary and an outlook for future work.

2 Representation of Geometric Entities and Transformations

A necessary step of developing a system for geometric reasoning is the choice of the representation of the geometric objects that are going to be used. In section 1.3 we restricted the possible shapes of objects to polytopes. Polyhedra and polygons may be described by linear subspaces namely points, (infinite) lines and planes. In this chapter we want to explore a mathematical description of points, lines and planes in 2D and 3D, which from now on are referenced as *basic geometric entities* or just *geometric entities.*

We use results from projective geometry, which have been successfully applied to Computer Vision in the last decades. This chapter summarizes some of the aspects of projective geometry that are important for our task. Note however that the notation is simplified compared to a rigorous treatment as in FAUGERAS AND LUONG 2001, that is too verbose for our purposes.

Geometric entities will be represented as projective subspaces, which in turn can be parametrized by homogeneous vectors. An important observation in this chapter is that all homogeneous vectors can be consistently subdivided into a homogeneous and a Euclidean part, as proposed in BRAND 1966 who described the distinction without the explicit context of projective geometry.

After discussing the homogeneous representation of geometric entities, some task-relevant transformations are reviewed, such as homographies and pinhole-camera transformations. Finally the duality-principle of projective geometry, that will be used in the subsequent chapters, is introduced.

2.1 Projective Geometry

When dealing with images of the real world, we are naturally exposed to elements at infinity: for example the horizon is an image of infinitely distant space points[1], which can be easily seen by images taken in flat environments: parallel roadsides meet at infinity. Unfortunately Euclidean geometry is not sufficient to represent infinite elements and therefore it is hard to deal with

[1] To be precise, this would be true if the world was planar. But the real, almost spherical world can be locally approximated (or modeled) by a plane if we measure buildings and garages within a small region.

S. Heuel: Uncertain Projective Geometry, LNCS 3008, pp. 19-45, 2004.

them within the Euclidean framework. Therefore we need a more general framework which enables us to deal with infinite points, this mathematical framework is known as *projective geometry.*

In this and the following chapters, we will see that projective geometry provides a simple and consistent way to represent and transform geometric entities, that are important for Euclidean polyhedral reconstruction. One drawback though is the fact that properties such as parallelity or orthogonality can not be directly used; more precisely, they are not invariant with respect to projective transformations. The study of invariance of properties with respect to certain groups of transformation may be regarded as a definition of geometry, as KLEIN 1939 mentioned in his Erlanger Programm. We will come back to the issue of invariance in section 3.3 on page 70, where we investigate possible relations between geometric entities and their invariance properties with respect to Euclidean, similar, affine and projective geometry. We will see that for our application it is possible to infer parallelity and orthogonality of lines and planes from homogeneous vectors by using the Euclidean interpretation thereof.

For a detailed and more precise introduction to projective geometry in the context of Computer Vision, see FAUGERAS AND LUONG 2001 or chapter 23 in MUNDY 1992.

2.1.1 Projective Space and Homogeneous Coordinates

Generally, a projective space $\mathbb{P}^n$ is the set of one-dimensional vector subspaces of a given vector space $\mathbb{R}^{n+1}$. Each one-dimensional vector subspace represents a projective point $\mathbf{x}$ with the *projective* dimension 0.

Assuming $n = 1$, the projective space is the projective line $\mathbb{P}^1$ and is visualized in figure 2.1(a) on the facing page: the projective points $\mathbf{x}_{1/2} \in \mathbb{P}^1$ can be identified with lines going through the origin of the vector space $\mathbb{R}^2$. Thus these lines are one-dimensional subspaces of $\mathbb{R}^2$. Note that no coordinate system but only the origin is defined here.

A definition of a projective point $\mathbf{x} \in \mathbb{P}^n$ may be given as follows:

Definition 2 (Projective point) *A projective point* $\mathbf{x} \in \mathbb{P}^n$ *of dimension* n *is an element of the quotient of* $\mathbb{R}^{n+1} \setminus \mathbf{0}$ *and the equivalence relation* $\cong$*:*

$$\mathbf{x} \cong \mathbf{y} \quad \Leftrightarrow \quad \exists \lambda \neq 0, \mathbf{x} = \lambda \mathbf{y} \tag{2.1}$$

In other words: all vectors $\lambda \mathbf{x}^h = (\lambda x_1, \dots, \lambda x_n, \lambda)^\mathsf{T} \in \mathbb{R}^{n+1}$ $\lambda \neq 0$ are equivalent with respect to $\cong$ and define a projective point $\mathbf{x}$. In other words, a projective point $\mathbf{x} \in \mathbb{P}^n$ is defined by representing it with a $n+1$ vector $\mathbf{x}^h \in \mathbb{R}^{n+1}$, the so-called *homogeneous vector* representation. This representation is overdetermined: the equivalence relation states, that the lengths of the vectors $\mathbf{x}^h, \mathbf{y}^h \in \mathbb{R}^{n+1}$ do not matter, only their directions.

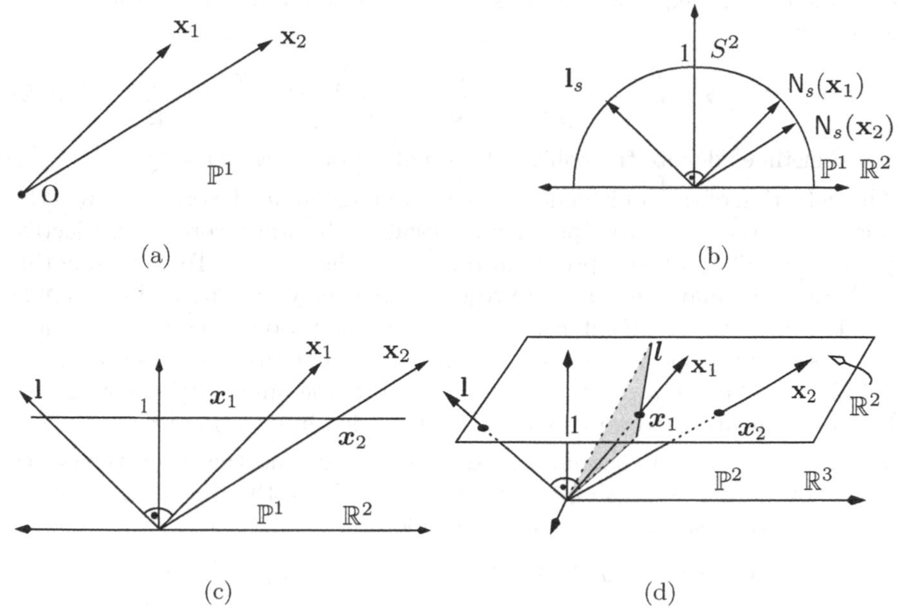

Fig. 2.1. Different interpretations of the projective line $\mathbb{P}^1$: in (a), projective points $\mathbf{x}_1, \mathbf{x}_2$ are interpreted as lines going through the origin without defining any coordinate system. In (b), the projective points are homogeneous vectors, that are normalized to the sphere S^2. In (c) the homogeneous vectors intersect the Euclidean axis and define the points $\boldsymbol{x}_1$ and $\boldsymbol{x}_2$ on the Euclidean axis. The role of a hyperplane $\mathbf{l}$ can be visualized by considering the projective plane $\mathbb{P}^2$ in (d): two projective points $\mathbf{x}_1$ and $\mathbf{x}_2$ define two points $\boldsymbol{x}_1, \boldsymbol{x}_2$ on the Euclidean plane $\mathbb{R}^2$. The first point $\boldsymbol{x}_1$ is incident to a line $\boldsymbol{l}$ on the Euclidean plane. The line $\boldsymbol{l}$ in turn is defined by the homogeneous vector $\mathbf{l} \in \mathbb{R}^3 \simeq \mathbb{P}^2$, which is the normal of the plane that contains the origin of $\mathbb{R}^3$ and the line on the Euclidean plane $\mathbb{R}^2$. As the first Euclidean point $\boldsymbol{x}_1$ is incident to the Euclidean line, the homogeneous vector $\mathbf{l}$ is orthogonal to the homogeneous vector $\mathbf{x}_1$, or $\mathbf{l}^\mathsf{T}\mathbf{x}_1 = 0$.

To simplify the notation, we identify projective points $\mathbf{x} \in \mathbb{P}^n$ with their (non-unique) coordinate vector representation based on $\mathbf{x}^h \in \mathbb{R}^{n+1}$, i.e. $\mathbf{x} = \mathbf{x}^h$. Therefore we are not strict with respect to mathematical notation and do *not* distinguish between projective points and its coordinate vectors as in FAUGERAS AND LUONG 2001, p. 78. HARTLEY AND ZISSERMAN 2000 take a similar approach, see p. 3f. In the following, we assume that the correct interpretation is clear from the context.

One can *normalize* a homogeneous vector $\mathbf{x}_s := \frac{\mathbf{x}}{|\mathbf{x}|}$ to avoid ambiguities, see fig.2.1(b). The index s in the homogeneous vector $\mathbf{x}_s$ indicates that the vector is normalized and thus lies on the unit sphere. We could use the normalization of a homogeneous vector with its length for an alternative definition of the equivalence class $\cong$ for two homogeneous vectors $\mathbf{x}, \mathbf{y} \in \mathbb{R}^{n+1} \setminus \mathbf{0}$:

$$\mathbf{x} \cong \mathbf{y} \quad \Leftrightarrow \quad \left(\frac{\mathbf{x}}{|\mathbf{x}|} = \frac{\mathbf{y}}{|\mathbf{y}|} \text{ or } \frac{\mathbf{x}}{|\mathbf{x}|} = -\frac{\mathbf{y}}{|\mathbf{y}|} \right) \quad \Leftrightarrow \quad \frac{\mathbf{x}}{|\mathbf{x}|} = \pm \frac{\mathbf{y}}{|\mathbf{y}|} \tag{2.2}$$

We sometimes denote the spherical normalization by $\mathsf{N}_s(\mathbf{x}) = \frac{\mathbf{x}}{|\mathbf{x}|}$.

The definition of the equivalence class $\cong$ is ambiguous as the orientation of the normalized vectors can be positive or negative. In other words, a projective point from $\mathbb{P}^n$ has two representations on a sphere S^{n+1}. To overcome this ambiguity one may use *oriented* projective geometry, as proposed by STOLFI 1991, where every element has an intrinsic orientation. This is analytically expressed by the specialized definition, that homogeneous vectors are identical if they differ by a *positive* scale factor. But the ambiguity is not critical for our application, so we will keep the definition in equation (2.2).

The link between an original Euclidean space $\mathbb{R}^n$ and the projective space $\mathbb{P}^n$ is now quite easy: a simple mapping $\mathbb{R}^n \to \mathbb{P}^n$ of the Euclidean space to the projective space can be written as follows:

$$\boldsymbol{x} = (x_1, \ldots, x_n)^\mathsf{T} \in \mathbb{R}^n \to \mathbf{x}_e = (x_1, \ldots, x_n, 1)^\mathsf{T} \in \mathbb{P}^n \tag{2.3}$$

We will denote vectors from $\mathbb{R}^n$ with italic bold letters such as $\boldsymbol{x}, \boldsymbol{y}$, whereas homogeneous vectors from $\mathbb{P}^n \backsimeq \mathbb{R}^{(n+1)}$ are denoted with upright bold letters, $\mathbf{x}, \mathbf{y}$.

The index e in the homogeneous vector $\mathbf{x}_e$ denotes that we can directly extract the Euclidean information from the homogeneous vector. It is a unique member of the equivalence class defined by $\mathbf{x} \cong \mathbf{x}_e$. In figure 2.1(c) one may interpret the fixing of the last coordinate of the homogeneous vector to 1 as follows: take the line parallel to the first axis, intersecting the second axis in 1. This line represents the real axis $\mathbb{R}^1$, on which the Euclidean numbers $\boldsymbol{x}_{1/2}$ are fixed. Connecting the origin with $\boldsymbol{x}_{1/2}$ on the constructed line yield the projective points.

For the inverse direction of (2.3), any projective point $\mathbf{x} \in \mathbb{P}^n$ (except of ideal points) can be mapped to an euclidized version, see also figure 2.2:

$$\mathbf{x} = (x_1, \ldots, x_n, x_{n+1})^\mathsf{T} \in \mathbb{P}^n \to \mathbf{x}_e = 1/x_{n+1}\,(x_1, \ldots, x_n, 1)^\mathsf{T} \in \mathbb{P}^n \tag{2.4}$$

For lower dimensions, we will replace the coordinates x_1, x_2, x_3 with the letters u, v, w to avoid overloaded use of indices.

Linear Transformations of Projective Points A linear transformation H from $\mathbb{P}^n$ to $\mathbb{P}^m$ can be represented as a $(m+1) \times (n+1)$, such that for $\mathbf{x} \in \mathbb{P}^n$ and $\mathbf{x}' \in \mathbb{P}^m$:

$$\mathbf{x}' = \mathsf{H}\mathbf{x} \tag{2.5}$$

If $n = m$ and the matrix H is regular, we call H a *homography.*

Properties that remain invariant under a transformation H are called projectively invariant. Examples for such properties are collinearity and incidence.

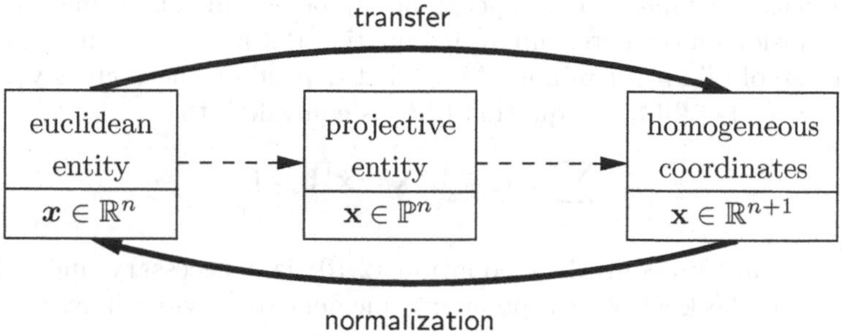

Fig. 2.2. Transfer between Euclidean and Homogeneous Coordinates: giving a Euclidean point $\boldsymbol{x} \in \mathbb{R}^n$ we can derive a projective point $\mathbf{x} \in \mathbb{P}^n$ which we represent as a homogeneous vector $\mathbf{x} \in \mathbb{R}^{n+1}$, see also equation (2.3). Given a homogeneous vector $\mathbf{x} \in \mathbb{R}^{n+1}$, we may obtain a Euclidean representation by normalization, yielding $\boldsymbol{x} \in \mathbb{R}^m$, equation (2.4). In general, the normalization is done by dividing the homogeneous vector by its homogeneous part as described in section 2.2. Note that we use the symbol $\mathbf{x}$ both for a projective point and for homogeneous vectors representing the projective points.

2.1.2 Hyperplanes

A point in projective space can be regarded as the smallest linear subspace that is non-zero. Opposed to a point, a *hyperplane* is the largest linear subspace that is not equal to the complete space. Examples for hyperplanes are lines in 2D and planes in 3D.

In general, a hyperplane $\mathbf{h}$ in $\mathbb{P}^n$ is defined by n linearly independent points $\mathbf{y}_1, \cdots, \mathbf{y}_n$: for example in $\mathbb{P}^2$, two points $\mathbf{y}_1, \mathbf{y}_2$ define a line and in $\mathbb{P}^3$ three points $\mathbf{Y}_1, \mathbf{Y}_2, \mathbf{Y}_3$ define a plane. Any additional point $\mathbf{x}$ is incident to the a hyperplane $\mathbf{h}(\mathbf{y}_1, \cdots, \mathbf{y}_n)$ if the determinant of all concatenated homogeneous vectors is zero

$$|\mathbf{y}_1\, \mathbf{y}_2 \cdots \mathbf{y}_n\; \mathbf{x}| = 0 \tag{2.6}$$

This is a generalization of the well-known case when checking whether three 2D points are collinear BRONSTEIN *et al.* 1996:

$$\boldsymbol{y}_1, \boldsymbol{y}_2 \text{ and } \boldsymbol{x} \text{ are collinear} \quad \Leftrightarrow \quad \begin{vmatrix} \boldsymbol{y}_1 & \boldsymbol{y}_2 & \boldsymbol{x} \\ 1 & 1 & 1 \end{vmatrix} = 0 \tag{2.7}$$

which can be generalized to the to the 3D case when checking whether four points are coplanar:

$$\boldsymbol{Y}_1, \boldsymbol{Y}_2, \boldsymbol{Y}_3 \text{ and } \boldsymbol{X} \text{ are coplanar} \quad \Leftrightarrow \quad \begin{vmatrix} \boldsymbol{Y}_1 & \boldsymbol{Y}_2 & \boldsymbol{Y}_3 & \boldsymbol{X} \\ 1 & 1 & 1 & 1 \end{vmatrix} = 0 \tag{2.8}$$

(2.9)

An alternative definition of a hyperplane can be obtained by using the Laplacian expansion on the determinant in equation (2.6): if we expand by the last column, we obtain $n-1$ minors $|\mathsf{M}_{i,n}|$ that depend on the vectors $\mathbf{y}_1, \cdots \mathbf{y}_n$. With $h_i = (-1)^{i+n}|\mathsf{M}_{i,n}|$, equation (2.6) is equivalent to

$$\sum_i h_i x_i = \mathbf{h}^\mathsf{T}\mathbf{x} = \mathbf{x}^\mathsf{T}\mathbf{h} = 0 \tag{2.10}$$

Both (2.6) and its equivalent equation (2.10) is a necessary and sufficient condition to check whether a point $\mathbf{x}$ is incident to a hyperplane, $\mathbf{x} \in \mathbf{h}$.

Since the minors h_i with $i = 1, \cdots, n$ uniquely define a hyperplane, we can identify them with the coordinates of a hyperplane, defining the vector $\mathbf{h} = (h_1, \cdots, h_n)^\mathsf{T}$. The coordinates h_i are called Plücker coordinates, see also section 2.2.4. The vector $\mathbf{h}$ is a homogeneous vector as (2.10) is a homogeneous equation. As for projective points, it is assumed that it is clear from the context whether a homogeneous vector is interpreted as a projective point or a hyperplane: in the following lines in 2D are usually denoted by $\mathbf{l}, \mathbf{m}$ and planes in 3D are denoted by $\mathbf{A}, \mathbf{B}$ opposed to points denoted by $\mathbf{x}, \mathbf{y}$ resp. $\mathbf{X}, \mathbf{Y}$.

It is important to note that $\mathbf{h}$ is actually a linear form of $\mathbb{R}^{n+1}$, since each coordinate is a function of the form $\mathbb{R}^{n+1} \to \mathbb{R}$. Thus the vector $\mathbf{h}$ represents an element of the dual space $(\mathbb{R}^{n+1})^*$ or equivalently $(\mathbb{P}^n)^*$.

Given a linear transformation $\mathbf{x}' = \mathsf{H}\mathbf{x}$ for projective points as in (2.5), one may be interested in the corresponding transformation for hyperplanes. H is a linear mapping from $\mathbb{P}^n$ to $\mathbb{P}^m$, its transpose H^T is defined as the mapping from $(\mathbb{P}^m)^*$ to $(\mathbb{P}^n)^*$. With $\mathbf{h}' \in (\mathbb{P}^m)^*$ being the imaged hyperplane of $\mathbf{h} \in (\mathbb{P}^n)^*$ with respect to H, one obtains

$$\mathsf{H}^\mathsf{T}\mathbf{h}' = \mathbf{h} \tag{2.11}$$

If H is a homography, thus $n = m$ and H being regular, one can prove (2.11) as follows: given $\mathbf{x}' = \mathsf{H}\mathbf{x}$, we are looking for the corresponding transformation $\mathbf{h}' = \mathsf{J}\mathbf{h}$, such that for all $\mathbf{y} \in \mathbf{h}$ and $\mathbf{y}' \in \mathbf{h}'$: $\mathbf{y}' = \mathsf{H}\mathbf{y}$. Thus the following equation must hold

$$\mathbf{y'}^\mathsf{T}\mathbf{h}' = \mathbf{x'}^\mathsf{T}\mathsf{J}\mathbf{h} = \mathsf{x}^\mathsf{T}\mathsf{H}^\mathsf{T}\mathsf{J}\mathbf{h} = 0$$

This can be true only if $\mathsf{J} = \mathsf{H}^{-\mathsf{T}}$.

2.2 Representation of Geometric Entities

This section introduces the basic geometric entities, that will be used throughout the rest of this text. As we want to reconstruct polyhedral objects, it is sufficient to deal with geometric entities that can be modeled as linear subspaces, cf. section 1.3.2 on page 6. Therefore only points, lines and planes in 2D and 3D are covered. They are represented using homogeneous vectors, but we will give a Euclidean interpretation for the vectors of every entity. At the end of this section, we will see that the chosen representations are special cases of the Plücker coordinates, a representation of subspaces within a projective space.

2.2.1 Points and Lines in 2D

Given a *2D point* $\boldsymbol{x} = (x, y)^\mathsf{T} \in \mathbb{R}^2$ on the Euclidean plane, one may obtain a homogeneous representation using equation (2.3):

$$\mathbf{x} = \begin{pmatrix} x \\ y \\ 1 \end{pmatrix} \cong \begin{pmatrix} wx \\ wy \\ w \end{pmatrix} \cong \begin{pmatrix} u \\ v \\ w \end{pmatrix}$$

see also figure 2.1(d) on page 21.

A Euclidean representation of a 2D point is straight forward by specifying two coordinates, but the Euclidean representation of a *2D line* can be manifold:

slope-intercept form	$y = m\,x + b$
axis-intercept form	$\frac{x}{a} + \frac{y}{b} = 1$
angle-distance form	$x\,cos(\varphi) + y\,sin(\varphi) = d$
homogeneous form	$a\,x + b\,y + c = 0$
determinant form	$\begin{vmatrix} x_1 & x_2 & x \\ y_1 & y_2 & y \\ 1 & 1 & 1 \end{vmatrix} = 0$
point-angle form	$(x - x_0)\,cos(\varphi) + (y - y_0)\,sin(\varphi) = 0$

Some of these representations do not cover all possible cases (e.g. vertical lines for the slope-intercept form) and some require more parameters than necessary: the determinant form requires two points $\boldsymbol{x}_1$, $\boldsymbol{x}_2$ hence four parameters opposed to the fact that only two parameters are necessary, for example angle and distance.

We will start with the homogeneous form $a\,x + b\,y + c = 0$, then the parameters can be written in a homogeneous vector

$$\mathbf{l} = \begin{pmatrix} a \\ b \\ c \end{pmatrix}$$

If we multiply the homogeneous form with a scalar w and use the homogeneous vector $\mathbf{x} = (wx, wy, w)^\mathsf{T} \cong (u, v, w)^\mathsf{T}$ for a point, we obtain

$$a\,u + b\,v + c\,w = \mathbf{l}^\mathsf{T}\mathbf{x} = \mathbf{x}^\mathsf{T}\mathbf{l} = 0 \tag{2.12}$$

which is the 2D case for equation (2.10) for hyperplanes.

A *line segment* $\mathbf{s}_l$ is a finite segment of an (infinite) line. It can be represented by a triple $(\mathbf{l}_{fs}, \mathbf{x}_{fs}, l)$, the infinite line $\mathbf{l}_{fs}$, a mid-point $\mathbf{x}_{fs}$ of the line segment and a length l. This is equivalent compared to defining two enclosing points on the line, cf. definition 1 on page 6.

2.2.1.1 Homogeneous and Euclidean Parts From the homogeneous vectors for points it is easy to determine the Euclidean distance from the point to the origin $\boldsymbol{O}$: for a projective point $\mathbf{x} = (u, v, w)^\mathsf{T}$, the Euclidean distance to the origin is given by $d(\mathbf{x}, \boldsymbol{O}) = \sqrt{(u/w)^2 + (v/w)^2} = \sqrt{u^2 + v^2}/|w|$.
The coordinates of the 2D line $\mathbf{l} = (a, b, c)^\mathsf{T}$ may be interpreted by comparing equation (2.12) with the angle-distance form: they are essentially the same up to a multiplication factor $\lambda_l = 1/\sqrt{a^2 + b^2}$:

$$\mathbf{l} = \begin{pmatrix} a \\ b \\ c \end{pmatrix} \cong 1/\sqrt{a^2 + b^2} \begin{pmatrix} a \\ b \\ c \end{pmatrix} = \begin{pmatrix} cos(\varphi) \\ sin(\varphi) \\ -d \end{pmatrix} \tag{2.13}$$

As the scalar d refers to the distance of the line to the origin, we obtain $d(\mathbf{l}, \boldsymbol{O}) = |c|/\sqrt{a^2 + b^2}$. Note that the angle φ is the angle of the normal of the line $\mathbf{l}$.
Motivated by the above observation of computing the distances to the origin and as suggested by BRAND 1966 p. 51ff., we will divide the homogeneous vectors for 2D points and lines into two parts:

- the *Euclidean part* depends upon the position of the Euclidean origin $\boldsymbol{O}$. This part of the homogeneous vector will be indexed with O and is the enumerator of the fractions $d(\mathbf{x}, \boldsymbol{O})$ and $d(\mathbf{l}, \boldsymbol{O})$ as above.
- the *homogeneous part* is independent of the position of the origin $\boldsymbol{O}$. It refers to the denominator of the fractions $d(\mathbf{x}, \boldsymbol{O})$ and $d(\mathbf{l}, \boldsymbol{O})$.

For a 2D point $\mathbf{x}$ and for a 2D line $\mathbf{l}$ the partitioning yields

$$\mathbf{x} = \begin{pmatrix} u \\ v \\ \hline w \end{pmatrix} = \begin{pmatrix} \boldsymbol{x}_O \\ x_h \end{pmatrix} \quad ; \quad \mathbf{l} = \begin{pmatrix} a \\ b \\ \hline c \end{pmatrix} = \begin{pmatrix} \boldsymbol{l}_h \\ l_O \end{pmatrix}$$

2.2.1.2 Normalization to Euclidean Coordinates Dividing the homogeneous vectors by the norm of the homogeneous parts, $\mathbf{x}/|x_h|$ and $\mathbf{l}/|\boldsymbol{l}_h|$, we obtain the Euclidean representation of the 2D entities: the Euclidean coordinates of a 2D point and the angle-distance form of a 2D line. We call this division *Euclidean normalization*: for any homogeneous vector $\mathbf{x}$ representing a geometric entity, the Euclidean normalization $\mathsf{N}_O(\cdot)$ is defined as

$$\mathsf{N}_O(\mathbf{x}) := \mathbf{x}/|\boldsymbol{x}_h| \tag{2.14}$$

where $\boldsymbol{x}_h$ is the homogeneous part of the entity $\mathbf{x}$.

Note that the proposed normalization $\mathsf{N}_O(\cdot)$ is ambiguous with respect to its sign: a Euclidean point $\boldsymbol{x} = (2,3)^\mathsf{T}$ may be represented by $\mathbf{x}_1 = (4,6,2)^\mathsf{T}$ and $\mathbf{x}_2 = (-4,-6,-2)^\mathsf{T}$. This leads to $\mathsf{N}_O(\mathbf{x}_1) = -\mathsf{N}_O(\mathbf{x}_2)$. The same applies to a 2D line with $\varphi = 90°$ and $d = 3$, where two representations $\mathbf{m}_1 = (2,0,6)^\mathsf{T}$ and $\mathbf{m}_2 = (-2,0,-6)^\mathsf{T}$ yield $\mathsf{N}_O(\mathbf{m}_1) = -\mathsf{N}_O(\mathbf{m}_1)$, see figure 2.3.

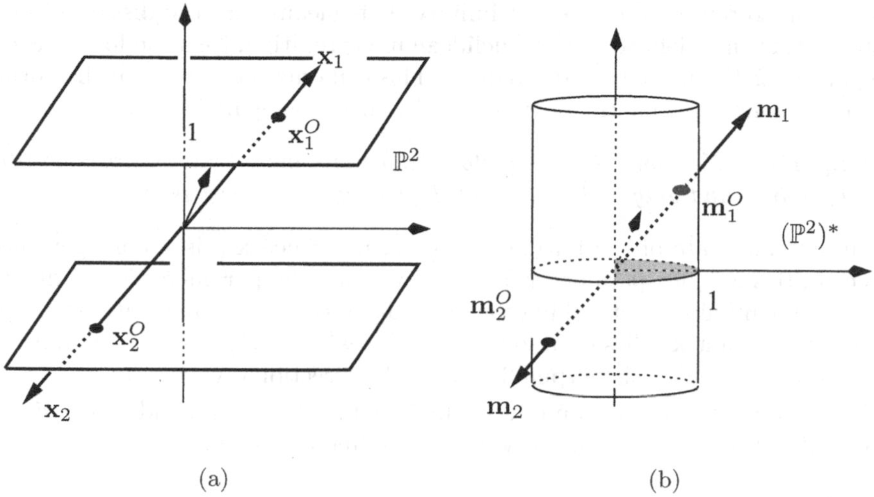

Fig. 2.3. Ambiguous Euclidean normalization of a 2D point $\boldsymbol{x}$ and a 2D line $\boldsymbol{m}$, which can be represented by $\mathbf{x}_1, \mathbf{x}_2$ resp. $\mathbf{m}_1, \mathbf{m}_2$: the normalizations $\mathsf{N}_O(\mathbf{x}_1)$ and $\mathsf{N}_O(\mathbf{x}_2)$ resp. $\mathsf{N}_O(\mathbf{m}_1)$ and $\mathsf{N}_O(\mathbf{m}_2)$ yield different results. We may choose one representation where the homogeneous part is positive oriented, see text for details.

To reach a unique representation, it is possible to demand that the homogeneous part $(\boldsymbol{x}_e)_h$ of any *normalized* homogeneous vector $\mathbf{x}_e = \mathsf{N}_O(\mathbf{x})$ should be *positive oriented*: for points, this means that the homogeneous scale factor x_h is positive, thus for a 2D point $\mathbf{x}$, we skip the norm by $\mathsf{N}_O(\mathbf{x}) = \mathbf{x}/x_h$. For lines, the direction $\boldsymbol{l}_h$ should be on the upper half of the circle S^2, thus $\langle (0,1)^\mathsf{T}, \boldsymbol{l}_h \rangle > 0$, see figure 2.3(b) As for the spherical normalization, the ambiguity may be resolved using oriented projective geometry, STOLFI 1991, but the ambiguity only affects the Euclidean output of the reasoning process: as long as only homogeneous coordinates are used, the sign of the vectors is not important. We will not use Euclidean coordinates directly during the complete reasoning, see chapter 3, thus the ambiguity can be disregarded in our case.

To summarize, we are now able to transfer back and forth between Euclidean and homogeneous representation as in figure 2.2 on page 23. Furthermore, we can easily obtain the distances to the origin:

$$d(\mathbf{x}, \boldsymbol{O}) = \frac{|\boldsymbol{x_O}|}{|x_h|} \quad ; \quad d(\mathbf{l}, \boldsymbol{O}) = \frac{|l_O|}{|\boldsymbol{l_h}|} \tag{2.15}$$

In the next sections, we will define Euclidean and homogeneous parts for 3D entities, too, and we will see that there both parts of the vector play the same role as in the 2D case.

2.2.1.3 Infinite Points and Line As stated earlier, projective geometry allows us to represent entities at infinity. This means that the distance from the origin is not defined in the Euclidean interpretation, thus the formulas in equation (2.15) are not valid anymore. This can only happen when the norm of the enumerator, i.e. the norm of the homogeneous part is zero.

Property 1 *If a homogeneous vector is non-zero, then it represents an entity at infinity if and only if the norm of the homogeneous part is zero.*

Infinite points are of the form $\mathbf{x}_\infty = (u, v, 0)^\mathsf{T}$. Since $\mathbf{x}_\infty$ is a homogeneous vector, this is a one-parameter family of vectors; the parameter is the direction of the infinite point: taking the angle-distance form from equation (2.13), an infinite point $\mathbf{x}_\infty$ lies on a 2D line $\mathbf{l} = \lambda(cos(\varphi), sin(\varphi), -d)$ if and only if the direction angle φ is perpendicular to the directional vector $(u, v)^\mathsf{T}$.
The (unique) 2D line at infinity is of the form $\mathbf{l}_\infty = (0, 0, c)^\mathsf{T}$, and it is obvious that all 2D points $\mathbf{x}_\infty$ at infinity are on the line at infinity.

Table 2.1. Interpretation of the canonical basis $\mathbf{e}_i$ as 2D points and 2D lines.

homogeneous vector	2D point interpretation	2D line interpretation
$\mathbf{e}_1 = (1, 0, 0)^\mathsf{T}$	infinite point in x-direction	y-axis
$\mathbf{e}_2 = (0, 1, 0)^\mathsf{T}$	infinite point in y-direction	x-axis
$\mathbf{e}_3 = (0, 0, 1)^\mathsf{T}$	origin	line at infinity

Canonical Examples As examples for points and lines, we look at the canonical basis vectors $\mathbf{e}_i = (0, \ldots, \underset{i}{1}, \ldots, 0)$ and interpret them as 2D points or lines, see table 2.1. This will be useful when analyzing homogeneous transformations, section 2.3 on page 34 and construction of new entities, section 3.1.1 on page 49.

2.2.2 Points and Planes in 3D

Points in 3D are represented analogously to points in 2D: given a *3D point* $\boldsymbol{X} = (X, Y, Z)^\mathsf{T} \in \mathbb{R}^3$, the homogeneous representation is according to equation (2.3):

$$\mathbf{X} = \begin{pmatrix} X \\ Y \\ Z \\ 1 \end{pmatrix} \cong \begin{pmatrix} TX \\ TY \\ TZ \\ T \end{pmatrix} \cong \begin{pmatrix} U \\ V \\ W \\ T \end{pmatrix}$$

A *plane* in 3D can be represented in several ways, similar to 2D line representations, see table 2.2.1 on page 25. Since 3D planes are hyperplanes for $\mathbb{P}^3$ as 2D lines are hyperplanes for $\mathbb{P}^2$, a plane is represented as a homogeneous vector $\mathbf{A} = (A, B, C, D)^\mathsf{T}$, resembling the homogeneous form $AX + BY + CZ + D = 0$ with a Euclidean point $\boldsymbol{X} = (X, Y, Z)$. Again the homogeneous form can be rewritten with a homogeneous point vector $\mathbf{X}$:

$$AU + BV + CW + D = \mathbf{A}^\mathsf{T}\mathbf{X} = \mathbf{X}^\mathsf{T}\mathbf{A}$$

Since we deal with polyhedral objects, a *plane segment* or a *planar patch* can be defined by its polygonal outline: a planar patch $\mathbf{P} = (\mathbf{A}_P, \mathbf{l}_{P1}, \ldots, \mathbf{l}_{Pn})$ is a planar region enclosed by n 2D lines $\mathbf{l}_{Pi}, i = 0 \ldots n$ which lie on the plane $\mathbf{A}_P$, cf. definition 1 on page 6.

2.2.2.1 Homogeneous and Euclidean Parts We can transfer the definition from section 2.2.1 and define the homogeneous and Euclidean parts of a point $\mathbf{X}$ and a plane $\mathbf{A}$

$$\mathbf{X} = \begin{pmatrix} U \\ V \\ W \\ \hline T \end{pmatrix} = \begin{pmatrix} \boldsymbol{X}_O \\ \hline X_h \end{pmatrix} \quad ; \quad \mathbf{A} = \begin{pmatrix} A \\ B \\ C \\ \hline D \end{pmatrix} = \begin{pmatrix} \boldsymbol{A}_h \\ \hline A_O \end{pmatrix}$$

The Euclidean normalization of $\mathbf{X}$ is then $\mathsf{N}_O(\mathbf{X}) = \mathbf{X}/X_h$ and $\mathsf{N}_O(\mathbf{A}) = \mathbf{A}/|\boldsymbol{A}_h|$. Similar to the 2D case, the ambiguity in the sign of $\mathbf{A}_e = \mathsf{N}_O(\mathbf{A})$ is resolved by testing whether the normalized direction $(\boldsymbol{A}_e)_h$ is on the upper half of the sphere S^3, thus $\langle (0,0,1)^\mathsf{T}, (\boldsymbol{A}_e)_h \rangle > 0$. The Euclidean distance $d(\cdot, \boldsymbol{O})$ to the origin is

$$d(\mathbf{X}, \boldsymbol{O}) = \frac{|\boldsymbol{X}_O|}{|X_h|} = |\mathsf{N}_O(\mathbf{X})| \quad ; \quad d(\mathbf{A}, \boldsymbol{O}) = \frac{|A_O|}{|\boldsymbol{A}_h|} = |\mathsf{N}_O(\mathbf{A})| \qquad (2.16)$$

2.2.2.2 Ideal Points and Plane and Canonical Examples With property 1 on page 28 it is clear that the infinite point $\mathbf{X}_\infty = (U, V, W, 0)^\mathsf{T}$ lies on the infinite plane $\mathbf{A}_\infty = (0, 0, 0, 1)$. The direction of an infinite point $\mathbf{X}_\infty$ is determined by Euclidean part $(\boldsymbol{X}_\infty)_O$.

The interpretation for the canonical vectors $\mathbf{E}_i, i = 1, \ldots, 4$ are listed in table 2.2, it is a direct generalization of table 2.1 on the preceding page.

Table 2.2. Interpretation of the canonical basis $\mathbf{E}_i$ as 3D points and 3D planes.

homogeneous vector	3D point interpretation	3D line interpretation
$\mathbf{E}_1 = (1,0,0,0)^\mathsf{T}$	infinite point in x-direction	yz-plane
$\mathbf{E}_2 = (0,1,0,0)^\mathsf{T}$	infinite point in y-direction	xz-plane
$\mathbf{E}_3 = (0,0,1,0)^\mathsf{T}$	infinite point in z-direction	xy-plane
$\mathbf{E}_4 = (0,0,0,1)^\mathsf{T}$	origin	plane at infinity

2.2.3 Lines in 3D

A *3D line* is different from the previous entities since it is neither a point nor a hyperplane. It is rather a one-dimensional subspace in 3D which can be represented by two (Euclidean) points $\boldsymbol{X}_0$ and $\boldsymbol{Y}_0$, cf. figure 2.4(a). One needs four parameters to represent a line in 3D, as shown in figure 2.4(b): assume two planes have been chosen in advance such as the coordinate planes on the XY respectively XZ axes. Then the two intersection points of the 3D line and the planes can be described by two parameters each.

There exists a number of other representations for 3D lines, but in the following we will derive a homogeneous vector representation based on two given points, see figure 2.4(a): a 3D line refers to a homogeneous 6-vector

$$\mathbf{L}(\boldsymbol{X}_0, \boldsymbol{Y}_0) = \begin{pmatrix} \boldsymbol{Y}_0 - \boldsymbol{X}_0 \\ \hline \boldsymbol{X}_0 \times \boldsymbol{Y}_0 \end{pmatrix} = \begin{pmatrix} \boldsymbol{L}_h \\ \hline \boldsymbol{L}_O \end{pmatrix}$$

which consists of two parts:

- the first part $\boldsymbol{L}_h$ contains the direction $\boldsymbol{L}_h = \boldsymbol{Y}_0 - \boldsymbol{X}_0$ of the line $\mathbf{L}$ and has 2 degrees of freedom. Note that this part is the *homogeneous part* of the 3D line vector as it is independent from the origin.
- the two points $\boldsymbol{X}_0, \boldsymbol{Y}_0$ and the origin $\boldsymbol{O}$ define a plane, whose normal is given by the cross-product $\boldsymbol{L}_O = \boldsymbol{X}_0 \times \boldsymbol{Y}_0$. With the direction of $\boldsymbol{L}_O$ we have fixed the line $\mathbf{L}(\boldsymbol{X}_0, \boldsymbol{Y}_0)$ in a specific plane. Fixing the plane adds only one additional independent parameter to the existing two of the first part $\boldsymbol{L}_h$, since $\boldsymbol{L}_O$ and $\boldsymbol{L}_h$ are constrained to be perpendicular to each other,

  $$\boldsymbol{L}_O^\mathsf{T} \boldsymbol{L}_h = 0 \tag{2.17}$$

  This constraint is called the *Plücker constraint* of a 3D line, see below for more details.

  In order to uniquely fix the 3D line we need the minimal distance $d(\mathbf{L}, \boldsymbol{O})$ of the line to the origin as the fourth parameter of the 3D-line. $d(\mathbf{L}, \boldsymbol{O})$ is equivalent to the altitude h of the triangle $(\boldsymbol{O}, \boldsymbol{X}_0, \boldsymbol{Y}_0)$. Because the norm $|\boldsymbol{L}_O|$ of the cross-product $\boldsymbol{X}_0 \times \boldsymbol{Y}_0$ yields twice the area of the triangle and $|\boldsymbol{L}_h|$ is the length of the base of the triangle, we can write

$$d(\mathbf{L}, \boldsymbol{O}) = \frac{|\boldsymbol{L}_O|}{|\boldsymbol{L}_h|} \tag{2.18}$$

Thus the distance as the fourth parameter the 3D line is implicitly given by $\boldsymbol{L}_O$. The 3-vector $\boldsymbol{L}_O = \boldsymbol{X}_0 \times \boldsymbol{Y}_0$ completely depends on the distance of the origin and thus $\boldsymbol{L}_O$ is the *Euclidean part* of the 6-vector $\mathbf{L} = (\boldsymbol{L}_h, \boldsymbol{L}_O)^\mathsf{T}$. The Euclidean part vanishes, i.e. $\boldsymbol{L}_O = \mathbf{0}$, if and only if the line intersects the origin.

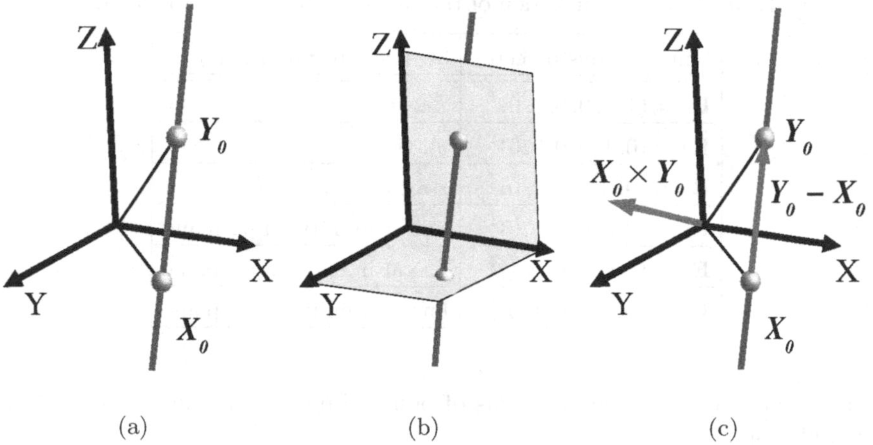

Fig. 2.4. A 3D line may be determined by two points which intersect with two planes of the coordinate system; the two points have 2 degrees of freedom since they are fixed to planes (left). Two Euclidean points $\boldsymbol{X}_0$ and $\boldsymbol{Y}_0$ define a 3D line (middle). In this work we choose two 3-vectors for the representation of $\mathbf{L}$: the direction of the line $\boldsymbol{L}_h = \boldsymbol{X}_0 - \boldsymbol{Y}_0$ and the 3-vector $\boldsymbol{L}_O = \boldsymbol{X}_0 \times \boldsymbol{Y}_0$. We assume $\boldsymbol{L}_O$ to have the length equal to the distance of the 3D line $\mathbf{L}$ to the origin (right).

To summarize the above analysis, a *3D line* is represented by a homogeneous 6-vector $\mathbf{L} = (L_1, L_2, L_3, L_4, L_5, L_6)^\mathsf{T}$, where the homogeneous part $\boldsymbol{L}_h = (L_1, L_2, L_3)^\mathsf{T}$ is constrained to be orthogonal to the Euclidean part $\boldsymbol{L}_O = (L_4, L_5, L_6)^\mathsf{T}$, i.e. $\boldsymbol{L}_h^\mathsf{T}\boldsymbol{L}_O = 0$. Thus the 6-vector $\mathbf{L}$ has 4 degrees of freedom, considering both the orthogonality and homogeneous constraint. As in 2D, a line segment $\mathbf{S}_L$ is a finite fragment of $\mathbf{L}$ and is represented by a triple $(\mathbf{L}, \mathbf{X}_{fs}, L)$, where $\mathbf{X}_{fs}$ is the mid-point of the line segment and L is the length of the segment.

The Euclidean normalization of a 3D line $\mathsf{N}_O(\mathbf{L}) = \mathbf{L}/|\boldsymbol{L}_h|$ uses the formula already known from points and hyperplanes. The ambiguity in the sign of the homogeneous part $(\boldsymbol{L}_e)_h$ of $\mathbf{L}_e = \mathsf{N}_O(\mathbf{L})$ is resolved by deciding whether the normalized direction $(\boldsymbol{L}_e)_h$ is on the upper half of the sphere S^3: $\langle (0,0,1)^\mathsf{T}, (\boldsymbol{L}_e)_h \rangle > 0$.

Some Examples Again, we look at some canonical examples. A line $\mathbf{L} = (\boldsymbol{L}_h, \boldsymbol{L}_O)^\mathsf{T}$ lies at infinity if the homogeneous part vanishes, $\boldsymbol{L}_h = \mathbf{0}$. Thus a line at infinity is defined by the normal direction $\boldsymbol{L}_O$ and has 2 degrees of freedom. Note that a line at infinity can only be computed using equation(2.19).
Taking the canonical 6-vectors $\mathbf{E}_i$, we obtain the coordinate axis and three distinct lines at infinity, namely the lines lying on the canonical planes of the coordinate system, cf. table 2.3.

Table 2.3. Interpretation of the canonical basis $\mathbf{E}_i$ as 3D lines.

homogeneous 6-vector	3D line interpretation
$\mathbf{E}_1 = (1,0,0,0,0,0)^\mathsf{T}$	x-axis
$\mathbf{E}_2 = (0,1,0,0,0,0)^\mathsf{T}$	y-axis
$\mathbf{E}_3 = (0,0,1,0,0,0)^\mathsf{T}$	z-axis
$\mathbf{E}_4 = (0,0,0,1,0,0)^\mathsf{T}$	line at infinity on yz-plane
$\mathbf{E}_6 = (0,0,0,0,0,1)^\mathsf{T}$	line at infinity on xy-plane
$\mathbf{E}_5 = (0,0,0,0,1,0)^\mathsf{T}$	line at infinity on xz-plane

We summarize the representations of points, lines and planes in 2D and 3D in table 2.4

Table 2.4. Homogeneous representation of points, lines and planes in 2D and 3D. A line in 3D is represented with Plücker coordinates, for which the Plücker constraint $\boldsymbol{L}^\mathsf{T}\boldsymbol{L}_O = 0$ holds.

	2D	3D
point	$\mathbf{x}^\mathsf{T} = (u, v; w) = (\boldsymbol{x}_O^\mathsf{T}, x_h)$	$\mathbf{X}^\mathsf{T} = (U, V, W; T) = (\boldsymbol{X}_O^\mathsf{T}, X_h)$
line	$\mathbf{l}^\mathsf{T} = (a, b; c) = (\boldsymbol{l}_h^\mathsf{T}, l_O)$	$\mathbf{L}^\mathsf{T} = (L_1, L_2, L_3; L_4, L_5, L_6) = (\boldsymbol{L}_h^\mathsf{T}, \boldsymbol{L}_O^\mathsf{T})$
plane	-	$\mathbf{A}^\mathsf{T} = (A, B, C; D) = (\boldsymbol{A}_h^\mathsf{T}, A_O)$

2.2.4 Plücker Coordinates

The previous representation of the 3D line is compatible with the representation using the so-called Plücker coordinates. The geometrician Julius Plücker[2] introduced a representation of k-dimensional subspaces within an

[2] Julius Plücker, born 1801 in Elberfeld, Germany, died 22 May 1868 in Bonn, Germany. Professor of mathematics at Halle in 1834, at Bonn since 1836.

n-dimensional projective space based on $\binom{n+1}{k+1}$ minors of a $(n+1) \times (k+1)$ matrix with given coordinates, cf. FAUGERAS AND LUONG 2001.

An example for Plücker coordinates was already introduced in section 2.1.2, where three points $\mathbf{X}, \mathbf{Y}, \mathbf{Z}$ form a 2-dimensional subspace, a hyperplane. From the 4×3 matrix $\mathsf{M}_A = (\mathbf{X}\,\mathbf{Y}\,\mathbf{Z})$ one obtains 4 minors or 4 Plücker coordinates.

Assume now a 3D line $\mathbf{L}$ which was formed by two points $\mathbf{X}, \mathbf{Y}$. It is a one-dimensional subspace with $\mathbb{P}^3$ and there are $\binom{n+1}{k+1} = 6$ coordinates, which are derived from 2×2 minors l_{ij} of a 4×2 matrix $\mathsf{M}_L = (\mathbf{X}, \mathbf{Y})$. This matrix consists of two homogeneous vectors of 3D points $\mathbf{X}, \mathbf{Y}$ - we can also use 3D planes, as we will see in section 2.5 on page 40. For M_L we obtain

$$\mathsf{M}_L = \begin{pmatrix} X_1 & Y_1 \\ X_2 & Y_2 \\ X_3 & Y_3 \\ X_4 & Y_4 \end{pmatrix}; \quad \begin{array}{lll} l_{12} = \begin{vmatrix} X_1 & Y_1 \\ X_2 & Y_2 \end{vmatrix} & l_{23} = \begin{vmatrix} X_2 & Y_2 \\ X_3 & Y_3 \end{vmatrix} & l_{34} = \begin{vmatrix} X_3 & Y_3 \\ X_4 & Y_4 \end{vmatrix} \\ l_{14} = \begin{vmatrix} X_1 & Y_1 \\ X_4 & Y_4 \end{vmatrix} & l_{24} = \begin{vmatrix} X_2 & Y_2 \\ X_4 & Y_4 \end{vmatrix} & l_{13} = \begin{vmatrix} X_1 & Y_1 \\ X_3 & Y_3 \end{vmatrix} \end{array} \tag{2.19}$$

If we assume the homogeneous part of $\mathbf{X}, \mathbf{Y}$ to be fixed as $X_4 = Y_4 = 1$, it is easy to see that the Plücker coordinates l_{ij} directly relate to our 6-vector $\mathbf{L} = (\boldsymbol{L}_h, \boldsymbol{L}_O)^\mathsf{T}$: $\boldsymbol{L}_h = -(l_{14}, l_{24}, l_{34})^\mathsf{T} = (l_{41}, l_{42}, l_{43})^\mathsf{T}$ and $\boldsymbol{L}_O = (l_{23}, -l_{13}, l_{12})^\mathsf{T} = (l_{23}, l_{31}, l_{12})^\mathsf{T}$. Scaling $\mathbf{X}, \mathbf{Y}$ with homogeneous factor λ resp. μ scales the Plücker coordinates by $\lambda \cdot \mu$.

Note that the Plücker coordinates l_{ij} have been expressed according to a particular order of the index pair i, j. We are free though to choose other index pairs, that means any other combination of 2×2 minors. For our choice of $\mathbf{L} = (\boldsymbol{L}_h, \boldsymbol{L}_O)$ we have $(i, j) \in \{(4,1), (4,2), (4,3), (2,3), (3,1), (1,2)\}$. This has only the effect on the ordering and the signs of the Plücker coordinates within a vector, but makes it possible to distinguish clearly between a homogeneous and a Euclidean part. Note that other authors choose different Plücker coordinates for lines, cf. HARTLEY AND ZISSERMAN 2000, STOLFI 1991.

Plücker Constraint In general, the Plücker coordinates are not independent from each other, in fact there is a set of quadratic relations between them. For the cases in 2D and 3D space, a quadratic relation of Plücker coordinates only occurs for line in 3D, we obtain only one condition:

$$l_{41}l_{23} + l_{42}l_{31} + l_{43}l_{12} = 0 \tag{2.20}$$

All 6-vectors $\mathbf{L} = (l_{41}, l_{42}, l_{43}, l_{23}, l_{31}, l_{12})$ satisfying equation (2.20) correspond to valid 3D lines. As mentioned above the Plücker constraint (or Plücker condition) is obviously satisfied with the proposed construction as $\boldsymbol{L}_h^\mathsf{T} \boldsymbol{L}_O = 0$. Note that the constraint (2.20) defines a manifold in $\mathbb{R}^6$, containing all valid 3D-lines. If the vector $\mathbf{L}$ is known to be spherically normalized, the constraint defines a manifold on the sphere S^5.

Thus the vector notation for lines appears to have a disadvantage compared to the vector representation for points and planes: in addition to the homogeneous factor it has another degree of freedom which is constraint by the Plücker condition. This situation can be compared to the fundamental matrix, see section 2.3.3 on page 36, which is a homogeneous 3×3 matrix with 7 degrees of freedom. In chapter 3 and 4 we will see that it is easy to cope with the over-parameterization in the context of geometric reasoning.

2.3 Basic Geometric Transformations

After introducing the basic geometric entities, we now review possible geometric transformations on them. One property of these transformations is that they at least preserve straight lines and incidence relations. An in-depth treatment of such geometric transformations on projective entities can be found in HARTLEY AND ZISSERMAN 2000.

Transformations may be applied to points, lines or planes, though usually one refers to point transformations. Reformulating point transformations to transformations of 2D lines and 3D planes is mentioned in section 2.5.2 on page 44.

In the following, we concentrate on the following types of transformations:

- Linear mappings within image space resp. object space, called homographies in 2D and 3D with affine transformations and similarity transformations as special cases
- Linear mapping from object space to image space, called projective camera transformations with affine camera transformations as a special case
- Linear mapping between two different images, called fundamental matrix.

A more detailed analysis and the application of the transformations are introduced in section 3.1.2 on page 56.

Matrix and Vector Representation The transformations will generally be represented by a $n \times m$ matrix M, such that $\boldsymbol{x}' = \mathit{M}\boldsymbol{x}$. If a matrix is a transformation of homogeneous vectors, then it is a homogeneous transformation $\mathsf{M} \cong \lambda\mathsf{M}$ as it is defined up to scale, denoted be upright sans-serif letters.

Alternatively, we will sometimes represent the matrix as a vector $\boldsymbol{m} = \text{vec}(\mathit{M}^{\mathsf{T}}))$ where the transposed rows of M are stacked up in a mn-vector, see definition B.2 on page 184.

2.3.1 Homography

Homographies are projective transformations, that linearly map points within a projective space $\mathbb{P}^n$. Thus it can be written as a simple matrix-vector multiplication

$$\mathsf{x}' = \mathsf{H}\mathbf{x}$$

with $\mathbf{x}', \mathbf{x} \in \mathbb{R}^{n+1}$ and the $(n+1) \times (n+1)$ regular matrix H.

An affine transformation H_a is a special case of a homography and combines translation T, rotation R, the scaling of both axes M and skew W of the original space $\mathbb{P}^2$ resp. $\mathbb{P}^3$. For 2D, we obtain

$$\begin{aligned}\mathsf{H}_a &= \mathsf{W}\,\mathsf{M}\,\mathsf{R}\,\mathsf{T} \\ &= \begin{pmatrix} 1 & s/2 & 0 \\ s/2 & 1 & 0 \\ 0 & 0 & 1 \end{pmatrix} \begin{pmatrix} m_1 & 0 & 0 \\ 0 & m_2 & 0 \\ 0 & 0 & 1 \end{pmatrix} \begin{pmatrix} \cos(\alpha) & \sin(\alpha) & 0 \\ -\sin(\alpha) & \cos(\alpha) & 0 \\ 0 & 0 & 1 \end{pmatrix} \begin{pmatrix} 1 & 0 & -t_x \\ 0 & 1 & -t_y \\ 0 & 0 & 1 \end{pmatrix} \\ &= \begin{pmatrix} h_{11} & h_{12} & h_{13} \\ h_{21} & h_{22} & h_{23} \\ 0 & 0 & 1 \end{pmatrix} \end{aligned} \tag{2.21}$$

As an affine transformation in 2D has six degrees of freedom, all six entries in H_a are not constrained. We may constrain the upper left 2×2 matrix $H'_a = \mathsf{H}_a^{(1,2),(1,2)}$ to obtain a similarity transformation by $H'^{\mathsf{T}}_a H'_a = m^2 I$. This leads to two independent quadratic constraints in 2D. In a similar way, we obtain five independent quadratic constraints in 3D.

In general, the entries of a homography H can be interpreted using the canonical examples from section 2.2.1.3 on page 28: the multiplication $\mathsf{H}\,\mathbf{e}_i$ with the canonical vectors $\mathbf{e}_i$ yields the i-th column of the homography H. Thus for example the first column $(h_{11}, h_{21}, h_{31})^{\mathsf{T}}$ is the image of the infinite point in the direction of the x-axis. It is obvious that for an affine transformations H_a infinite points remain at infinity.

The first row $(h_{12}, h_{12}, h_{13})^{\mathsf{T}}$ of a homography H can be interpreted using lines: as $\mathsf{H}^{\mathsf{T}}\mathbf{l}' = \mathbf{l}$, cf. equation (2.11), the first row is the line which is mapped to $\mathbf{e}_1$. Thus it is easy to see that for affine transformations, the line at infinity remains at infinity.

2.3.2 Projective Camera

The most common model for cameras is the pinhole camera model: a point $\mathbf{X}$ in 3D space is projected into an image by computing a viewing ray $\mathbf{L}'(\mathbf{x}')$ from the unique projection center $\mathbf{X}_0$ to $\mathbf{X}$ and intersecting this viewing ray with a unique image plane, see figure 2.5. The image formation process described above can also be modeled as a general linear transformation of $\mathbb{P}^3$ to $\mathbb{P}^2$, which again can be written as a matrix-vector multiplication $\mathbf{x}' = \mathsf{P}\mathbf{X}$, where P is a 3×4 matrix.

This time we interpret the *rows* $\mathbf{A}^{\mathsf{T}}$, $\mathbf{B}^{\mathsf{T}}$ and $\mathbf{C}^{\mathsf{T}}$ of the matrix $\mathsf{P} = \begin{pmatrix} \mathbf{A}^{\mathsf{T}} \\ \mathbf{B}^{\mathsf{T}} \\ \mathbf{C}^{\mathsf{T}} \end{pmatrix}$ as planes $\mathbf{A}, \mathbf{B}, \mathbf{C}$ and the multiplication $\mathbf{x}' = \mathsf{P}\,\mathbf{X}$ contains three equations

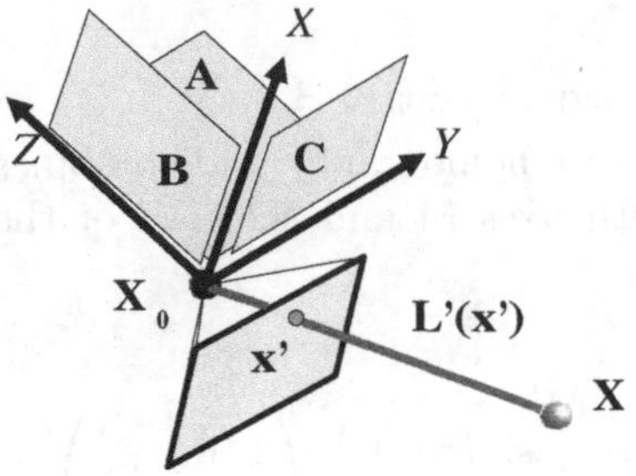

Fig. 2.5. Pinhole camera model and the three camera planes **A**, **B** and **C**.

of the form (2.10) on page 24. If $\mathbf{x}' = (u, v, w)^\mathsf{T}$, then $u = 0$ if and only if $\mathbf{X} \in \mathbf{A}$. Similarly, $v = 0 \Leftrightarrow \mathbf{X} \in \mathbf{B}$ and $w = 0 \Leftrightarrow \mathbf{X} \in \mathbf{C}$, thus a point $\mathbf{x}'$ lies at infinity if $\mathbf{X} \in \mathbf{C}$. The three planes intersect in the projection center $\boldsymbol{X}_0$, as the image point is undefined if $\mathbf{X} = (\boldsymbol{X}_0, 1)^\mathsf{T}$, already observed by DAS 1949.

Given the exterior orientation of a camera with rotation R and projection center $\boldsymbol{X}_0$ and the inner orientation with the calibration matrix K, a Euclidean interpretation of the projection matrix can be formulated as follows:

$$\mathsf{P} = \mathsf{K}R[I| - \boldsymbol{X}_0] \quad \text{with } \mathsf{K} = \begin{pmatrix} c & cs & x_H \\ 0 & c(1+m) & y_h \\ 0 & 0 & 1 \end{pmatrix} \tag{2.22}$$

with skew s, focal length c, principle point $(x_H, y_H)^\mathsf{T}$ and scale difference m between the first and second coordinate axis in the image.

Finally, as in the last section we can fix the last row of the transformation matrix with $\mathbf{C}^\mathsf{T} = (0, 0, 0, 1)^\mathsf{T}$ to obtain a special subclass of projective cameras: the plane $\mathbf{C}$ then lies at infinity and therefore the projection center $\boldsymbol{X}_0$, too. This means that all viewing rays are parallel to each other and we obtain the affine camera

$$\mathsf{P}^a = \begin{pmatrix} p_{11} & p_{12} & p_{13} & p_{14} \\ p_{21} & p_{22} & p_{23} & p_{24} \\ 0 & 0 & 0 & 1 \end{pmatrix}$$

A more detailed description of various camera models can be found in HARTLEY AND ZISSERMAN 2000 or in FÖRSTNER 2001B.

2.3.3 Fundamental Matrix

Assume there are two images given, taken with different cameras P_1 and P_2. Suppose now a point $\mathbf{X}$ in object space is projected in the first and second image, $\mathbf{x}' = \mathsf{P}_1\mathbf{X}$ and $\mathbf{x}'' = \mathsf{P}_2\mathbf{X}$, cf. figure 2.6. It is clear that the projection center $\boldsymbol{X}_{01}$, $\boldsymbol{X}_{02}$ and the point $\mathbf{X}$ define a plane, the so-called epipolar plane.

Now suppose we take two arbitrarily chosen image points $\mathbf{x}'$ and $\mathbf{x}''$ from the first resp. second image. If we want to test whether they are caused by the same object point, one has to test whether the two viewing rays $\mathbf{L}'$ and $\mathbf{L}''$ are coplanar.

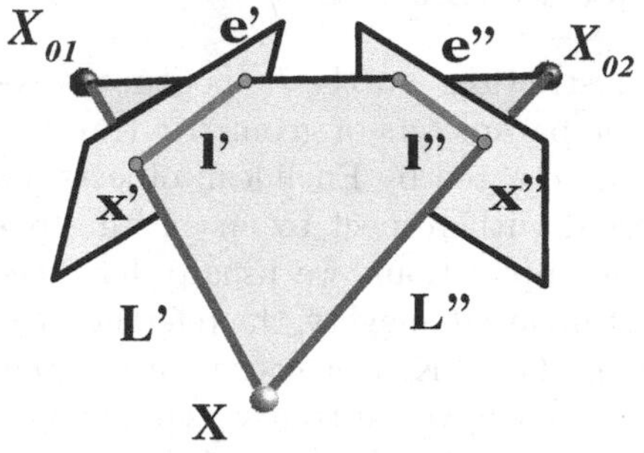

Fig. 2.6. Epipolar geometry: two projection centers $\boldsymbol{X}_{01}$ and $\boldsymbol{X}_{02}$ are connected by an epipolar line, which intersects the image planes at $\mathbf{e}'$ and $\mathbf{e}''$. The two viewing rays $\mathbf{L}'$ and $\mathbf{L}''$ point to the object point $\mathbf{X}$.

But one can also apply this test in 2D by projecting the viewing-ray $\mathbf{L}'$ into the second image, thus obtaining an image line $\mathbf{l}''$, the so-called epipolar line of $\mathbf{x}'$ which in turn has to be incident to the second image point $\mathbf{x}''$. This 2D relationship between the first point $\mathbf{x}'$ and the second point $\mathbf{x}''$ can be expressed as a coplanarity condition with the 3×3 so-called fundamental matrix F:

$$\mathbf{x}'^{\mathsf{T}} \mathsf{F} \mathbf{x}'' = 0 \tag{2.23}$$

Thus the fundamental matrix relates one image to another image. There are many ways to derive an algebraic expression for the fundamental matrix, for a review see HARTLEY AND ZISSERMAN 2000 or FAUGERAS AND LUONG 2001[3]. In section 3.1.3 we will give an explicit formula for computing F given two cameras $\mathsf{P}_1, \mathsf{P}_2$.

The vector defined by $\mathsf{F}\mathbf{x}''$ is actually the homogeneous vector for the epipolar line $\mathbf{l}''$ and by transposing equation 2.23, one also obtains the epipolar line $\mathbf{l}'$, in summary:

$$\mathbf{l}' = \mathbf{l}'(\mathbf{x}'') = \mathsf{F}\mathbf{x}'' \qquad \mathbf{l}'' = \mathbf{l}''(\mathbf{x}') = \mathsf{F}^{\mathsf{T}}\mathbf{x}'$$

Note that the projection of camera center $\boldsymbol{X}_{01}$ into the second image results in a point $\mathbf{e}'$, which is called epipole; likewise we can obtain the second epipole $\mathbf{e}''$. For epipoles, there is no epipolar line defined, thus e.g. $\mathsf{F}\mathbf{e}'' = \mathbf{0}$ and $\mathbf{e}''$ being the nullvector of F.

The fundamental matrix has 7 degrees of freedom: this is due to the fact that the 3×3 matrix is homogeneous and singular, i.e. $\det(\mathsf{F}) = 0$. Thus there are two constraints for the 9 matrix elements.

[3] Note that they use the transpose F^{T} as a definition for the fundamental matrix.

If the internal camera parameters K_1 and K_2 are known, only 5 parameters are unknown and the matrix is called *essential matrix*, LONGUET-HIGGINS 1981 and has 5 degrees of freedom.

2.4 Conditioning of Homogeneous Entities

Up to now we have not imposed any restrictions on the homogeneous coordinates of geometric entities nor on the parameters of geometric transformations. As the homogeneous entities are created by Euclidean objects, cf. figure 2.2, their coordinates are expressed with respect to an existing reference coordinate system. In close-range applications, we usually have the reference system nearby the entities. But in aerial imagery, the reference system is usually a geodetic system, like the Gauß-Krüger coordinate system or its international variant, the UTM system (universal-transversal-mercator system), cf. WITTE AND SCHMIDT 1995. For example, the tip of the dome of Aachen has the coordinates $x = 2505940.53[m], y = 5626590.37[m]$. Embedding this point in homogeneous coordinates as in equation (2.3) would yield a difference of seven orders of magnitudes between the Euclidean part $\boldsymbol{X}_O$ and the homogeneous part X_h. When considering subsequent computations with these coordinates, such a coordinate representation is undesirable in terms of numerical stability.

A solution is to define a local coordinate system that is close to the set of given object points, which is advisable to do prior to all methods that are discussed in this work. But if the distances between different points are large with respect to the units, an additional scaling of the coordinates is needed. For homogeneous vectors of geometric entities this means to either choose a larger X_h instead of 1 or equivalently to scale down the Euclidean coordinates by some factor. Both solutions can be achieved by scaling one part appropriately without changing the other part. We call this scaling operation *conditioning* of a geometric entity.

One may choose the second option for conditioning, i.e. scaling the Euclidean part of a geometric entity without scaling its homogeneous part. This has the effect, that the geometric entities of the vectors move closer to the Euclidean origin, cf. section 2.2. The conditioning operation can be represented by a matrix multiplication $\mathbf{x}' = \mathsf{W}(f)\mathbf{x}$, where f is the conditioning factor and W is the identity matrix except of the diagonal elements, which correspond to the Homogeneous part of the vector. The following matrices for 2D and 3D entities are obtained:

$$\begin{gathered}\mathsf{W}_x(f) = \begin{pmatrix} f I_2 & \mathbf{0} \\ \mathbf{0}^\mathsf{T} & 1 \end{pmatrix} ; \mathsf{W}_l(f) = \begin{pmatrix} I_2 & \mathbf{0} \\ \mathbf{0}^\mathsf{T} & f \end{pmatrix} ; \\ \mathsf{W}_X(f) = \begin{pmatrix} f I_3 & \mathbf{0} \\ \mathbf{0}^\mathsf{T} & 1 \end{pmatrix} ; \mathsf{W}_L(f) = \begin{pmatrix} I & O_3 \\ O_3 & f I_3 \end{pmatrix} ; \mathsf{W}_A(f) = \begin{pmatrix} f I_3 & \mathbf{0} \\ \mathbf{0}^\mathsf{T} & 1 \end{pmatrix}\end{gathered} \tag{2.24}$$

For scaling down the Euclidean part, one has to Choose $f < 1$. To go back to the original coordinates, one simply applies the inverse of the conditioning factor, $\mathbf{x} = \mathsf{W}(\frac{1}{f})$. Given a set of geometric entities, one has to find the entity

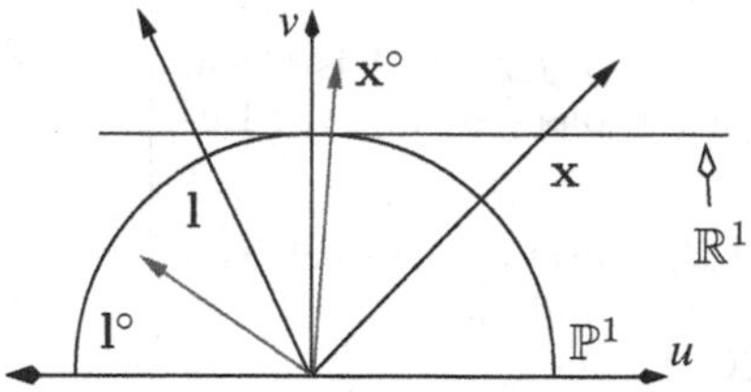

Fig. 2.7. Effect of conditioning of homogeneous vectors: a point $\mathbf{x}$ moves closer to the u-axis, a line $\mathbf{l}$ moves closer to the v-axis, but the angle between the two new vectors remains invariant.

$\mathbf{x}$ that is furthest away from the origin, and choose a conditioning factor f_{min}, such that $\mathbf{x}' = \mathsf{W}(f_{min})\mathbf{x}$ is sufficiently close to the origin. Then apply the conditioning factor to all other entities.

For the projective axis, the effect of conditioning to the homogeneous vectors is depicted in figure 2.7: for a point $\mathbf{x} = (u, v)^\mathsf{T}$, the first coordinate is scaled by a factor f, the second coordinate remains invariant. The vector is now closer to the v-axis, a Euclidean normalization onto $\mathbb{R}^1$ would yield a scalar closer to zero. For a hyperplane $\mathbf{l} = (a, b)^\mathsf{T}$, the second coordinate is assumed to be the Euclidean part and thus the first coordinate is scaled by f, yielding a vector closer to the u-axis. Note that the dot-product between $\mathbf{x}$ and $\mathbf{l}$ remains invariant when scaling with the same conditioning factor.

For a transformation, we generally have two different sets of entities $\mathbf{x}_i$ and $\mathbf{x}'_i$, for which two different conditioning factors f and f' have to be computed. A conditioned transformation matrix H° can be obtained by

$$\mathsf{H}^\circ = \mathsf{W}(f')\mathsf{H}\mathsf{W}^{-1}(f) \tag{2.25}$$

The reconditioning of the estimated transformation H° is then given by

$$\mathsf{H} = \mathsf{W}^{-1}(f')\mathsf{H}^\circ\mathsf{W}(f) \tag{2.26}$$

The conditioning operation is not only useful to obtain small coordinates: as we are going to see in the next two chapters, conditioning becomes important when

- computing optimal estimations from a set of observations, cf. section 3.4
- reducing the bias of the statistical representation of uncertain geometric entities, cf. section 4.3.

Algorithm 2.1 shows a method for conditioning of a set of geometric entities. The algorithm is used in section 3.4 on page 80 when constructing an unknown entity by a set of observations and will turn out especially useful

Conditioning a set of geometric entities

Objective: Transform a set of lines, points and planes in 3D such that they move closer to the Euclidean origin.

Given are $n = n_X + n_L + n_A$ entities with n_X 3D points $\mathbf{X}_k$, n_L lines $\mathbf{L}_l$ and n_A planes $\mathbf{A}_m$.

Additionally, the required conditioning factor $f_{min} < 1$ is given, i.e. the minimal ratio between the homogeneous part and the Euclidean part of a geometric entity.

1. find the maximum ratio max_{hO} of the norm of the Euclidean and homogeneous parts for points, lines and planes
   $max_{hO} := \max_{i,j,k} \frac{|\mathbf{v}_h|}{|\mathbf{v}_O|}$ for $\mathbf{v} \in \{\mathbf{X}_i, \mathbf{L}_j, \mathbf{A}_k\}$
2. determine whether it is necessary to scale the Euclidean parts.
   if $\max_{hO} < f_{min}$ then
   $f := f_{min} \cdot max_{hO}$
   if $f < \epsilon_M$ then $f := f_{min}/max_O$
   with $max_O = \max_{i,j,k} |\mathbf{v}_O|$ for $\mathbf{v} \in \{\mathbf{X}_i, \mathbf{L}_j, \mathbf{A}_k\}$
   $\mathbf{v}^\circ := \mathsf{W}(f)\mathbf{v}$ for $\mathbf{v} \in \{\mathbf{X}_i, \mathbf{L}_j, \mathbf{A}_k\}$

Algorithm 2.1: Algorithm for conditioning a set of geometric entities. The conditioning changes the ratio between the norms of the Euclidean and homogeneous part of a vector, which should be smaller than f_{min}. The machine accuracy ϵ_M is the smallest number such that $1.0 + \epsilon_M > 1.0$. The conditioned entity is denoted by $\mathbf{v}^\circ$, for a definition of the matrices $\mathsf{W}(f)$, see equation (2.24).

when discussing uncertainty representations of the homogeneous vectors, see section 4.3.3 on page 117.

Note that because we have subdivided all homogeneous vectors of geometric entities into a homogeneous and a Euclidean part, the conditioning of geometric entities is very simple and works consistently for homogeneous vectors for points, lines and planes within a coordinate frame.

2.5 Duality Principle

The duality principle is an important feature of projective geometry as all propositions occur twice: taking the dual of a true proposition yields a new proposition, possibly with a different meaning. As an example for duality, consider the property from section 2.1.2: three 2D points $\mathbf{x}_1, \mathbf{x}_2, \mathbf{x}_3 \in \mathbb{P}^2$ are collinear iff $|\mathbf{x}_1\ \mathbf{x}_2\ \mathbf{x}_3| = 0$. It is possible to replace the points with lines, which are elements of the dual space $(\mathbb{P}^2)^*$. Then we obtain the dual property: three 2D lines $\mathbf{l}_1, \mathbf{l}_2, \mathbf{l}_3$ intersect in one point iff $|\mathbf{l}_1\ \mathbf{l}_2\ \mathbf{l}_3| = 0$, see figure 2.8.

In the following we will discuss dual entities and their transformations. They become important for the geometric constructions described in chapter 3, where we introduce the duality between the join and intersection of entities.

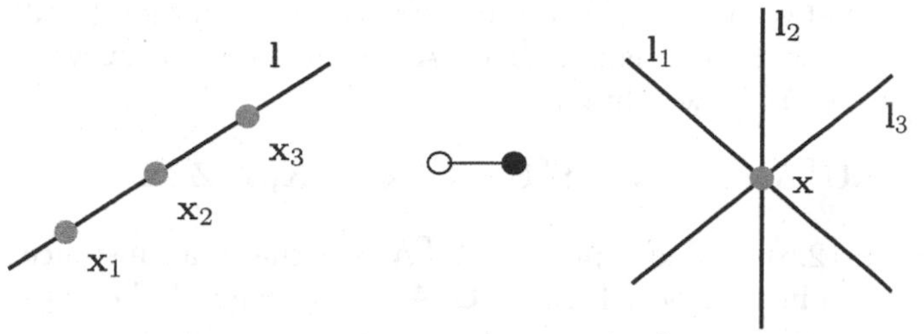

Fig. 2.8. Checking whether three points are collinear *(left)* is dual to checking whether three lines intersect in one point *(right)*, see text for details. The sign o—• is used here to express the duality between the two geometric situations.

2.5.1 Dual Entities

Dual Points and Hyperplanes As demonstrated by the last example and mentioned in section 2.1.2, points and hyperplanes in 2D and 3D are called dual to each other. As they use the same kind of homogeneous vectors, one can interpret a homogeneous point vector $\mathbf{x}$ as a hyperplane: we denote this change of interpretation by $\overline{\mathbf{x}}$ and call $\overline{\mathbf{x}}$ the dual of $\mathbf{x}$. The vector $\overline{\mathbf{x}}$ has the same coordinates, but now represents a hyperplane instead of a point.

For example, given a 2D point $\mathbf{x} = (x, y, 1)^\mathsf{T}$ its dual $\overline{\mathbf{x}}$ is a 2D line. Setting $\mathbf{l} = \overline{\mathbf{x}}$, we obtain the normalized line $\mathsf{N}_O(\mathbf{l}) = \frac{1}{x^2+y^2}(x, y, 1)^\mathsf{T}$. Thus the distance of the line $\mathbf{l} = \overline{\mathbf{x}}$ is the inverse distance of the point $\mathbf{x}$ to the origin. Additionally, the point and its dual line lie on opposite sides of the origin, see figure 2.9(left). The same reasoning applies to 3D points $\mathbf{X}$, where its dual is a plane $\overline{\mathbf{X}}$, see figure 2.9(middle).

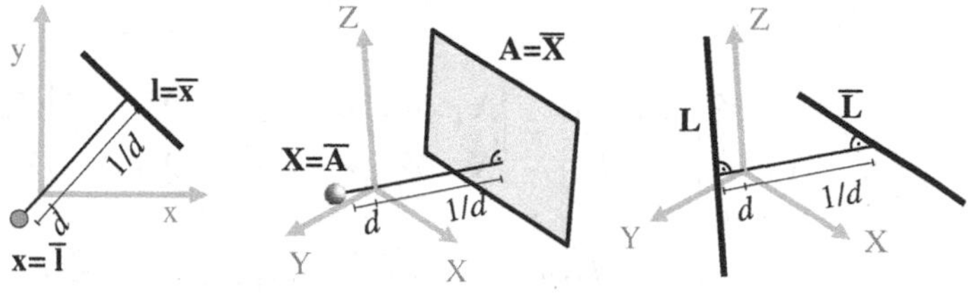

Fig. 2.9. Visualization of dual entities. *Left*: the dual $\overline{\mathbf{x}}$ of a 2D point $\mathbf{x}$ is a 2D line: *middle*: the dual $\overline{\mathbf{X}}$ of a 3D point $\mathbf{X}$ is a plane; *right*: the dual $\overline{\mathbf{L}}$ of a 3D line $\mathbf{L}$ is again a 3D line. the distance d of the dual entities is inverse to the distance of the original entities.

Dual Lines One question has not been answered yet: what is the dual of a 3D line $\mathbf{L}$? To answer this question, we consider again the relationship that was derived between a point and a hyperplane, see equation (2.10). In 3D, a point $\mathbf{U}$ is incident to a plane $\mathbf{A}$ if $\mathbf{U}^\mathsf{T}\mathbf{A} = 0$. If the plane $\mathbf{A}$ was derived by three points $\mathbf{X}, \mathbf{Y}, \mathbf{Z}$, we obtain

$$\mathbf{U}^\mathsf{T}\mathbf{A} = 0 \quad \Leftrightarrow \quad \mathbf{A}^\mathsf{T}\mathbf{U} = 0 \quad \Leftrightarrow \quad |\mathbf{X}, \mathbf{Y}, \mathbf{Z}\ ;\ \mathbf{U}| = 0 \tag{2.27}$$

see equation (2.6). The dot-product $\mathbf{U}^\mathsf{T}\mathbf{A}$ is actually an instance of the so-called cap-product, denoted by $< \mathbf{U}, \mathbf{A} >$: in general the cap-product is defined as a product of elements of two dual projective subspaces onto $\mathbb{R}$. More precisely, the cap-product is defined for the two corresponding *vector* spaces where the sum of the dimensions of the underlying *vector* subspaces is equal to the dimension of the vector space. For example, a 3D point with projective dimension 0 has a vector dimension of 1 as the projective point in $\mathbb{P}^3$ is represented by a line in $\mathbb{R}^4$, cf. figure 2.1(d) for the 2D case. The dual of a 3D point is a 3D plane with projective dimension 2 resp. vector dimension 3. Summing up the vector dimensions yields $1 + 3 = 4$ the dimension of $\mathbb{R}^4$. To derive a cap-product involving 3D lines, we can re-interpret the determinant $|\mathbf{X}, \mathbf{Y}, \mathbf{Z}\ ;\ \mathbf{U}|$ by defining two lines $\mathbf{L}(\mathbf{X}, \mathbf{Y})$ and $\mathbf{M}(\mathbf{Z}, \mathbf{U})$:

$$|\underbrace{\mathbf{X}\,\mathbf{Y}}_{\mathbf{L}}\ ;\ \underbrace{\mathbf{Z}\,\mathbf{U}}_{\mathbf{M}}| = 0 \tag{2.28}$$

This equation holds if the four points are coplanar or equivalently, if the lines $\mathbf{L}$ and $\mathbf{M}$ intersect in one point. We can expand the determinant in (2.28) to hold the Plücker coordinates of the lines $\mathbf{L}$, $\mathbf{M}$:

$$\begin{vmatrix} X_1 & Y_1 & Z_1 & U_1 \\ X_2 & Y_2 & Z_2 & U_2 \\ X_3 & Y_3 & Z_3 & U_3 \\ X_4 & Y_4 & Z_4 & U_4 \end{vmatrix} = \begin{vmatrix} X_1 & Y_1 \\ X_4 & Y_4 \end{vmatrix}\begin{vmatrix} Z_2 & U_2 \\ Z_3 & U_3 \end{vmatrix} - \begin{vmatrix} X_2 & Y_2 \\ X_4 & Y_4 \end{vmatrix}\begin{vmatrix} Z_1 & U_1 \\ Z_3 & U_3 \end{vmatrix} + \begin{vmatrix} X_3 & Y_3 \\ X_4 & Y_4 \end{vmatrix}\begin{vmatrix} Z_1 & U_1 \\ Z_2 & U_2 \end{vmatrix}$$

$$+ \begin{vmatrix} X_2 & Y_2 \\ X_3 & Y_3 \end{vmatrix}\begin{vmatrix} Z_1 & U_1 \\ Z_4 & U_4 \end{vmatrix} - \begin{vmatrix} X_1 & Y_1 \\ X_3 & Y_3 \end{vmatrix}\begin{vmatrix} Z_2 & U_2 \\ Z_4 & U_4 \end{vmatrix} + \begin{vmatrix} X_1 & Y_1 \\ X_2 & Y_2 \end{vmatrix}\begin{vmatrix} Z_3 & U_3 \\ Z_4 & U_4 \end{vmatrix} \tag{2.29}$$

$$= +l_{14}m_{23} - l_{24}m_{13} + l_{34}m_{12} + l_{23}m_{14} - l_{13}m_{24} + l_{12}m_{34}$$
$$= -l_{41}m_{23} - l_{42}m_{31} - l_{43}m_{12} - l_{23}m_{41} - l_{31}m_{42} - l_{12}m_{43}$$

The 2×2 minors are the Plücker coordinates for the lines $\mathbf{L}$ and $\mathbf{M}$, see also section 2.2.4 on page 32. One can check that equation (2.28) is equivalent to

$$-\boldsymbol{L}_h^\mathsf{T}\boldsymbol{M}_O - \boldsymbol{L}_O^\mathsf{T}\boldsymbol{M}_h = 0 \tag{2.30}$$

Thus we have derived a condition for the incidence of two lines $\mathbf{L}$ and $\mathbf{M}$ from the incidence of a plane $\mathbf{A}$ and a point $\mathbf{U}$.

Equation (2.30) defines the cap-product $< \mathbf{L}, \mathbf{M} >$ of two lines in a similar way as $< \mathbf{A}, \mathbf{U} >$ with equation (2.10); the dual of a 3D line is again a 3D line. Interpreting the cap-product by a dot-product of dual entities, we obtain

$$< \mathbf{L}, \mathbf{M} > := -\boldsymbol{L}_h^{\mathsf{T}} \boldsymbol{M}_O - \boldsymbol{L}_O^{\mathsf{T}} \boldsymbol{M}_h = -\mathbf{L}^{\mathsf{T}} \overline{\mathbf{M}} \quad \text{with } \overline{\mathbf{M}} := \begin{pmatrix} \boldsymbol{M}_O \\ \boldsymbol{M}_h \end{pmatrix}$$

where $\overline{\mathbf{M}}$ is the dual of the line $\mathbf{M}$ and is obtained by switching the homogeneous and Euclidean part of the vector $\mathbf{M}$. The visualization of a line and its dual is shown in figure 2.9(left). Note that again the point and its dual line are on opposite sides of the origin and the dual has an inverse distance to the origin with respect to the original entity.

To summarize, the possible cap-products in 2D and 3D are:

entity 1	dim. of vector subspace	dual entity 2	dim. of vector subspace	dim. of vector space	cap-product
point $\mathbf{x}$	1	line $\mathbf{l}$	2	1+2=3	$< \mathbf{x}, \mathbf{l} > \;:= \mathbf{x}^{\mathsf{T}} \mathbf{l}$
point $\mathbf{X}$	1	plane $\mathbf{A}$	3	1+3=4	$< \mathbf{X}, \mathbf{A} > \;:= \mathbf{X}^{\mathsf{T}} \mathbf{A}$
line $\mathbf{L}$	2	line $\mathbf{M}$	2	2+2=4	$< \mathbf{L}, \mathbf{M} > \;:= \mathbf{L}^{\mathsf{T}} \overline{\mathbf{M}}$

Usually the cap-product is defined in the context of the so-called double or Grassmann-Cayley algebra, where operations such as join and intersection form an algebra on projective subspaces. In the next chapter we introduce these operations on the geometric entities defined. Again, for a rigorous treatment of this subject, see FAUGERAS AND LUONG 2001.

An alternative proof that two lines are incident if equation (2.30) does not explicitly use determinants but only analytical geometry and follows BRAND 1966, p. 54: assume a point $(\boldsymbol{X}^{\mathsf{T}}, 1) \in \mathbf{L}^{\mathsf{T}} = (\boldsymbol{L}_h^{\mathsf{T}}, \boldsymbol{L}_O^{\mathsf{T}})$ and $(\boldsymbol{Y}^{\mathsf{T}}, 1) \in \mathbf{M}^{\mathsf{T}} = (\boldsymbol{M}_h^{\mathsf{T}}, \boldsymbol{M}_O^{\mathsf{T}})$. The lines are coplanar if and only if the vectors $(\boldsymbol{X} - \boldsymbol{Y})$, $\boldsymbol{L}_h$ and $\boldsymbol{M}_h$ are coplanar, or algebraically $(\boldsymbol{X} - \boldsymbol{Y})^{\mathsf{T}} (\boldsymbol{L}_h \times \boldsymbol{M}_h) = 0$. The assertion follows now from the fact that $\boldsymbol{L}_O = \boldsymbol{X} \times \boldsymbol{L}_h$ and $\boldsymbol{M}_O = \boldsymbol{Y} \times \boldsymbol{M}_h$.

Algebraic Expression for Dual Entities As we want to express all operations on entities within linear algebra, we define the dual operator as a linear transformation from one projective subspace to another. For points and lines in 2D and for points and planes in 3D, the dual operator is the identity matrix I_3 resp. I_4:

$$\overline{\mathbf{x}} := I_3 \, \mathbf{x} \quad , \; \overline{\mathbf{l}} := I_3 \, \mathbf{l} \qquad ; \qquad \overline{\mathbf{X}} := I_4 \, \mathbf{X} \quad , \; \overline{\mathbf{A}} := I_4 \, \mathbf{A}$$

For lines in 3D, we define the dual operator as follows, considering our choice for the Plücker-coordinates:

$$\overline{\mathbf{L}} := C\mathbf{L} \quad \text{with } C := \begin{pmatrix} 0&0&0&1&0&0\\ 0&0&0&0&1&0\\ 0&0&0&0&0&1\\ 1&0&0&0&0&0\\ 0&1&0&0&0&0\\ 0&0&1&0&0&0 \end{pmatrix} = \begin{pmatrix} 0 & I_3 \\ I_3 & 0 \end{pmatrix} \tag{2.31}$$

Note that applying the dual operators I_3, I_4 and C changes the interpretation of a homogeneous vector from a projective entity to its dual.

2.5.2 Dual of Point Transformations

The existence of dual entities also has an effect on the transformations mentioned in section 2.3 on page 34, namely homographies, projective transformations and fundamental matrices. There exists a theorem to derive a duality property for linear mappings of all geometric entities. In section 2.1.2 it was already mentioned that given any linear map H from $\mathbb{P}^n$ to $\mathbb{P}^m$, the dual mapping is the transpose H^T that maps the dual spaces $(\mathbb{P}^m)^*$ to $(\mathbb{P}^n)^*$. This behavior can be transferred to any of the proposed geometric entities, see Faugeras and Luong 2001, p. 154.

For example assume a projective transformation P of a 3D-point $\mathbf{X}$ to 2D point $\mathbf{x}'$, the duality relation for $\mathbf{x}' = \mathsf{P}\mathbf{X}$ can be written as

$$\mathbf{x}' = \mathsf{P}\mathbf{X} \quad \circ\!\!-\!\!\bullet \quad \mathsf{P}^\mathsf{T}\mathbf{l}' = \mathbf{A} \tag{2.32}$$

where again the simplified notation $\overline{\mathbf{x}'} = \mathbf{l}'$ and $\overline{\mathbf{X}} = \mathbf{A}$ holds. The sign $\circ\!\!-\!\!\bullet$ is used here to express the duality between the two expressions.

Since we know how to interpret the elements of the dual spaces $(\mathbb{P}^m)^*$ and $(\mathbb{P}^n)^*$ we can directly apply this to 2D and 3D. In equation (2.32), a 2D image line may be back-projected in 3D using a projective camera matrix P and results in a projective plane $\mathbf{A}$, which intersects the focal point and the 2D line on the image plane.

Similarly, the rules of duality for homographies H, projective transformations P and fundamental matrices F are

$$\begin{aligned} \mathbf{x}' = \mathsf{H}\mathbf{x} \quad &\circ\!\!-\!\!\bullet \quad \mathsf{H}^\mathsf{T}\mathbf{l}' = \mathbf{l} \\ \mathbf{X}' = \mathsf{H}\mathbf{X} \quad &\circ\!\!-\!\!\bullet \quad \mathsf{H}^\mathsf{T}\mathbf{A}' = \mathbf{A} \\ \mathbf{L}' = \mathsf{H}_L\mathbf{L} \quad &\circ\!\!-\!\!\bullet \quad \mathsf{H}_L^\mathsf{T}\overline{\mathbf{L}'} = \overline{\mathbf{L}} \\ \mathbf{x}' = \mathsf{P}\mathbf{X} \quad &\circ\!\!-\!\!\bullet \quad \mathsf{P}^\mathsf{T}\mathbf{l}' = \mathbf{A} \\ \mathbf{l}' = \mathsf{F}\mathbf{x}'' \quad &\circ\!\!-\!\!\bullet \quad \mathsf{F}^\mathsf{T}\mathbf{x}' = \mathbf{l}'' \\ \mathbf{l}' = \mathsf{Q}\mathbf{L} \quad &\circ\!\!-\!\!\bullet \quad \mathsf{Q}^\mathsf{T}\mathbf{x}' = \overline{\mathbf{L}} \end{aligned}$$

The last duality relation involves the projective camera transformation Q for 2D lines, which will be covered in section 3.1.2.2 on page 57.

Right now, the representation of the line transformation $\mathbf{L}' = \mathsf{H}_L\mathbf{L}$ is not obvious as not all homographies $\mathsf{H}_L : \mathbb{P}^5 \to \mathbb{P}^5$ within the five-dimensional projective space correspond to valid homographies in $\mathbb{P}^3$. We will deal with line transformations H_L in section 3.1.2 on page 56: as soon as we have introduced algebraic expressions for constructions join and intersection, we are able to derive a construction of H_L given a 3D homography H. The same argument applies to projective camera transformations Q for 3D lines, see section 3.1.2.2.

3 Geometric Reasoning Using Projective Geometry

After we have introduced a representation of points, lines and planes within the framework of projective geometry, we now turn our attention to the question of how to perform reasoning on these entities. Geometric reasoning involves the manipulation of given entities and the derivation of new knowledge by checking relations between entities.

The manipulation of geometric entities usually means to *construct* new entities – either from given ones, such as a line from two points or by transformation, such as the projection of an object point to an image. For the latter type of construction, we can use the basic projective transformation from the last chapter, though they have not yet been explicitly given for all types of transformations such as 3D lines to 2D lines, which will be worked out in this chapter. The construction of new entities from given ones will first be described by the dual operations join and intersection, for which we will give simple algebraic expressions using three types of matrix functions for these operations. The three matrix functions in turn can also be used for expressing incidence and identity relations between the geometric entities and thus we gain algebraic expressions for *testing relations* between objects. The relations between objects finally enables us to express any over-constrained construction problem as for example any forward intersection problems involving points and lines.

Additionally to construction and testing of geometric entities, the process of geometric reasoning using projective geometry involves the transformation between the Euclidean coordinates to homogeneous coordinates, see figure 2.2 in the last chapter. Thus we seperate the geometric reasoning process into three parts, cf. fig. 3.1:

- the transfer of Euclidean observations to a projective entity,
- the construction and/or tests using simplified formulas from projective geometry and
- the normalization of the projective entities back to a Euclidean representation

An important aspect here is that we don't explicitly work with projective entities in $\mathbb{P}^n$ (which is not a vector space) but with their homogeneous –

S. Heuel: Uncertain Projective Geometry, LNCS 3008, pp. 47-95, 2004.

and thus over-parametrized – representation in the vector space $\mathbb{R}^{n+1}$. This will allow us to use algebraic operations for vector spaces.

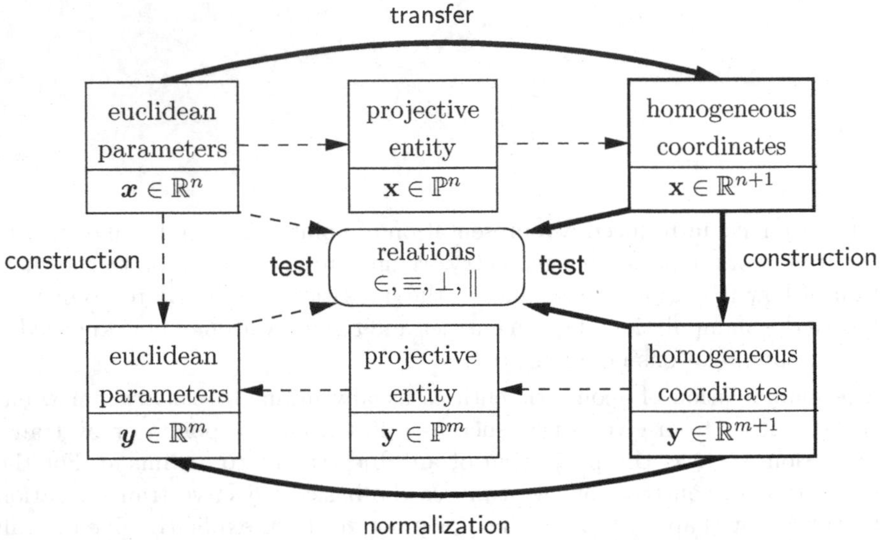

Fig. 3.1. This diagram illustrates the process of geometric reasoning using projective geometry: starting with a Euclidean entity $\boldsymbol{x} \in \mathbb{R}^n$ we can derive a projective entity $\mathbf{x} \in \mathbb{P}^n$ which we can represent as a homogeneous vector $\mathbf{x} \in \mathbb{R}^{n+1}$, see also (2.3). We may now construct a new homogeneous vector $\mathbf{y} \in \mathbb{R}^{m+1}$ using homogeneous vectors $\mathbf{x}_i$ from $\mathbb{R}^{n_i+1}$. If we are interested in the Euclidean representation, we apply a normalization as in (2.3), yielding $\boldsymbol{y} \in \mathbb{R}^m$. For testing relations between objects, we also use the homogeneous representation instead of the Euclidean.

3.1 Unique Constructions of Entities

In this section we describe how to algebraically compute the unique construction of new entities from given ones. There are two types of unique constructions that are discussed in this work: the first type involves the computation using the operators join and intersection. The proposed algebraic expressions are equivalent to the determinant operations of the Grassmann-Cayley algebra, yet are solely based on a matrix-vector notation.

The second type of unique constructions use the projective transformation matrices from section 2.3. Using join and intersection, we will introduce algebraic expression for projective transformations involving 3D lines. Additionally, yet different transformations can be easily defined such as the inverse projective camera, which projects image points onto arbitrary planes.

3.1.1 Join and Intersection

Constructions using join and intersection are natural ways of defining new points, lines and planes, see figure 3.2. The join operator takes given entities and extend them to the smallest projective subspace that contains the given entities: two (linearly independent) points form a line, three (linearly independent) points form a plane. The intersection operator does the opposite: it takes the largest projective subspace such that it is incident to the given entities: two 2D-lines intersect in a 2D point, three planes intersect in a 3D point.

Join and intersection yield only a geometric entity if the resulting entity is always non-empty; for example: in general two 3D lines are not coplanar, in other words they do not intersect in a point and thus we don't define an intersection operator with two 3D lines. Note that the resulting entity of a valid intersection may lie at infinity: two parallel but different 3D planes intersect in a line at infinity.

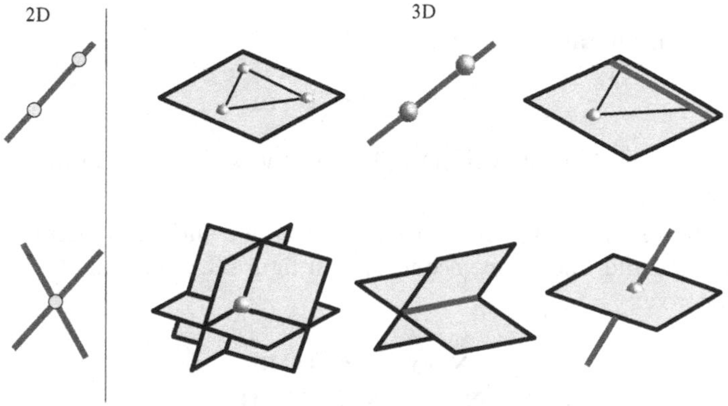

Fig. 3.2. Eight unique constructions, the upper row displays joins, the lower row intersections. From top left to lower right: two 2D points define one 2D line; three 3D points define a plane; two 3D points define a 3D line; a 3D point and a line define a plane; two 2D lines intersect in a 2D point; three planes intersect in a point; two planes intersect in a line; a 3D point as the intersection of a 3D line and a plane

The join operator is denoted with $\wedge$ and the intersection or meet operator with $\cap$. With the linear subspaces in projective space, $\wedge$ and $\cap$ form a so-called *double algebra* or *Grassmann-Cayley algebra.* In this algebra join and intersection but also incidence of entities are operations represented by determinants of their coordinates. We will give an equivalent description of these operations in terms of matrix-vector multiplication in the next chapter. For a good overview of the Grassmann-Cayley algebra, see FAUGERAS AND LUONG 2001 or CARLSSON 1994.

3.1.1.1 Duality of Join and Intersection Implicitly we have already used a join operation in the last chapter, e.g. when mentioning the test on collinearity of three 2D points $\mathbf{x}_1, \mathbf{x}_2, \mathbf{x}_2$, cf. section 2.5: they are collinear iff the determinant $|\mathbf{x}_1\mathbf{x}_2\mathbf{x}_3| = 0$, see also figure 2.8. We can infer from the collinearity property that the point $\mathbf{x}_3$ is incident to the line $\mathbf{l} = \mathbf{x}_1 \wedge \mathbf{x}_2$ that is defined by the join of $\mathbf{x}_1$ and $\mathbf{x}_2$: $\mathbf{x}_3 \in (\mathbf{x}_1 \wedge \mathbf{x}_2)$ Note that the 2D line may be expressed in Plücker coordinates as it was done for the 3D line. The dual case to the collinearity of three points is the common intersection of three 2D lines $\mathbf{l}_1, \mathbf{l}_2, \mathbf{l}_3$ which is checked by $|\mathbf{l}_1, \mathbf{l}_2, \mathbf{l}_3| = 0$. As before we can construct a point $\mathbf{x} = \mathbf{l}_1 \cap \mathbf{l}_2$ which is the unique intersection of the first two lines and the property is reformulated as $\mathbf{l}_3 \ni (\mathbf{l}_1 \cap \mathbf{l}_2)$.
From the above example it can bee seen that there exists a duality property between join and intersection which can be characterized by

$$\overline{\mathbf{x} \wedge \mathbf{y}} \equiv \overline{\mathbf{x}} \cap \overline{\mathbf{y}} \equiv \mathbf{l} \cap \mathbf{m} \tag{3.1}$$

or with another notation

$$\mathbf{x} \wedge \mathbf{y} \quad \circ\!\!-\!\!\bullet \quad \mathbf{l} \cap \mathbf{m} \tag{3.2}$$

meaning that the join of two 2D points $\mathbf{x} \wedge \mathbf{y}$ is dual to the intersection of two 2D lines $\mathbf{l} \cap \mathbf{m}$.
This can be extended to the 3D case with join and intersection involving points, planes and lines: the second row in figure 3.2 contains the dual cases of the first row:

$$\begin{aligned}
\mathbf{x} \wedge \mathbf{y} \quad &\circ\!\!-\!\!\bullet \quad \mathbf{l} \cap \mathbf{m} \\
\mathbf{X} \wedge \mathbf{Y} \quad &\circ\!\!-\!\!\bullet \quad \mathbf{A} \cap \mathbf{B} \\
\mathbf{X} \wedge \mathbf{Y} \wedge \mathbf{Z} \quad &\circ\!\!-\!\!\bullet \quad \mathbf{A} \cap \mathbf{B} \cap \mathbf{C} \\
\mathbf{X} \wedge \mathbf{L} \quad &\circ\!\!-\!\!\bullet \quad \mathbf{A} \cap \overline{\mathbf{L}}
\end{aligned}$$

The actual existence of the duality property of join and intersection can be shown within the Grassmann–Cayley algebra using the Hodge-operator, again see FAUGERAS AND LUONG 2001 for more details.

In the following, we will explicitly develop algebraic expressions for the join of two points in 2D and 3D and the join of a point and a line in 3D. From these two constructions, we can infer algebraic expressions for the other cases depicted in figure 3.2.

3.1.1.2 Join of Two Points in 2D and 3D We start with the join of two 2D points $\mathbf{x}, \mathbf{y}$ as in figure 3.2, top left. We require that the constructing points must lie in the resulting line, $\mathbf{x}, \mathbf{y} \in \mathbf{l}$. From equation (2.10) on page 24 we know that points lie on a line if and only if $\mathbf{x}^\mathsf{T}\mathbf{l} = 0$ and $\mathbf{y}^\mathsf{T}\mathbf{l} = 0$. Thus the join of two points $\mathbf{x} \wedge \mathbf{y}$ is defined using the vector (or cross) product

$$\mathbf{x} \wedge \mathbf{y} = \mathbf{x} \times \mathbf{y} = -\mathbf{y} \times \mathbf{x} \tag{3.3}$$
$$= \mathsf{S}(\mathbf{x})\mathbf{y} = -\mathsf{S}(\mathbf{y})\mathbf{x} \tag{3.4}$$

The matrix $\mathsf{S}(\mathbf{x})$, sometimes denoted as $[\mathbf{x}]_\times$ is a skew-symmetric matrix:

$$\mathsf{S}(\mathbf{x}) := \frac{\partial \mathbf{x} \wedge \mathbf{y}}{\partial \mathbf{y}} = \begin{pmatrix} 0 & -x_3 & x_2 \\ x_3 & 0 & -x_1 \\ -x_2 & x_1 & 0 \end{pmatrix} \tag{3.5}$$

The matrix $\mathsf{S}(\mathbf{x})$ induces the cross-product and at the same time is the Jacobian $\partial \mathbf{x} \wedge \mathbf{y} / \partial \mathbf{y}$ of the join with respect to $\mathbf{x}$. The rank of S equals two. Note that the join is undefined for $\mathbf{x} \wedge \mathbf{x} = \mathsf{S}(\mathbf{x})\mathbf{x} = \mathbf{0}$.

The join $\mathbf{x} \wedge \mathbf{y}$ is a bilinear operation as it can be expressed linearly in $\mathbf{x}$ and $\mathbf{y}$. Still, this is *not* a linear operation as products of two variables occur.

The join of two Euclidean points $\boldsymbol{X}, \boldsymbol{Y}$ in 3D has already been introduced in section 2.2.3 on page 30, where the resulting 3D line consists of a homogeneous part $\boldsymbol{L}_h = \boldsymbol{Y} - \boldsymbol{X}$ and a Euclidean part $\boldsymbol{L}_O = \boldsymbol{X} \times \boldsymbol{Y}$. For homogeneous coordinates $\mathbf{X} = (\boldsymbol{X}_O, X_h)^\mathsf{T} = (U_1, V_1, W_1, T_1)^\mathsf{T}$ and $\mathbf{Y} = (\boldsymbol{Y}_O, Y_h)^\mathsf{T} = (U_2, V_2, W_2, T_2)^\mathsf{T}$ the Euclidean points are obtained by $\boldsymbol{X} = \boldsymbol{X}_O / X_h$ and $\boldsymbol{Y} = \boldsymbol{Y}_O / Y_h$. Thus

$$\mathbf{L} = \mathbf{X} \wedge \mathbf{Y} = \begin{pmatrix} \boldsymbol{L}_h \\ \boldsymbol{L}_O \end{pmatrix} = \begin{pmatrix} \boldsymbol{Y} - \boldsymbol{X} \\ \boldsymbol{X} \times \boldsymbol{Y} \end{pmatrix} \tag{3.6}$$
$$\cong X_h Y_h \begin{pmatrix} \boldsymbol{Y} - \boldsymbol{X} \\ \boldsymbol{X} \times \boldsymbol{Y} \end{pmatrix} \tag{3.7}$$
$$= \begin{pmatrix} X_h \boldsymbol{Y}_O - Y_h \boldsymbol{X}_O \\ S(\boldsymbol{X}_O) \boldsymbol{Y}_O \end{pmatrix} \tag{3.8}$$

The homogeneous part $\boldsymbol{L}_h$ of $\mathbf{L}$ can be written in matrix-vector form and equation (3.8) becomes

$$\mathbf{L} = \mathbf{X} \wedge \mathbf{Y} \tag{3.9}$$
$$= \begin{pmatrix} X_h \mathsf{I} & -\boldsymbol{X}_O \\ \mathsf{S}(\boldsymbol{X}_O) & \mathbf{0} \end{pmatrix} \mathbf{Y} \tag{3.10}$$
$$= \Pi(\mathbf{X}) \mathbf{Y} \tag{3.11}$$

$$\Pi(\mathbf{X}) := \frac{\partial \mathbf{X} \wedge \mathbf{Y}}{\partial \mathbf{Y}} = \begin{pmatrix} T_1 & 0 & 0 & -U_1 \\ 0 & T_1 & 0 & -V_1 \\ 0 & 0 & T_1 & -W_1 \\ 0 & -W_1 & V_1 & 0 \\ W_1 & 0 & -U_1 & 0 \\ -V_1 & U_1 & 0 & 0 \end{pmatrix} \tag{3.12}$$

Thus the join $\mathbf{X} \wedge \mathbf{Y}$ can be expressed by a simple matrix-vector multiplication. The join in 3D is an antisymmetric operation as $\mathbf{X} \wedge \mathbf{Y} = -\mathbf{Y} \wedge \mathbf{X}$, cf. FAUGERAS AND LUONG 2001,p. 133, which is reflected in the matrix-vector multiplication:

$$\mathbf{X} \wedge \mathbf{Y} = \Pi(\mathbf{X})\mathbf{Y} = -\Pi(\mathbf{Y})\mathbf{X} = -\mathbf{Y} \wedge \mathbf{X} \tag{3.13}$$

The rank of the matrix $\Pi(\mathbf{X})$ equals 3. The join $\mathbf{X} \wedge \mathbf{X}$ is undefined as $\Pi(\mathbf{X})\mathbf{X} = \mathbf{0}$.

3.1.1.3 Join of Point and Line (3D) From the construction in equation (3.12) it is possible to derive all other operations shown in figure 3.2 on page 49. First we examine the join of a point and a line in 3D, which also leads to a bilinear expression, depending on the point and the line. The bilinear expression reveal the so-called Plücker matrix as the third important matrix that we use for constructions besides S and Π.

Linear Expression with Respect to the Line-vector **L** We start from the join of two points in 3D, $\mathbf{L} = \mathbf{X} \wedge \mathbf{y} = \mathbf{L} = \Pi(\mathbf{X})\mathbf{Y}$, and fix the point $\mathbf{X}$, which is emphasized by defining a matrix $\mathsf{M} := \Pi(\mathbf{X})$. The matrix M can be regarded as a linear transformation from $\mathbb{P}^4$ to $\mathbb{P}^6$. Using the duality property for linear mappings from section 2.5.2, we obtain

$$\mathbf{L} = \mathsf{M}\mathbf{Y} \quad \circ\!\!-\!\!\bullet \quad \mathsf{M}^{\mathsf{T}}\overline{\mathbf{L}} = \mathbf{A} \tag{3.14}$$

Note that the dual of a point is a plane with the notation $\overline{\mathbf{Y}} = \mathbf{A}$. As M is a matrix depending on $\mathbf{X}$, we have just derived a linear expression for the join of a fixed point $\mathbf{X}$ and a line $\mathbf{L}$, resulting in a plane $\mathbf{A}$:

$$\mathbf{A} = \mathbf{X} \wedge \mathbf{L} = \Pi^{\mathsf{T}}(\mathbf{X})\overline{\mathbf{L}} \tag{3.15}$$

$$= \overline{\Pi}^{\mathsf{T}}(\mathbf{X})\mathbf{L} \tag{3.16}$$

$$\text{with } \overline{\Pi}(\mathbf{X}) := \frac{\partial \mathbf{X} \wedge \mathbf{L}}{\partial \mathbf{L}} = C\Pi(\mathbf{X}) = \Pi(\mathbf{X}) \tag{3.17}$$

where C is the dual-operator for 3D lines[1], see equation (2.31) on page 44. Note that only bilinear scalar multiplications occur in the matrix products $\overline{\Pi}^{\mathsf{T}}(\mathbf{X})\mathbf{L} = \Pi^{\mathsf{T}}(\mathbf{X})\overline{\mathbf{L}}$ of the join $\mathbf{X} \wedge \mathbf{L}$.

Linear Expression with Respect to the Point-vector $\mathbf{X}$ Up to now we assumed the point $\mathbf{X}$ to be fixed. If we now switch the roles of $\mathbf{L}$ and $\mathbf{X}$ and assume $\mathbf{L}$ to be fixed and $\mathbf{X}$ to be a vector variable, we are able to reformulate the join $\mathbf{X} \wedge \mathbf{L}$ as a matrix-vector product with a vector $\mathbf{X}$ and using a 4×4 matrix O:

$$\mathbf{A} = \mathbf{X} \wedge \mathbf{L} = \Pi^{\mathsf{T}}(\mathbf{X})\overline{\mathbf{L}} = \begin{pmatrix} -S(\boldsymbol{L}_h) & -\boldsymbol{L}_O \\ \boldsymbol{L}_O^{\mathsf{T}} & 0 \end{pmatrix} \mathbf{X} \tag{3.18}$$

$$= \mathsf{O}(\mathbf{L})\,\mathbf{X} \tag{3.19}$$

$$\mathsf{O}(\mathbf{L}) := \frac{(\partial \mathbf{X} \wedge \mathbf{L})}{\partial \mathbf{X}} = \begin{pmatrix} 0 & L_3 & -L_2 & -L_4 \\ -L_3 & 0 & L_1 & -L_5 \\ L_2 & -L_1 & 0 & -L_6 \\ L_4 & L_5 & L_6 & 0 \end{pmatrix} \tag{3.20}$$

with $\mathbf{L} = (L_1, L_2, L_3, L_4, L_5, L_6)^{\mathsf{T}}$ and $\mathbf{X} = (X_1, X_2, X_3, X_4)^{\mathsf{T}}$. The matrix $\mathsf{O}(\mathbf{L})$ has the form of a Plücker matrix of a 3D line, cf. HARTLEY AND ZISSERMAN 2000.

We are now able to express the join $\mathbf{A} = \mathbf{X} \wedge \mathbf{L}$ as a bilinear matrix-vector multiplication

$$\mathbf{A} = \mathbf{X} \wedge \mathbf{L} = \overline{\Pi}^{\mathsf{T}}(\mathbf{X})\mathbf{L} = \mathsf{O}(\mathbf{L})\mathbf{X} \tag{3.21}$$

For the remainder of this work it turns out that it is notationally advantageous to replace the matrix $\mathsf{O}(\mathbf{L})$ by an alternative definition $\Gamma(\mathbf{L}) := -\mathsf{O}(\overline{\mathbf{L}})$. The matrix $\Gamma(\mathbf{L})$ is identical to the original 4×4-matrix, but only differs in the sign and the permutation of the Plücker coordinates $L_1, \cdots, L_6$, see also section 2.2.4 on page 32. Thus (3.18) becomes:

$$\mathbf{A} = \mathbf{X} \wedge \mathbf{L} = \overline{\Gamma}^{\mathsf{T}}(\mathbf{L})\mathbf{X} \tag{3.22}$$

$$\text{with} \quad \Gamma(\mathbf{L}) := -\mathsf{O}(\overline{\mathbf{L}}) = \mathsf{O}^{\mathsf{T}}(\overline{\mathbf{L}}) = \begin{pmatrix} 0 & L_6 & -L_5 & -L_1 \\ -L_6 & 0 & L_4 & -L_2 \\ L_5 & -L_4 & 0 & -L_3 \\ L_1 & L_2 & L_3 & 0 \end{pmatrix} \tag{3.23}$$

$$\text{and} \quad \overline{\Gamma}(\mathbf{L}) := \Gamma(\overline{\mathbf{L}}) \tag{3.24}$$

[1] Note that the dual sign in $\overline{\Pi}^{\mathsf{T}}(\mathbf{X})$ is a bit misleading as the dual of a linear mapping is its transpose. In this case, the dual sign is a notational abbreviation such that we can multiply $\overline{\Pi}^{\mathsf{T}}(\mathbf{X})$ with a line $\mathbf{L}$ and not with its dual $\overline{\mathbf{L}}$, cf. equation (3.28) on the following page.

As only the order and the sign of the Plücker coordinates changes with respect to $\mathsf{O}(\mathbf{L})$, the matrix $\mathsf{\Gamma}(\mathbf{L})$ is also a Plücker matrix of a 3D line $\mathbf{L}$, cf. HARTLEY AND ZISSERMAN 2000. The matrix $\overline{\mathsf{\Gamma}}(\mathbf{L}) = \mathsf{\Gamma}(\overline{\mathbf{L}})$ is called the Plücker matrix for the dual line or just dual Plücker matrix[2]. Note that $\overline{\overline{\mathsf{\Gamma}}}(\mathbf{L}) = \mathsf{\Gamma}(\mathbf{L})$.
The expression in equation (3.21) on the page before now becomes

$$\mathbf{A} = \mathbf{X} \wedge \mathbf{L} = \overline{\mathsf{\Pi}}^{\mathsf{T}}(\mathbf{X})\mathbf{L} = \overline{\mathsf{\Gamma}}^{\mathsf{T}}(\mathbf{L})\mathbf{X} \tag{3.25}$$

Note that equation (3.25) is written such that it resembles the fact that the join of a point $\mathbf{X}$ and a line $\mathbf{L}$ is a symmetrical operation in the Grassmann-Cayley algebra.
Summarizing the last pages on join operations, we observe that the matrices $\mathsf{S}(\cdot)$, $\mathsf{\Pi}(\cdot)$ and $\mathsf{\Gamma}(\cdot)$ play similar roles for calculation: they are the only matrices that are necessary to express join and intersection in matrix-vector notation. We call these matrices construction matrices:

Definition 3 (Construction Matrices) *The matrices* $\mathsf{S}(\cdot)$, $\mathsf{\Pi}(\cdot)$ *and* $\mathsf{\Gamma}(\cdot)$ *defined in equations* (3.5), (3.12) *and* (3.23) *are called* construction matrices *of their geometric entities.*
The matrix $\mathsf{S}(\cdot)$ *is a construction matrix of a 2D point* $\mathbf{x}$ *or a 2D line* $\mathbf{l}$. $\mathsf{\Pi}(\cdot)$ *is a construction matrix of a 3D point* $\mathbf{X}$ *or a plane* $\mathbf{A}$. *Finally* $\mathsf{\Gamma}(\cdot)$ *is a construction matrix of a 3D line* $\mathbf{L}$ *or its dual* $\overline{\mathbf{L}}$.

3.1.1.4 Dual Operations: Intersections in 2D and 3D We are now ready to derive the intersection operations: using the three join constructions in equations (3.4), (3.9) and (3.25) we can take advantage of the duality of join and intersection to obtain three intersection constructions:

$$\begin{aligned} \mathbf{l} = \mathbf{x} \wedge \mathbf{y} &= \mathsf{S}(\mathbf{x})\mathbf{y} = -\mathsf{S}(\mathbf{y})\mathbf{x} \\ &\multimap \mathbf{x} = \mathbf{l} \cap \mathbf{m} = \mathsf{S}(\mathbf{l})\mathbf{m} = -\mathsf{S}(\mathbf{m})\mathbf{l} \end{aligned} \tag{3.26}$$

$$\begin{aligned} \mathbf{L} = \mathbf{X} \wedge \mathbf{Y} &= \mathsf{\Pi}(\mathbf{X})\mathbf{X} = -\mathsf{\Pi}(\mathbf{Y})\mathbf{X} \\ &\multimap \overline{\mathbf{L}} = \mathbf{A} \cap \mathbf{B} = \mathsf{\Pi}(\mathbf{A})\mathbf{B} = -\mathsf{\Pi}(\mathbf{B})\mathbf{A} \end{aligned} \tag{3.27}$$

$$\begin{aligned} \mathbf{A} = \mathbf{X} \wedge \mathbf{L} &= \overline{\mathsf{\Pi}}^{\mathsf{T}}(\mathbf{X})\mathbf{L} = \overline{\mathsf{\Gamma}}^{\mathsf{T}}(\mathbf{L})\mathbf{X} \\ &\multimap \mathbf{X} = \mathbf{A} \cap \overline{\mathbf{L}} = \overline{\mathsf{\Pi}}^{\mathsf{T}}(\mathbf{A})\overline{\mathbf{L}} = \overline{\mathsf{\Gamma}}^{\mathsf{T}}(\overline{\mathbf{L}})\mathbf{A} \end{aligned} \tag{3.28}$$

With the definition

$$\overline{\mathsf{\Gamma}}(\mathbf{L}) := \mathsf{\Gamma}(\overline{\mathbf{L}}) = \mathsf{\Gamma}(C\mathbf{L}) \quad \text{and} \quad \overline{\mathsf{\Pi}}(\mathbf{A}) := C\mathsf{\Pi}(\mathbf{A}) \tag{3.29}$$

and C defined in (2.31) the intersection operators in 3D can be reformulated:

[2] In earlier work (FÖRSTNER *et al.* 2000, HEUEL 2001, HEUEL AND FÖRSTNER 2001) the definition of the matrices $\mathsf{\Gamma}(\mathbf{L})$ and $\overline{\mathsf{\Gamma}}(\mathbf{L})$ is permuted. The current definition though provides a more consistent notation of the algebraic expressions, see table 3.1.

$$\mathbf{L} = \mathbf{A} \cap \mathbf{B} = \overline{\Pi}(\mathbf{A})\mathbf{B} = -\overline{\Pi}(\mathbf{B})\mathbf{A} \tag{3.30}$$

$$\mathbf{X} = \mathbf{A} \cap \mathbf{L} = \Pi^{\mathsf{T}}(\mathbf{A})\mathbf{L} = \Gamma^{\mathsf{T}}(\mathbf{L})\mathbf{A} \tag{3.31}$$

It is important to note that with equations (3.30) and (3.31) we use the join and intersection as *operations on geometric objects* and not as algebraic operations of the Grassmann-Cayley algebra. For example, the *algebraic* operation of the intersection of two planes yields the dual line $\mathbf{L}$, see equation (3.27). In practice though, we are interested in the line $\mathbf{L}$ and not in its dual, therefore we "hide" the duality in the matrix-vector multiplication with $\overline{\Pi}(\mathbf{A}) = C\Pi(\mathbf{A})$, thus we obtain a *geometric* operation $\mathbf{L} = \mathbf{A} \cap \mathbf{B}$. Also note that the expression $\mathbf{X}\mathbf{Y}^{\mathsf{T}} - \mathbf{Y}\mathbf{X}^{\mathsf{T}}$ with two points $\mathbf{X}, \mathbf{Y}$ may serve as an alternative definition of the Plücker matrix Γ as

$$\Gamma(\mathbf{X} \wedge \mathbf{Y}) = \mathbf{X}\mathbf{Y}^{\mathsf{T}} - \mathbf{Y}\mathbf{X}^{\mathsf{T}} \tag{3.32}$$

Equation (3.32) can be proven by comparing the matrix elements of $\mathbf{X}\mathbf{Y}^{\mathsf{T}} - \mathbf{Y}\mathbf{X}^{\mathsf{T}}$ and $\Gamma(\Pi(\mathbf{X})\mathbf{Y})$. Note that with two planes $\mathbf{A}, \mathbf{B}$ we obtain

$$\overline{\Gamma}(\mathbf{A} \cap \mathbf{B}) = \mathbf{A}\mathbf{B}^{\mathsf{T}} - \mathbf{B}\mathbf{A}^{\mathsf{T}} \tag{3.33}$$

as $\overline{\Gamma}(\mathbf{A} \cap \mathbf{B}) = \overline{\Gamma}(\overline{\Pi}(\mathbf{A})\mathbf{B}) = \Gamma(\Pi(\mathbf{A})\mathbf{B}) = \mathbf{A}\mathbf{B}^{\mathsf{T}} - \mathbf{B}\mathbf{A}^{\mathsf{T}}$.

3.1.1.5 Trilinear Constructions Two more unique constructions are shown in figure 3.2: the join of three points and dually the intersection of three planes. The join of three points $\mathbf{A} = \mathbf{X} \wedge \mathbf{Y} \wedge \mathbf{Z}$ results in a planc $\mathbf{A}$. Since the join operation $\wedge$ is associative, we may write

$$\mathbf{A} = (\mathbf{X} \wedge \mathbf{Y}) \wedge \mathbf{Z} \tag{3.34}$$

$$= \overline{\Gamma}^{\mathsf{T}}(\mathbf{X} \wedge \mathbf{Y})\mathbf{Z} \tag{3.35}$$

$$= \overline{\Gamma}^{\mathsf{T}}\left(\Pi(\mathbf{X})\mathbf{Y}\right)\mathbf{Z} \tag{3.36}$$

which can be written as a trilinear expression, since

$$\mathbf{X} \wedge \mathbf{Y} \wedge \mathbf{Z} = \mathbf{Z} \wedge \mathbf{X} \wedge \mathbf{Y} = \mathbf{Y} \wedge \mathbf{Z} \wedge \mathbf{X}$$

Dually one obtains identical expressions $\mathbf{X} = \mathbf{A} \cap \mathbf{B} \cap \mathbf{C}$ for planes:

$$\mathbf{X} = (\mathbf{A} \cap \mathbf{B}) \cap \mathbf{C} \tag{3.37}$$

$$= \Gamma^{\mathsf{T}}(\mathbf{A} \cap \mathbf{B})\mathbf{C} \tag{3.38}$$

$$= \Gamma^{\mathsf{T}}\left(\overline{\Pi}(\mathbf{A})\mathbf{B}\right)\mathbf{C} \tag{3.39}$$

$$= \overline{\Gamma}^{\mathsf{T}}\left(\Pi(\mathbf{A})\mathbf{B}\right)\mathbf{C} \tag{3.40}$$

and

$$\mathbf{A} \cap \mathbf{B} \cap \mathbf{C} = \mathbf{C} \cap \mathbf{A} \cap \mathbf{B} = \mathbf{B} \cap \mathbf{C} \cap \mathbf{A}$$

We summarize the algebraic expression for all eight unique constructions in table 3.1.

Table 3.1. Unique Construction using join $\wedge$ and intersection $\cap$ of new points lines and planes. The matrices S, Π, $\overline{\Pi}$, Γ and $\overline{\Gamma}$ are defined in equations (3.5), (3.12), (3.17), (3.23) and (3.28). All forms are linear in the coordinates of the given coordinate vectors. Note that join $\wedge$ and intersection $\cap$ are used as *operations on geometric objects* and not strictly as algebraic operations of the Grassmann-Cayley algebra: the duality operation in the dual line $\overline{\mathbf{L}}$ as in $\overline{\mathbf{L}} = \mathbf{A} \cap \mathbf{B}$ is hidden in the algebraic expressions, see equations (3.27)-(3.31).

entities	construction	expression
points $\mathbf{x}, \mathbf{y}$	$\mathbf{l} = \mathbf{x} \wedge \mathbf{y}$	$\mathbf{l} = \mathsf{S}(\mathbf{x})\mathbf{y} = -\mathsf{S}(\mathbf{y})\mathbf{x}$
lines $\mathbf{l}, \mathbf{m}$	$\mathbf{x} = \mathbf{l} \cap \mathbf{m}$	$\mathbf{x} = \mathsf{S}(\mathbf{l})\mathbf{m} = -\mathsf{S}(\mathbf{m})\mathbf{l}$
points $\mathbf{X}, \mathbf{Y}$	$\mathbf{L} = \mathbf{X} \wedge \mathbf{Y}$	$\mathbf{L} = \Pi(\mathbf{X})\mathbf{Y} = -\Pi(\mathbf{Y})\mathbf{X}$
planes $\mathbf{A}, \mathbf{B}$	$\mathbf{L} = \mathbf{A} \cap \mathbf{B}$	$\mathbf{L} = \overline{\Pi}(\mathbf{A})\mathbf{B} = -\overline{\Pi}(\mathbf{B})\mathbf{A}$
point $\mathbf{X}$, line $\mathbf{L}$	$\mathbf{A} = \mathbf{X} \wedge \mathbf{L}$	$\mathbf{A} = \overline{\Pi}^{\mathsf{T}}(\mathbf{X})\mathbf{L} = \overline{\Gamma}^{\mathsf{T}}(\mathbf{L})\mathbf{X}$
plane $\mathbf{A}$, line $\mathbf{L}$	$\mathbf{X} = \mathbf{A} \cap \mathbf{L}$	$\mathbf{X} = \Pi^{\mathsf{T}}(\mathbf{A})\mathbf{L} = \Gamma^{\mathsf{T}}(\mathbf{L})\mathbf{A}$
points $\mathbf{X}, \mathbf{Y}, \mathbf{Z}$	$\mathbf{A} = \mathbf{X} \wedge \mathbf{Y} \wedge \mathbf{Z}$	$\overline{\Gamma}^{\mathsf{T}}(\Pi(\mathbf{X})\mathbf{Y})\mathbf{Z} = \overline{\Gamma}^{\mathsf{T}}(\Pi(\mathbf{Y})\mathbf{Z})\mathbf{X} = \overline{\Gamma}^{\mathsf{T}}(\Pi(\mathbf{Z})\mathbf{X})\mathbf{Y}$
planes $\mathbf{A}, \mathbf{B}, \mathbf{C}$	$\mathbf{X} = \mathbf{A} \cap \mathbf{B} \cap \mathbf{C}$	$\overline{\Gamma}^{\mathsf{T}}(\Pi(\mathbf{A})\mathbf{B})\mathbf{C} = \overline{\Gamma}^{\mathsf{T}}(\Pi(\mathbf{B})\mathbf{C})\mathbf{A} = \overline{\Gamma}^{\mathsf{T}}(\Pi(\mathbf{C})\mathbf{A})\mathbf{B}$

3.1.2 Transformation of Points, Lines, and Planes

We have already seen ways to construct points using homogeneous transformations, cf. 2.3 on page 34. Taking advantage of the duality property, it is easy to find the algebraic expression for transforming hyperplanes given a point transformation, cf. 2.5 on page 40.

In this section we especially deal with the transformation of 3D lines given a 3D point transformation. In particular we derive algebraic expressions for homographies and projective cameras using join and intersection. As an example for an application, we will derive an algebraic expression for a transformation from image entities to 3D entities that are incident to a given plane.

3.1.2.1 Homography Given a 4×4 matrix H which describes a point-to-point homography in 3D, it is now easy to derive an expression for 3D lines, which is a 6×6 matrix H_L: the i-th column $\mathsf{H}_L^{(i)}$ of H_L is the image of the canonical 6-vector ${}^6\mathbf{E}_i = (0, \cdots, \underset{i}{1}, \cdots, 0)^{\mathsf{T}}$. With our choice of Plücker coordinates, the canonical lines ${}^6\mathbf{E}_i = ({}^4\mathbf{E}_j) \wedge ({}^4\mathbf{E}_k)$ can be expressed by two canonical points, represented by the 4-vectors ${}^4\mathbf{E}_j$ and ${}^4\mathbf{E}_k$, where $(j, k) = (4,1), (4,2), (4,3), (2,3), (3,1), (1,2)$ for $i = 1, \ldots, 6$, see section 2.2.4 on page 32. Thus we can express the i-th column $\mathsf{H}_L^{(i)}$ of the line homography by using the canonical points ${}^4\mathbf{E}_j$ and ${}^4\mathbf{E}_k$:

$$\begin{aligned}\mathsf{H}_L^{(i)} &= \mathsf{H}^{(j)} \wedge \mathsf{H}^{(k)} \\ &= (\mathsf{H} \cdot ({}^4\mathbf{E}_j)) \wedge (\mathsf{H} \cdot ({}^4\mathbf{E}_k)) \\ &= \overline{\overline{\sqcap}}(\mathsf{H}^{(j)})\, \mathsf{H}^{(k)}\end{aligned} \tag{3.41}$$

with the choice of j, k as above. For example, assume a Euclidean rotation $R = (\boldsymbol{r}_1, \boldsymbol{r}_2, \boldsymbol{r}_3)$ and a translation by $\boldsymbol{X}$. The Euclidean motion ${}^m\mathsf{H}$ is a special homography and is given by

$${}^m\mathsf{H} = \begin{pmatrix} \boldsymbol{r}_1 & \boldsymbol{r}_2 & \boldsymbol{r}_3 & \boldsymbol{X} \\ 0 & 0 & 0 & 1 \end{pmatrix}$$

For the sake of simplicity, we only consider $i = 1$ and $i = 4$, thus $(j, k) = (4, 1)$ and $(2, 3)$. All other cases are similar to the chosen two:

$${}^m\mathsf{H}_L^{(1)} = \overline{\overline{\sqcap}}\begin{pmatrix} \boldsymbol{X} \\ 1 \end{pmatrix}\begin{pmatrix} \boldsymbol{r}_1 \\ 0 \end{pmatrix} = \begin{pmatrix} I_3 & \boldsymbol{X} \\ S(\boldsymbol{X}) & \mathbf{0} \end{pmatrix}\begin{pmatrix} \boldsymbol{r}_1 \\ 0 \end{pmatrix} = \begin{pmatrix} \boldsymbol{r}_1 \\ S(\boldsymbol{X})\boldsymbol{r}_1 \end{pmatrix}$$

$${}^m\mathsf{H}_L^{(4)} = \overline{\overline{\sqcap}}\begin{pmatrix} \boldsymbol{r}_2 \\ 1 \end{pmatrix}\begin{pmatrix} \boldsymbol{r}_3 \\ 0 \end{pmatrix} = \begin{pmatrix} 0 & \boldsymbol{r}_2 \\ S(\boldsymbol{r}_2) & \mathbf{0} \end{pmatrix}\begin{pmatrix} \boldsymbol{r}_3 \\ 0 \end{pmatrix} = \begin{pmatrix} \mathbf{0} \\ S(\boldsymbol{r}_2)\boldsymbol{r}_3 \end{pmatrix} = \begin{pmatrix} \mathbf{0} \\ \boldsymbol{r}_1 \end{pmatrix}$$

We obtain

$${}^m\mathsf{H}_L = \begin{pmatrix} R & 0 \\ S(\boldsymbol{X})R & R \end{pmatrix} \tag{3.42}$$

see also NAVAB *et al.* 1993.

The computation of H_L given a plane-to-plane homography $\mathsf{H}_A = \mathsf{H}^{-\mathsf{T}}$ is dually given by

$$\mathsf{H}_L^{(i)} = \overline{\overline{\sqcap}}(\mathsf{H}_A^{(j)})\, \mathsf{H}_A^{(k)}$$

Note that the derivation of the general homogeneous line-homography H_L is simpler than the one of BARTOLI AND STURM 2001

3.1.2.2 Projective Pinhole Camera In section 2.3.2 on page 35 the transformation matrix for a projective camera P was introduced for points only, $\mathbf{x}' = \mathsf{P}\mathbf{X}$. With the new constructors we are now able to construct a projection matrix Q that can be applied to 3D lines, yielding 2D image lines,

$$\mathbf{l}' = \mathsf{Q}\mathbf{L} \tag{3.43}$$

In the following, we will describe the derivation of the matrix Q in detail, since some of the previously introduced notions can be applied.

Suppose we have the projective camera matrix $\mathsf{P} = \begin{pmatrix} \mathbf{A}^{\mathsf{T}} \\ \mathbf{B}^{\mathsf{T}} \\ \mathbf{C}^{\mathsf{T}} \end{pmatrix}$, where the three rows $\mathbf{A}, \mathbf{B}$ and $\mathbf{C}$ may be interpreted as three distinct planes, cf. section 2.3.2 on page 35, intersecting in the projection center. Then the projective camera matrix can be derived by intersections of the three planes:

Proposition 1 (Projective camera matrix for lines) *Given a projective camera matrix for points,* $\mathsf{P} = \begin{pmatrix} \mathbf{A}^\mathsf{T} \\ \mathbf{B}^\mathsf{T} \\ \mathbf{C}^\mathsf{T} \end{pmatrix}$ *where* $\mathbf{x}' = \mathsf{P}\mathbf{X}$*, the projective camera matrix for lines is given as* $\mathsf{Q}(\mathsf{P}) = \mathsf{Q} = \begin{pmatrix} (\overline{\mathbf{B} \cap \mathbf{C}})^\mathsf{T} \\ (\overline{\mathbf{C} \cap \mathbf{A}})^\mathsf{T} \\ (\overline{\mathbf{A} \cap \mathbf{B}})^\mathsf{T} \end{pmatrix}$*, where* $\mathbf{l}' = \mathsf{Q}\mathbf{L}$.

Proof: Assume a 3D line $\mathbf{L}$ was constructed by two 3D points $\mathbf{X}$ and $\mathbf{Y}$, thus $\mathbf{L} = \mathbf{X} \wedge \mathbf{Y}$. Then the image line $\mathbf{l}'$ of the 3D line $\mathbf{L}$ is given as $\mathbf{l}' = \mathbf{x}' \wedge \mathbf{y}' = \mathsf{P}\mathbf{X} \wedge \mathsf{P}\mathbf{Y}$. Using $\mathbf{x}' \wedge \mathbf{y}' = \mathsf{S}(\mathbf{x}')\mathbf{y}'$, this can be written as

$$\mathbf{l}' = \begin{pmatrix} \mathbf{B}^\mathsf{T}\mathbf{X}\,\mathbf{C}^\mathsf{T}\mathbf{Y} - \mathbf{C}^\mathsf{T}\mathbf{X}\,\mathbf{B}^\mathsf{T}\mathbf{Y} \\ \mathbf{C}^\mathsf{T}\mathbf{X}\,\mathbf{A}^\mathsf{T}\mathbf{Y} - \mathbf{A}^\mathsf{T}\mathbf{X}\,\mathbf{C}^\mathsf{T}\mathbf{Y} \\ \mathbf{A}^\mathsf{T}\mathbf{X}\,\mathbf{B}^\mathsf{T}\mathbf{Y} - \mathbf{B}^\mathsf{T}\mathbf{X}\,\mathbf{A}^\mathsf{T}\mathbf{Y} \end{pmatrix} \tag{3.44}$$

We now have to prove that equation (3.44) is equivalent with

$$\mathbf{l}' = \mathsf{Q}(\mathbf{X} \wedge \mathbf{Y}) = \begin{pmatrix} (\overline{\mathbf{B} \cap \mathbf{C}})^\mathsf{T}\,(\mathbf{X} \wedge \mathbf{Y}) \\ (\overline{\mathbf{C} \cap \mathbf{A}})^\mathsf{T}\,(\mathbf{X} \wedge \mathbf{Y}) \\ (\overline{\mathbf{A} \cap \mathbf{B}})^\mathsf{T}\,(\mathbf{X} \wedge \mathbf{Y}) \end{pmatrix} = \begin{pmatrix} -<\mathbf{X} \wedge \mathbf{Y}\,,\, \mathbf{B} \cap \mathbf{C}> \\ -<\mathbf{X} \wedge \mathbf{Y}\,,\, \mathbf{C} \cap \mathbf{A}> \\ -<\mathbf{X} \wedge \mathbf{Y}\,,\, \mathbf{A} \cap \mathbf{B}> \end{pmatrix} \tag{3.45}$$

Without loss of generality we will only prove the first inner product:

$$\begin{aligned} \mathbf{B}^\mathsf{T}\mathbf{X}\,\mathbf{C}^\mathsf{T}\mathbf{Y} - \mathbf{C}^\mathsf{T}\mathbf{X}\,\mathbf{B}^\mathsf{T}\mathbf{Y} &= \mathbf{X}^\mathsf{T}\mathbf{B}^\mathsf{T}\,\mathbf{C}^\mathsf{T}\mathbf{Y} - \mathbf{X}^\mathsf{T}\mathbf{C}\,\mathbf{B}^\mathsf{T}\mathbf{Y} \\ &= \mathbf{X}^\mathsf{T}(\mathbf{B}\mathbf{C}^\mathsf{T} - \mathbf{C}\mathbf{B}^\mathsf{T})\mathbf{Y} \\ &= \mathbf{X}^\mathsf{T}(\overline{\Gamma}(\mathbf{B} \cap \mathbf{C}))\mathbf{Y} \\ &= \mathbf{X}^\mathsf{T}(-\overline{\Pi}^\mathsf{T}(\mathbf{Y}))\,(\mathbf{B} \cap \mathbf{C}) \\ &= -\,(\Pi(\mathbf{Y})\mathbf{X})^\mathsf{T}\,(\mathbf{B} \cap \mathbf{C}) \\ &= -\left(\Pi^\mathsf{T}(\mathbf{Y})\mathbf{X}\right)^\mathsf{T}\,(\overline{\mathbf{B} \cap \mathbf{C}}) \\ &= \left(\Pi^\mathsf{T}(\mathbf{X})\mathbf{Y}\right)^\mathsf{T}\,(\overline{\mathbf{B} \cap \mathbf{C}}) \\ &= (\mathbf{X} \wedge \mathbf{Y})^\mathsf{T}\,\overline{\mathbf{B} \cap \mathbf{C}} \;=\; -<\mathbf{X} \wedge \mathbf{Y}\,,\, \mathbf{B} \cap \mathbf{C}> \end{aligned}$$

Note that $\overline{\mathbf{B} \cap \mathbf{C}}$ is the dual of the intersection of the two planes $\mathbf{B}$ and $\mathbf{C}$. Thus we could also have written $\overline{\mathbf{B}} \wedge \overline{\mathbf{C}}$, meaning to take the join of the two points $\overline{\mathbf{B}}$ and $\overline{\mathbf{C}}$. □

Furthermore, the dual from equation (3.43) is

$$\mathbf{l}' = \mathsf{Q}\mathbf{L} \quad \circ\!\!-\!\!\bullet \quad \overline{\mathbf{l}' = \mathsf{Q}\mathbf{L}} \Leftrightarrow \mathsf{Q}^\mathsf{T}\overline{\mathbf{l}'} = \overline{\mathbf{L}} \tag{3.46}$$

$$\Leftrightarrow \mathcal{C}\mathsf{Q}^\mathsf{T}\mathbf{x}' = \mathbf{L} \tag{3.47}$$

$$\Leftrightarrow \overline{\mathsf{Q}}^\mathsf{T}\mathbf{x}' = \mathbf{L} \tag{3.48}$$

Here we assumed that the dual of an image line is an image point, $\overline{\mathbf{l}'} = \mathbf{x}'$ and used the convention $\overline{\mathsf{Q}} := \mathsf{Q}C$. Equation (3.48) reveals a constructor for a 3D line $\mathbf{L}$ given an image point $\mathbf{x}'$ and a projective camera, represented by Q. Thus the constructed 3D line is the viewing ray through the focal point, the image point $\mathbf{x}'$ and the object point $\mathbf{X}$, we sometimes write the viewing ray as $\mathbf{L}'$ to stress the fact, that the line is derived by an image point: $\mathbf{L}' = \overline{\mathsf{Q}}^\mathsf{T}\mathbf{x}'$. As in section 2.3 on page 34 we define a vector $\mathbf{q}$ containing the rows of the matrix Q

$$\mathbf{q} = \mathrm{vec}(\mathsf{Q}^\mathsf{T}) = \begin{pmatrix} \overline{\mathbf{B} \cap \mathbf{C}} \\ \overline{\mathbf{C} \cap \mathbf{A}} \\ \overline{\mathbf{A} \cap \mathbf{B}} \end{pmatrix} \tag{3.49}$$

Thus the projective camera vector $\mathbf{q}$ for lines depends on the projective camera vector $\mathbf{p}$ for points and we can write $\mathbf{q}$ as a function $\mathbf{q}(\mathbf{p})$. One can compute the projective camera vector $\mathbf{q}$ for lines as a matrix-vector multiplication

$$\mathbf{q} = \mathsf{J}_{\mathsf{qp}}(\mathbf{p})\mathbf{p} \tag{3.50}$$

$$\mathsf{J}_{\mathsf{qp}} := \frac{\partial \mathbf{q}(\mathbf{p})}{\partial \mathbf{p}} = \begin{pmatrix} 0 & -\overline{\Pi}(\mathbf{C}) & \overline{\Pi}(\mathbf{B}) \\ \overline{\Pi}(\mathbf{C}) & 0 & -\overline{\Pi}(\mathbf{A}) \\ -\overline{\Pi}(\mathbf{B}) & \overline{\Pi}(\mathbf{A}) & 0 \end{pmatrix} \tag{3.51}$$

Note that $\mathbf{q} = \mathsf{J}_{\mathsf{qp}}(\mathbf{p})\mathbf{p}$ is not linear in $\mathbf{p}$, as entries of $\mathbf{p}$ occur in J_{qp}.

3.1.3 Fundamental Matrix

As we now have an algebraic expression to construct a viewing ray $\mathbf{L}'$ from an image point $\mathbf{x}'$ by $\mathbf{L}' = \overline{\mathsf{Q}}^\mathsf{T}\mathbf{x}'$, we can now give a formula for computing the fundamental matrix from two given cameras P_1 and P_2, see section 3.1.3: first, compute the line projection matrix Q_1 and Q_2 as described above and obtain the first viewing ray $\mathbf{L}' = \overline{\mathsf{Q}}_1^\mathsf{T}\mathbf{x}'$. Projecting $\mathbf{L}'$ back to the second image yields

$$\mathbf{l}'' = \mathsf{Q}_2\overline{\mathsf{Q}}_1^\mathsf{T}\mathbf{x}' \quad \Leftrightarrow \quad \mathbf{l}'' = \mathsf{F}^\mathsf{T}\mathbf{x}' \qquad \text{with } \mathsf{F} := \overline{\mathsf{Q}}_1\mathsf{Q}_2 = \mathsf{Q}_1 C \mathsf{Q}_2 \tag{3.52}$$

3.1.4 Inverse Projective Camera with Respect to a Plane

As a demonstration of the capabilities of the described construction methods, we want to show a simple way to obtain the so-called inverse projective camera transformation which back-projects an image point or line onto a world plane (figure 3.3). This is a new transformation and can be expressed by combining a construction matrix with a projective camera matrix. The derivation

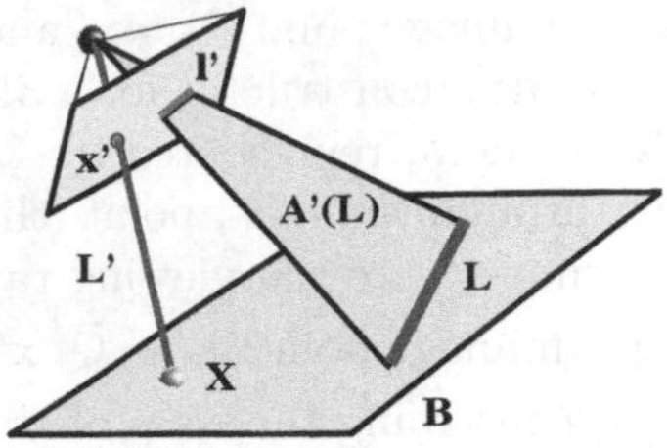

Fig. 3.3. Inverse projective camera: given a projective camera P and a 3D plane **B**, one can establish a one-to-one relation between the image entities $\mathbf{x}'$, $\mathbf{l}'$ and their corresponding 3D entities $\mathbf{X}, \mathbf{L}$ on the plane **B**.

of the transformation matrix is much simpler than the one in Faugeras and Luong 2001, p. 185f.

First, we demonstrate how to obtain backward projection from 2D points to 3D points by using the join operator: assume an arbitrary plane **B** in object space is given along with a projective camera P, see figure 3.3. A point $\mathbf{x}'$ induces a viewing ray $\mathbf{L}' = \overline{\mathsf{Q}}^{\mathsf{T}}\mathbf{x}'$ because of (3.48). We can now intersect the viewing ray $\mathbf{L}'$ with the given plane **B** to obtain the object point **X**:

$$\mathbf{X} = \mathbf{B} \cap \mathbf{L}' = \Pi^{\mathsf{T}}(\mathbf{B})\overline{\mathsf{Q}}^{\mathsf{T}}\mathbf{x}' = \overline{\Pi}^{\mathsf{T}}(\mathbf{B})\mathsf{Q}^{\mathsf{T}}\mathbf{x}' =: \mathsf{P}_B^{+}\mathbf{x}'$$

The 4×3 projection matrix P_B^{+} can be regarded as the inverse projection matrix to P with respect to a plane **B**: for any plane **B** not going through the projection center the equation $\mathbf{x}' = \mathsf{P}\mathsf{P}_B^{+}\mathbf{x}'$ holds. Thus the product is $\mathsf{P}\mathsf{P}_B^{+} = I_3$. and the rank of P_B^{+} has to be 3.

For image lines $\mathbf{l}'$, we obtain the following inverse projection matrix:

$$\mathbf{L} = \mathbf{B} \cap \mathbf{A}'(\mathbf{L}) = \overline{\Pi}(\mathbf{B})\mathsf{P}^{\mathsf{T}}\mathbf{l}' =: \mathsf{Q}_B^{+}\mathbf{l}'$$

Summarizing all transformations given a projective camera matrix P, we obtain table 3.2.

3.2 Construction Matrices and Their Interpretation

Before we explore relations between objects in section 3.3, we take a closer look at the construction matrices $\mathsf{S}(\cdot)$, $\Pi(\cdot)$ and $\Gamma(\cdot)$, defined in definition 3 on page 54. They have been used as linear mappings dependent on one entity and are relevant for algebraic expressions of join and intersection. In this section, we will investigate the nature of the construction matrices by interpreting their rows and the columns: it turns out that they can be seen as special geometric entities, that are incident to the entity that defines the matrix. This observation will help us to easily find algebraic expressions for all possible identity and incidence relations between entities in 2D and 3D.

Table 3.2. Possible constructions using a projective camera matrix P: forward projection (3D→2D) and backward projection (2D→3D). The derivation of the last two projection matrices are given in section 3.1.4.

dim	type	algebraic expression	matrix
3D→2D	point → point	$\mathbf{x}' = \mathsf{P}\mathbf{X}$	$\mathsf{P} = \mathsf{P}(\mathbf{p}) = \begin{pmatrix} \mathbf{A}^\mathsf{T} \\ \mathbf{B}^\mathsf{T} \\ \mathbf{C}^\mathsf{T} \end{pmatrix}$
3D→2D	line → line	$\mathbf{l}' = \mathsf{Q}\mathbf{L}$	$\mathsf{Q} = \mathsf{Q}(\mathbf{p}) = \begin{pmatrix} \overline{(\mathbf{B} \cap \mathbf{C})}^\mathsf{T} \\ \overline{(\mathbf{C} \cap \mathbf{A})}^\mathsf{T} \\ \overline{(\mathbf{A} \cap \mathbf{B})}^\mathsf{T} \end{pmatrix} = \begin{pmatrix} (\Pi(\mathbf{B})\mathbf{C})^\mathsf{T} \\ (\Pi(\mathbf{C})\mathbf{A})^\mathsf{T} \\ (\Pi(\mathbf{A})\mathbf{B})^\mathsf{T} \end{pmatrix}$
2D→3D	point → line	$\mathbf{L}' = \overline{\mathsf{Q}}^\mathsf{T}\mathbf{x}'$	$\overline{\mathsf{Q}}^\mathsf{T}(\mathbf{p}) = \begin{pmatrix} \mathbf{B} \cap \mathbf{C} & \mathbf{C} \cap \mathbf{A} & \mathbf{A} \cap \mathbf{B} \end{pmatrix} = \begin{pmatrix} \overline{\Pi}(\mathbf{B})\mathbf{C} & \overline{\Pi}(\mathbf{C})\mathbf{A} & \overline{\Pi}(\mathbf{A})\mathbf{B} \end{pmatrix}$
2D→3D	line → plane	$\mathbf{A}' = \mathsf{P}^\mathsf{T}\mathbf{l}'$	$\mathsf{P}^\mathsf{T} = \mathsf{P}^\mathsf{T}(\mathbf{p}) = \begin{pmatrix} \mathbf{A} & \mathbf{B} & \mathbf{C} \end{pmatrix}$
2D→3D	point → point	$\mathbf{X}'_B = \mathsf{P}^+_B\mathbf{x}'$	$\mathsf{P}^+_B = \mathsf{P}^+_B(\mathbf{B}, \mathbf{p}) = \overline{\Pi}^\mathsf{T}(\mathbf{B})\mathsf{Q}^\mathsf{T}(\mathbf{p})$
2D→3D	line → line	$\mathbf{L}'_B = \mathsf{Q}^+_B\mathbf{l}'$	$\mathsf{Q}^+_B = \mathsf{Q}^+_B(\mathbf{B}, \mathbf{p}) = \overline{\Pi}(\mathbf{B})\mathsf{P}^\mathsf{T}(\mathbf{p})$

3.2.1 Canonical Entities in Construction Matrices

In the following we will define special points, lines and planes, that are contained in the three defined construction matrices. For every point $\mathbf{X}$, line $\mathbf{L}$ or plane $\mathbf{A}$, there exists so-called canonical entities that relate $\mathbf{X}$, $\mathbf{L}$ resp. $\mathbf{A}$ with the given coordinate system in an obvious manner. We first explore the simpler 2D case before proceeding to 3D.

3.2.1.1 Points and Lines in 2D Given a 2D point $\mathbf{x}$, the construction matrix $\mathsf{S}(\mathbf{x})$ was used for constructions of lines passing through $\mathbf{x}$ using $\mathbf{l} = \mathsf{S}(\mathbf{x})\mathbf{y}$. The second point $\mathbf{y}$ may be anywhere on the projective plane besides on the location of $\mathbf{x}$. Choosing the three canonical points $\mathbf{e}_i$, see table 2.1 on page 28, we obtain three lines $\mathbf{l}_i$, $i = 1, \ldots, 3$, which are the columns of $\mathsf{S}(\mathbf{x})$, see figure 3.4(left): $\mathbf{l}_1$ is the join of $\mathbf{x}$ and the infinite point in the direction of the x-axis, thus being parallel to the x-axis and passing through point $\mathbf{x}$. Similarly, $\mathbf{l}_2$ is the line parallel to the y-axis passing through $\mathbf{x}$ and $\mathbf{l}_3$ passes the origin $\mathbf{e}_3$ and $\mathbf{x}$. Thus we can rewrite the skew-symmetric matrix as

$$\mathsf{S}(\mathbf{x}) = \begin{pmatrix} 0 & -w & v \\ w & 0 & -u \\ -v & u & 0 \end{pmatrix} = \begin{pmatrix} \mathbf{l}_1 & \mathbf{l}_2 & \mathbf{l}_3 \end{pmatrix}$$

$$= \begin{pmatrix} (\mathbf{x} \wedge \mathbf{e}_1) & (\mathbf{x} \wedge \mathbf{e}_2) & (\mathbf{x} \wedge \mathbf{e}_3) \end{pmatrix} = \begin{pmatrix} (\mathbf{e}_1 \wedge \mathbf{x})^\mathsf{T} \\ (\mathbf{e}_2 \wedge \mathbf{x})^\mathsf{T} \\ (\mathbf{e}_3 \wedge \mathbf{x})^\mathsf{T} \end{pmatrix}$$

The last equality stems from the fact that the rows of $\mathsf{S}(\mathbf{x})$ are the negated columns, thus also correspond to the three special lines $\mathbf{l}_i$.

The same argument applies for the matrix $\mathsf{S}(\mathbf{l})$ of a line $\mathbf{l} = (a, b, c)^\mathsf{T}$ leading to three special points $\mathbf{x}_i$ where two points $\mathbf{x}_1, \mathbf{x}_2$ are the intersections of the coordinate axes and $\mathbf{l}$ and the third point $\mathbf{x}_3$ lies at infinity in the direction of $\mathbf{l}$, thus

$$\mathsf{S}(\mathbf{l}) = \begin{pmatrix} 0 & -c & b \\ c & 0 & -a \\ -b & a & 0 \end{pmatrix} = \begin{pmatrix} \mathbf{x}_1 & \mathbf{x}_2 & \mathbf{x}_3 \end{pmatrix}$$

$$= \begin{pmatrix} (\mathbf{l} \cap \mathbf{e}_1) & (\mathbf{l} \cap \mathbf{e}_2) & (\mathbf{l} \cap \mathbf{e}_3) \end{pmatrix} = \begin{pmatrix} (\mathbf{e}_1 \cap \mathbf{l})^\mathsf{T} \\ (\mathbf{e}_2 \cap \mathbf{l})^\mathsf{T} \\ (\mathbf{e}_3 \cap \mathbf{l})^\mathsf{T} \end{pmatrix}$$

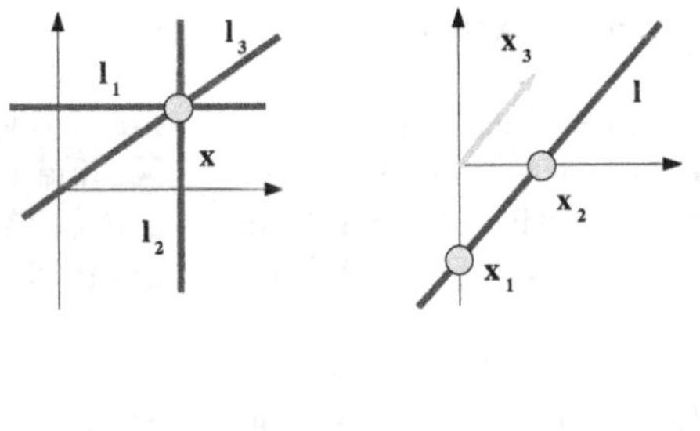

Fig. 3.4. Interpretation of construction matrix $\mathsf{S}(\cdot)$ for a point $\mathbf{x}$ (left) and a line $\mathbf{l}$ (right). A point $\mathbf{x}$ can be represented by two lines $\mathbf{l}_1, \mathbf{l}_2$ parallel to the coordinate axes and one line $\mathbf{l}_3$ passing the origin. A line $\mathbf{l}$ can be represented by two points $\mathbf{x}_1, \mathbf{x}_2$ on the coordinate axes and one point $\mathbf{x}_3$ at infinity.

Considering that there is a simple bijective relation between a point $\mathbf{x}$ and its construction matrix $\mathsf{S}(\mathbf{x})$, we will consider the matrix $\mathsf{S}(\mathbf{x})$ as the *matrix representation* of the point $\mathbf{x}$.

The entities that are contained in a construction matrix are subject of the following definition:

Definition 4 (Canonical entities in $\mathsf{S}(\cdot)$) *The lines* $(\mathbf{e}_i \wedge \mathbf{x})$ *contained in the construction matrix* $\mathsf{S}(\mathbf{x})$ *are called the* canonical lines *of a point* $\mathbf{x}$. *In* *the same manner, we call the points* $(\mathbf{e}_i \cap \mathbf{l})$ *contained in* $\mathsf{S}(\mathbf{l})$ canonical points *of the line* $\mathbf{l}$.

Choosing a Minimum Number of Canonical Entities In the case of a point $\mathbf{x}$, it is obvious that only two canonical lines—thus two columns resp. rows of $\mathsf{S}(\mathbf{x})$—are needed to completely determine the point $\mathbf{x}$. In the following we identify the canonical lines with the rows of $\mathsf{S}(\mathbf{x})$. As these rows are linearly dependent, the rank of the matrix equals the minimum number of canonical entities.

Sometimes it will be necessary to choose the two linearly independent canonical lines from the matrix $\mathsf{S}(\mathbf{x})$, as we will see when doing statistical tests of relations, section 4.5. But the choice of the rows depends on the position of $\mathbf{x}$: it may happen that two coordinates of $\mathbf{x} = (u, v, w)^\mathsf{T}$ are zero, then a row of $\mathsf{S}(\mathbf{x})$ is equal to the transposed zero-vector $\mathbf{0}^\mathsf{T}$ and the corresponding canonical line $(\mathbf{x} \wedge \mathbf{e}_i)$ is undefined; thus the other two rows of $\mathsf{S}(\mathbf{x})$have to be chosen.

Choosing the rows in a numerically stable manner is straight-forward: removing the row, which is closest to $\mathbf{0}^\mathsf{T}$ yields a 2×3 matrix $\mathsf{S}^{[2]}(\mathbf{x}) := \mathsf{J}^{[2]}\mathsf{S}(\mathbf{x})$, where $\mathsf{J}^{[2]}$ is the 2×3 reduction matrix, defined by keeping only two rows from the identity matrix I. The closest row $\mathbf{r}_i^\mathsf{T} = (\mathbf{x} \wedge \mathbf{e}_i)^\mathsf{T}$ to $\mathbf{0}^\mathsf{T}$ can easily be chosen by comparing the norms of the rows, for example the L-1 norm $||\mathbf{r}_i|| = \sum_j |(r_i)_j|$. Note that the reduced matrix $\mathsf{S}^{[2]}(\mathbf{x})$ still has rank 2, as the remaining rows are guaranteed to be linearly independent since the associated canonical lines are different.

Of course the same argumentation holds for lines $\mathbf{l}$ and their matrices $\mathsf{S}(\mathbf{l})$, yielding a reduced matrix $\mathsf{S}^{[2]}(\mathbf{l})$.

3.2.1.2 Points and Planes in 3D For a point $\mathbf{X}$ in 3D, the construction matrix Π may be interpreted in two ways: $\Pi(\mathbf{X})$ from $\mathbf{X} \wedge \mathbf{Y}$ and the dual transposed matrix $\overline{\Pi}^\mathsf{T}(\mathbf{X}) = \Pi^\mathsf{T}(\mathbf{X})C$ from $\mathbf{X} \wedge \mathbf{L}$, see table 3.1.. The matrix $\Pi(\mathbf{X})$ contains four columns, interpreted as lines $\mathbf{L}_i = \mathbf{X} \wedge \mathbf{E}_i$ with the four canonical points $\mathbf{E}_i$, see table 2.2 on page 30: three lines $\mathbf{L}_1, \mathbf{L}_2, \mathbf{L}_3$ are parallel to the coordinate axes and pass through the point $\mathbf{X}$, the fourth line $\mathbf{L}_4$ is incident to both the origin $\mathbf{E}_4$ and the point $\mathbf{X}$, see figure 3.5 (top left).

The matrix $\overline{\Pi}^\mathsf{T}(\mathbf{X})$ contains six columns, interpreted as planes $\mathbf{A}_i = \mathbf{X} \wedge \overline{\mathbf{E}}_i$ with the six canonical lines $\mathbf{E}_i$, see table 2.3 on page 32: three planes $\mathbf{A}_1, \mathbf{A}_2, \mathbf{A}_3$ are parallel to the coordinate system planes, e.g. $\mathbf{A}_1$ is parallel to the yz-plane and passes through $\mathbf{X}$. The other three planes $\mathbf{A}_4, \mathbf{A}_5, \mathbf{A}_6$ are constructed by the coordinate axes and the point $\mathbf{X}$, see figure 3.5 (bottom left).

Based on duality, we now know that for a plane $\mathbf{A}$ the matrices $\overline{\Pi}(\mathbf{A})$ and $\Pi^\mathsf{T}(\mathbf{A})$ contain four lines $\mathbf{L}_i = \mathbf{A} \cap \mathbf{E}_i$ resp. six points $\mathbf{X}_i = \mathbf{A} \cap \mathbf{E}_i$, see figure 3.5 (top and bottom right).

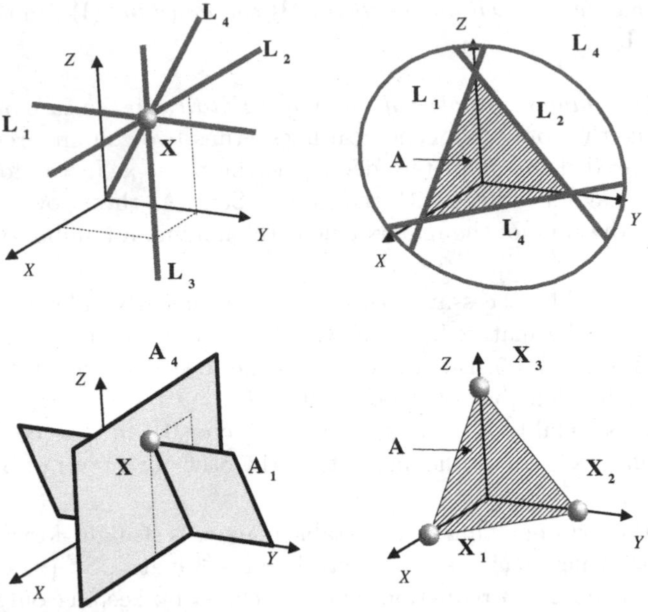

Fig. 3.5. Interpretation of construction matrices Π for a point $\mathbf{X}$ (left) and for a plane $\mathbf{A}$ (right). *Top left*: given $\Pi(\mathbf{X})$ a point $\mathbf{X}$ can be represented by three lines $\mathbf{L}_1, \mathbf{L}_2, \mathbf{L}_3$ parallel to the coordinate axes and one line $\mathbf{L}_4$ passing the origin. *Bottom left:* given $\overline{\Pi}^{\mathsf{T}}(\mathbf{X})$ a point $\mathbf{X}$ can be represented by six planes, which are either parallel to a coordinate system plane, e.g. $\mathbf{A}_4$ or intersecting a coordinate axis, e.g. $\mathbf{A}_1$. *Top right:* given $\overline{\Pi}(\mathbf{A})$ a plane $\mathbf{A}$ can be represented by three lines $\mathbf{L}_1, \mathbf{L}_2, \mathbf{L}_3$ incident to the coordinate system planes and one line $\mathbf{L}_4$ at infinity, incident to $\mathbf{A}$. *Bottom right:* given $\Pi^{\mathsf{T}}(\mathbf{A})$ a plane $\mathbf{A}$ can be represented by six points, three $\mathbf{X}_1, \mathbf{X}_2, \mathbf{X}_3$ on the coordinate axes and the plane, three $\mathbf{X}_4, \mathbf{X}_5, \mathbf{X}_6$ at infinity (not shown).

Analogous to definition 4 we can define canonical entities in the matrix $\Pi(\cdot)$ as follows:

Definition 5 (Canonical entities in $\Pi(\cdot)$) *The points, planes and lines contained in* $\Pi(\mathbf{X})$, $\overline{\Pi}(\mathbf{X})$, $\Pi^{\mathsf{T}}(\mathbf{A})$ *and* $\overline{\Pi}^{\mathsf{T}}(\mathbf{A})$ *are called the* canonical points, planes and lines *of* $\mathbf{X}$ *and* $\mathbf{A}$ *, see table 3.3 on page 66.*

Choosing a Minimum Number of Canonical Entities As a point $\mathbf{X}$ is fixed by three parameters, it is sufficient for the matrix $\Pi(\mathbf{X})$ to choose 3 out of 6 canonical planes, i.e. 3 out of 6 rows from $\Pi(\mathbf{X})$. Again the minimum number of canonical entities is equal to the rank of the construction matrix. Similar to the reduction of the matrix $\mathsf{S}(\cdot)$, we can remove the row which is closest to $\mathbf{0}^{\mathsf{T}}$, yielding a reduced matrix $\Pi^{[3]}(\mathbf{X}) = \mathsf{J}^{[3]}\Pi(\mathbf{X})$ where $\mathsf{J}^{[3]}$ is a 3×6 matrix and thus $\Pi^{[3]}(\mathbf{X})$ is a 3×4 matrix.

The rows of the matrix $\overline{\Pi}^{\mathsf{T}}(\mathbf{X})$ represent four canonical lines incident to the point $\mathbf{X}$, but again three are enough to define the point. So again we remove one row closest to $\mathbf{0}^{\mathsf{T}}$, the remaining rows are guaranteed to be linearly independent to each other as they represent three different lines. Thus the 3×6 matrix $\overline{\Pi}^{[3]\mathsf{T}}(\mathbf{X}) = \mathsf{J}^{[3]}\overline{\Pi}^{\mathsf{T}}(\mathbf{X})$ contains the minimum number of canonical lines necessary for representing the point.

Note that we could have selected rows of $\Pi(\mathbf{X})$ and $\overline{\Pi}^{\mathsf{T}}(\mathbf{X})$ instead of columns, but then the reduction matrix had to be multiplied from the right. The reduction of the construction matrices $\overline{\Pi}(\mathbf{A})$ and $\Pi^{\mathsf{T}}(\mathbf{A})$ for a plane $\mathbf{A}$ can be done in the same way as for the point matrices.

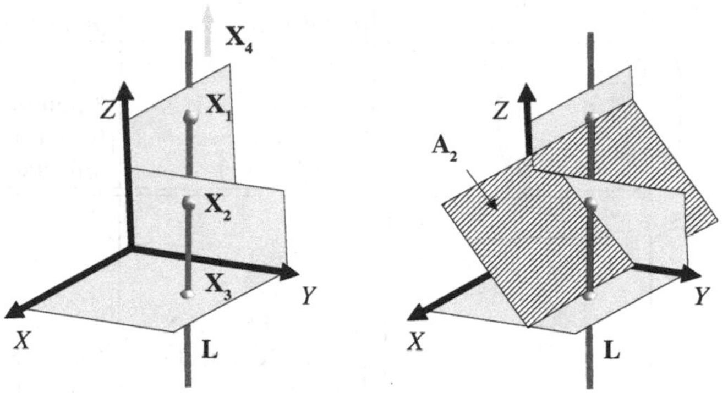

Fig. 3.6. Interpretation of construction- or Plücker-matrix Γ for a line $\mathbf{L}$. *Left:* given $\Gamma(\mathbf{L})$, a line $\mathbf{L}$ can be represented by three intersections $\mathbf{X}_1, \mathbf{X}_2, \mathbf{X}_3$ of the line $\mathbf{L}$ and the coordinate system planes. The fourth point $\mathbf{X}_4$ is equal to the intersection of $\mathbf{L}$ with the infinite plane. *Right:* given $\overline{\Gamma}(\mathbf{L})$, a line $\mathbf{L}$ can be represented by three planes $\mathbf{A}_1, \mathbf{A}_2, \mathbf{A}_3$ that are the join of the infinite points $\mathbf{E}_i$ and the line, the fourth plane $\mathbf{A}_4$ (not visible) is the join of the origin $\mathbf{E}_4$ and $\mathbf{L}$.

3.2.1.3 Lines in 3D Finally, for a line $\mathbf{L}$ and its dual $\overline{\mathbf{L}}$ the columns of the Plücker-matrices $\Gamma^{\mathsf{T}}(\mathbf{L})$ and the $\overline{\Gamma}^{\mathsf{T}}(\mathbf{L})$ can be interpreted as $\mathbf{L} \wedge \mathbf{E}_i$ resp. $\mathbf{L} \cap \mathbf{E}_i$: the first three columns of $\Gamma^{\mathsf{T}}(\mathbf{L})$ correspond to canonical points on the coordinate system plane, incident to the line $\mathbf{L}$. The fourth column is the intersection of $\mathbf{L}$ with the plane at infinite distance. Since the Plücker-matrix is skew-symmetric as S, the rows of $\Gamma^{\mathsf{T}}(\mathbf{X})$ correspond to the same points.

The Plücker matrix of the dual line, $\overline{\Gamma}^{\mathsf{T}}(\mathbf{L})$, contains the canonical planes, which are incident to the canonical points $\mathbf{E}_i$ and the line $\mathbf{L}$ itself

Note that the matrix representations S, Π play the same role as the Plücker matrix Γ for lines in 3D, but only S and Γ are usually used in Computer Vision and Graphics. The definition by applying canonical vectors follows

Table 3.3. Interpretation of rows and columns for the construction matrices of points, lines and planes. The skew-symmetric matrix $\mathsf{S}(\cdot)$ may represent a 2D point $\mathbf{x}$ resp. a 2D line $\mathbf{l}$. The rows and columns of the matrix are entities that intersect or are incident to $\mathbf{x}$ resp. $\mathbf{l}$. The Plücker-matrices $\Gamma(\mathbf{L}), \overline{\Gamma}(\mathbf{L})$ represent 3D lines $\mathbf{L}$, containing 4 incident points resp. 4 intersecting planes as rows and columns. The matrix $\Pi(\mathbf{X})$ represents a point $\mathbf{X}$, containing 6 planes or 4 lines and the matrix $\overline{\Pi}(\mathbf{A})$ represents a plane $\mathbf{A}$ containing 6 points or 4 lines,see also table 2.4 on page 32. Note that the canonical vectors $\mathbf{e}_i, \mathbf{E}_i$ have different interpretations, depending on the context: they may represent points, lines or planes.

dim	matrix	rows	columns
2D	$\mathsf{S}(\mathbf{x}) = \begin{pmatrix} 0 & -w & v \\ w & 0 & -u \\ -v & u & 0 \end{pmatrix}$	3 lines $\mathbf{e}_i \wedge \mathbf{x}$ (point $\mathbf{x}$ and unit point $\mathbf{e}_i$)	3 lines $\mathbf{x} \wedge \mathbf{e}_i$ (point $\mathbf{x}$ and unit point $\mathbf{e}_i$)
	$\mathsf{S}(\mathbf{l}) = \begin{pmatrix} 0 & -c & b \\ c & 0 & -a \\ -b & a & 0 \end{pmatrix}$	3 points $\mathbf{e}_i \cap \mathbf{l}$ (line $\mathbf{l}$ intersects unit line $\mathbf{e}_i$)	3 points $\mathbf{l} \cap \mathbf{e}_i$ (line $\mathbf{l}$ intersects unit line $\mathbf{e}_i$)
3D	$\Pi(\mathbf{X}) = \begin{pmatrix} T & 0 & 0 & -U \\ 0 & T & 0 & -V \\ 0 & 0 & T & -W \\ 0 & -W & V & 0 \\ W & 0 & -U & 0 \\ -V & U & 0 & 0 \end{pmatrix}$	6 planes ${}^6\overline{\mathbf{E}}_i \wedge \mathbf{X}$ (point $\mathbf{X}$ and dual unit line ${}^6\overline{\mathbf{E}}_i$)	4 lines $\mathbf{X} \wedge {}^4\mathbf{E}_i$ (point $\mathbf{X}$ and unit point ${}^4\mathbf{E}_i$)
	$\Gamma(\mathbf{L}) = \begin{pmatrix} 0 & L_6 & -L_5 & -L_1 \\ -L_6 & 0 & L_4 & -L_2 \\ L_5 & -L_4 & 0 & -L_3 \\ L_1 & L_2 & L_3 & 0 \end{pmatrix}$	4 points ${}^4\mathbf{E}_i \cap \mathbf{L}$ (line $\mathbf{L}$ intersects unit planes ${}^4\mathbf{E}_i$)	4 points $-(\mathbf{L} \cap \mathbf{E}_i)$ (line $\mathbf{L}$ intersects unit planes ${}^4\mathbf{E}_i$)
	$\overline{\Gamma}(\mathbf{L}) = \begin{pmatrix} 0 & L_3 & -L_2 & -L_4 \\ -L_3 & 0 & L_1 & -L_5 \\ L_2 & -L_1 & 0 & -L_6 \\ L_4 & L_5 & L_6 & 0 \end{pmatrix}$	4 planes ${}^4\mathbf{E}_i \wedge \mathbf{L}$ (line $\mathbf{L}$ and unit point ${}^4\mathbf{E}_i$)	4 planes $-(\mathbf{L} \wedge \mathbf{E}_i)$ (line $\mathbf{L}$ and unit point ${}^4\mathbf{E}_i$)
	$\overline{\Pi}(\mathbf{A}) = \begin{pmatrix} 0 & -C & B & 0 \\ C & 0 & -A & 0 \\ -B & A & 0 & 0 \\ D & 0 & 0 & -A \\ 0 & D & 0 & -B \\ 0 & 0 & D & -C \end{pmatrix}$	6 points ${}^6\overline{\mathbf{E}}_i \cap \mathbf{A}$ (plane $\mathbf{A}$ intersects dual unit lines ${}^6\overline{\mathbf{E}}_i$)	4 lines $\mathbf{A} \cap {}^4\mathbf{E}_i$ (plane $\mathbf{A}$ intersects unit planes ${}^4\mathbf{E}_i$)

equation (3.4) of Faugeras and Luong 2001, p. 135. Again, the entities in $\Gamma(\cdot)$ are defined as follows:

Definition 6 (Canonical entities in $\Gamma(\cdot)$) *The points and planes, that are contained in $\Gamma(\mathbf{L})$ and $\overline{\Gamma}(\mathbf{L})$, are called* canonical points and planes *of* $\mathbf{L}$.

Choosing a Minimum Number of Canonical Entities As a line $\mathbf{L}$ is already defined by two canonical points from the construction matrices $\Gamma(\mathbf{L})$ resp. two planes from $\overline{\Gamma}(\mathbf{L})$ we remove two rows from these matrices that are closest to $\mathbf{0}^\mathsf{T}$, thus $\Gamma^{[2]}(\mathbf{L}) = \mathsf{J}^{[2]}\Gamma(\mathbf{L})$ and $\overline{\Gamma}(\mathbf{L}) = \mathsf{J}^{[2]}\overline{\Gamma}(\mathbf{L})$.

3.2.2 Reduction of Construction Matrices

As seen before, we can reduce the construction matrices $\mathsf{S}(\cdot)$, $\Pi(\cdot)$ and $\Gamma(\cdot)$ to a minimum number of canonical entities or rows, which still uniquely define the underlying entity. The minimum number is equal to the rank of the construction matrices. A reduction of matrices will be especially important when we are doing statistical tests of relations and estimation of entities, see chapter 4; so the reader may skip this section at the first time.
For the reduction $\mathsf{U}^{[r]}(\mathbf{x})$ we delete those rows, which are closest to the zero-vector, thus closest to an undefined geometric entity. This way we obtain numerical stability for the remaining rows, which are guaranteed to be linearly independent. There are several criteria for closeness, e.g. we may take the L_1-norm.

3.2.2.1 Consistent Reduction of Matrices in Bilinear Forms Now assume the simultaneous reduction of two matrices, such as in bilinear forms $\mathsf{U}(\mathbf{x})\mathbf{y} = \mathsf{V}(\mathbf{y})\mathbf{x}$, where the reduction for $r = \min(r_U, r_V)$ is wanted (r_U and r_V are the minimum number of canonical elements for U and V). If the reduction of rows in $\mathbf{z}_1^{[r]} = \mathsf{U}^{[r]}(\mathbf{x})$ and $\mathbf{z}_2^{[r]} = \mathsf{V}^{[r]}(\mathbf{y})\mathbf{x}$ is done independently from each other, then it is possible that $\mathbf{z}_1^{[r]} \neq \mathbf{z}_2^{[r]}$. Therefore one always has to delete the same rows from $\mathsf{U}(\mathbf{x})$ and $\mathsf{V}(\mathbf{y})$ to ensure equality between the reduced vectors $\mathbf{z}_1^{[r]}$ and $\mathbf{z}_2^{[r]}$. With this constraint though, it may be impossible to avoid zero rows $\mathbf{0}^\mathsf{T}$ in one of the matrices $\mathsf{U}(\mathbf{x})$ and $\mathsf{V}(\mathbf{y})$, e.g. consider $\mathsf{S}(\mathbf{e}_1)$ and $-\mathsf{S}(\mathbf{e}_3)$, where only one common row is not $\mathbf{0}^\mathsf{T}$.
But under certain conditions, the simultaneous reduction of two construction matrices is indeed possible. We consider six compatible pairs, that are used in the bilinear forms of the constructions: $(\mathsf{S}(\mathbf{x}), \mathsf{S}(\mathbf{y}))$, $(\Pi(\mathbf{X}), \Pi(\mathbf{Y}))$ and $(\overline{\Pi}^\mathsf{T}(\mathbf{X}), \overline{\Gamma}^\mathsf{T}(\mathbf{L}))$ and their dual complements $(\mathsf{S}(\mathbf{l}), \mathsf{S}(\mathbf{m}))$, $(\overline{\Pi}(\mathbf{A}), \overline{\Pi}(\mathbf{B}))$ and $(\Pi^\mathsf{T}(\mathbf{A}), \Gamma^\mathsf{T}(\mathbf{L}))$. We will only cover the first three as the reasoning is equivalent for the dual pairs.

For the first pair we obtain

$$\mathsf{S}(\mathbf{x}) = \begin{pmatrix} 0 & -x_3 & x_2 \\ x_3 & 0 & -x_1 \\ -x_2 & x_1 & 0 \end{pmatrix} \quad \text{and} \quad \mathsf{S}(\mathbf{y}) = \begin{pmatrix} 0 & -y_3 & y_2 \\ y_3 & 0 & -y_1 \\ -y_2 & y_1 & 0 \end{pmatrix}$$

with $\mathbf{x} = (x_1, x_2, x_3)^\mathsf{T}$ and $\mathbf{y} = (y_1, y_2, y_3)^\mathsf{T}$. One can easily prove that there exists two common non-zero rows if and only if there exist an index i, such that $|x_i y_i| > 0$. A similar criteria exists for choosing three common non-zero rows from

$$\mathsf{\Pi}(\mathbf{X}) = \begin{pmatrix} X_4 & 0 & 0 & -X_1 \\ 0 & X_4 & 0 & -X_2 \\ 0 & 0 & X_4 & -X_3 \\ 0 & -X_3 & X_2 & 0 \\ X_3 & 0 & -X_1 & 0 \\ -X_2 & X_1 & 0 & 0 \end{pmatrix} \quad \text{and} \quad \mathsf{\Pi}(\mathbf{Y}) = \begin{pmatrix} Y_4 & 0 & 0 & -Y_1 \\ 0 & Y_4 & 0 & -Y_2 \\ 0 & 0 & Y_4 & -Y_3 \\ 0 & -Y_3 & Y_2 & 0 \\ Y_3 & 0 & -Y_1 & 0 \\ -Y_2 & Y_1 & 0 & 0 \end{pmatrix}$$

There exist three common non-zero rows if and only if there exists an index i, such that $|X_i Y_i| > 0$: either the first three rows are chosen ($i = 4$) or two of the last three and one of the first three rows ($i = 1, 2, 3$).

For the last pair of construction matrices, namely ($\overline{\overline{\mathsf{\Pi}}}^\mathsf{T}(\mathbf{X})$, $\overline{\mathsf{\Gamma}}^\mathsf{T}(\mathbf{L})$), the criteria is a bit more complicated. Given

$$\overline{\overline{\mathsf{\Pi}}}^\mathsf{T}(\mathbf{X}) = \begin{pmatrix} 0 & -X_3 & X_2 & X_4 & 0 & 0 \\ X_3 & 0 & -X_1 & 0 & X_4 & 0 \\ -X_2 & X_1 & 0 & 0 & 0 & X_4 \\ 0 & 0 & 0 & -X_1 & -X_2 & -X_3 \end{pmatrix} \quad \text{and} \quad \overline{\mathsf{\Gamma}}^\mathsf{T}(\mathbf{L}) = \begin{pmatrix} 0 & -L_3 & L_2 & -L_4 \\ L_3 & 0 & -L_1 & -L_5 \\ -L_2 & L_1 & 0 & -L_6 \\ L_4 & L_5 & L_6 & 0 \end{pmatrix}$$

we can only assume that at least one X_i and one L_j are not zero entries, for example $\mathbf{X} = \mathbf{E}_1$, $\mathbf{L} = \mathbf{E}_4$ ($i = 1, j = 4$). Checking all combinations of (i, j), only for the following pairs of (i, j), one can guarantee that at least two common rows are non-zero:

point index i \ line index j	1	2	3	4	5	6
1	•				•	•
2		•		•		•
3			•	•	•	
4	•	•	•			

The table above reminds of the pattern of the non-zero entries in the matrix $\mathsf{\Pi}(\mathbf{X})$ and we can formulate the condition for consistent reduction as follows: if there exists a pair of indices (i, j) such that $|\mathsf{\Pi}^\mathsf{T}_{i,j}(\mathbf{X})\, L_j| > 0$, a consistent reduction of both construction matrices is possible. Here, $\mathsf{\Pi}^\mathsf{T}_{i,j}(\mathbf{X})$ denotes the (i, j)-th entry of the matrix $\mathsf{\Pi}^\mathsf{T}(\mathbf{X})$.

Summarizing the criteria for a possible consistent reduction of the construction matrices, we obtain

Consistent reduction of construction matrices
Objective: Given two objects $\mathbf{x}$ and $\mathbf{y}$ (points, lines or planes in 2D and 3D) and the construction matrices $\mathsf{U}(\mathbf{x})$ and $\mathsf{V}(\mathbf{y})$ with r_U and r_V being the minimum number of canonical entities for U and V.
Return consistently reduced matrices $\mathsf{U}^{[r]}(\mathbf{x})$ and $\mathsf{V}^{[r]}(\mathbf{y})$ with $r = \min(r_U, r_V)$.

1. If the requirements in equation (3.53)-(3.55) are not fulfilled, rotate and possibly translate the coordinate system accordingly.
2. Choose r rows $\mathbf{u}_i^{\mathsf{T}}$, $\mathbf{v}_i^{\mathsf{T}}$ from $\mathsf{U}(\mathbf{x})$ and $\mathsf{V}(\mathbf{y})$ with the largest entries $|\mathbf{u}_i||\mathbf{v}_i|$. One obtains $\mathsf{U}^{[r]}(\mathbf{x})$ and $\mathsf{V}^{[r]}(\mathbf{y})$

Algorithm 3.1: Simultaneous and consistent reduction of pairs of construction matrices. It is guaranteed that the reduced matrices $\mathsf{U}^{[r]}(\mathbf{x})$ and $\mathsf{V}^{[r]}(\mathbf{y})$ have rank r.

$$(\mathbf{x}, \mathbf{y}): \quad \exists\, i: \quad |x_i y_i| > 0 \quad \text{with } \mathbf{x}, \mathbf{y} \text{ being points or hyperplanes} \tag{3.53}$$

$$(\mathbf{X}, \mathbf{L}): \quad \exists\, i,j: |\Pi^{\mathsf{T}}_{i,j}(\mathbf{X})\, L_j| > 0 \quad \text{with 3D line } \mathbf{L} \text{ and point } \mathbf{X} \tag{3.54}$$

$$(\mathbf{A}, \mathbf{L}): \quad \exists\, i,j: |\overline{\Pi}^{\mathsf{T}}_{i,j}(\mathbf{A})\, L_j| > 0 \quad \text{with 3D line } \mathbf{L} \text{ and plane } \mathbf{A} \tag{3.55}$$

Thus only in very rare cases a reduction of construction matrices is not possible, e.g. for the 3D plane $\mathbf{A} = \mathbf{E}_1$ and the line $\mathbf{L} = \mathbf{E}_3$. When checking geometric relations, see section 3.3.4 on page 79, these cases practically do not occur between entities.

If we still want to ensure that a consistent reduction is always possible, it is important to note that the consistent reduction depends on the choice of the coordinate system for the geometric entities. Therefore we may transform the coordinate system, e.g. by a rotation and a translation, such that the conditions in (3.53) - (3.55) are fulfilled.

Algorithm 3.1 shows how to reduce construction matrices for all possible cases discussed above.

3.2.3 Nullspaces of Construction Matrices

Nullspaces of construction matrices are of interest when introducing statistical notion to geometric reasoning, see chapter 4. One may skip this section for the current chapter.

The determination of the nullspace of a construction matrix can be accomplished using the interpretation of the matrix rows: for the "line-from-point" matrices $\mathsf{S}(\mathbf{x})$ and $\Pi(\mathbf{X})$, the nullspace is equal to the point $\mathbf{x}$ resp. $\mathbf{X}$, as this is the only point that is incident to the hyperplanes in the rows of $\mathsf{S}(\mathbf{x})$ resp. $\Pi(\mathbf{X})$. Similarly for the dual cases $\mathsf{S}(\mathbf{l})$ and $\overline{\Pi}(\mathbf{A})$, the nullspaces are determined by the homogeneous vectors.

Table 3.4. The two tables show the nullspaces of construction matrices for points and lines (left) and for planes and dual lines (right). The operator $\mathcal{I}(\cdot)$ denotes the image or column-space of a matrix, see text for details.

Construction matrix	Nullspace
$\mathsf{S}(\mathbf{x})$	$\{\mathbf{x}\}$
$\Pi(\mathbf{X})$	$\{\mathbf{X}\}$
$\Pi^{\mathsf{T}}(\mathbf{X})$	$\mathcal{I}(\overline{\Pi}(\mathbf{X}))$
$\Gamma(\mathbf{L})$	$\mathcal{I}(\overline{\Gamma}(\mathbf{L}))$

Construction matrix	Nullspace
$\mathsf{S}(\mathbf{l})$	$\{\mathbf{l}\}$
$\overline{\Pi}(\mathbf{A})$	$\{\mathbf{A}\}$
$\overline{\Pi}^{\mathsf{T}}(\mathbf{A})$	$\mathcal{I}(\Pi(\mathbf{A}))$
$\overline{\Gamma}(\mathbf{L}) = \Gamma(\overline{\mathbf{L}})$	$\mathcal{I}(\Gamma(\mathbf{L}))$

For $\Pi^{\mathsf{T}}(\mathbf{X})$, we are looking for all 3D lines $\mathbf{L}_i$, that are incident to the point $\mathbf{X}$, as then $\Pi^{\mathsf{T}}(\mathbf{A})\overline{\mathbf{L}_i} = \mathbf{0}$. But exactly these lines $\mathbf{L}_i$ are contained in the column-space or image $\mathcal{I}(\Pi(\mathbf{X}))$, thus $\Pi^{\mathsf{T}}(\mathbf{X})\overline{\Pi}(\mathbf{X}) = 0$. Note that the dimension of the column-space $\mathcal{I}(\Pi(\mathbf{X}))$ of $\Pi(\mathbf{X})$ is 3, although only two degrees of freedom are possible for a 3D-line incident to $\mathbf{X}$: the additional rank is due to the homogeneous scale.
Similarly one obtains the nullspace for the Plücker matrix: there, the column-space of its dual is its nullspace. All nullspaces are summarized in table 3.4.

Summary In section 3.2 we have shown how to interpret the rows and columns of a construction matrix of a geometric entity. A summary of the interpretation of all construction matrices is shown in table 3.3 on page 66. Thus any matrix (or linear map), which has the form of one of the homogeneous matrices $\mathsf{S}(\cdot)$, $\Pi(\cdot)$, $\overline{\Pi}(\cdot)$, $\Gamma(\cdot)$ or $\overline{\Gamma}(\cdot)$ can be interpreted as an over-parametrized representation for a geometric entity. One may reduce the number of rows in the matrices but still keep a complete representation for an entity.

3.3 Relations between Entities

In this section we investigate geometric relationships between entities constructed in the previous section, concentrating on pairwise relations. Especially we will cover four kinds of relations: incidence, identity, orthogonality and parallelity. Furthermore we provide simple computations for the Euclidean distances between various entities, such as the distance between a 3D line and a point.

3.3.1 Projective Relations

The first type of relations that we explore are projective relations, i.e. relations that are invariant with respect to a projective transformation. Examples of these relations are collinearity, coplanarity, incidence and identity.

3.3.1.1 Collinearity and Coplanarity In section 2.1.2 on page 23 we have found that we can use a determinant form to check the collinearity resp. coplanarity of three points in 2D resp. four points in 3D:

$$\mathbf{x}_1, \mathbf{x}_2 \text{ and } \mathbf{x}_3 \text{ are collinear} \Leftrightarrow |\, \mathbf{x}_1\ \mathbf{x}_2\, \mathbf{x}_3 \,| = 0 \tag{3.56}$$

$$\mathbf{X}_1, \mathbf{X}_2, \mathbf{X}_3 \text{ and } \mathbf{X}_4 \text{ are coplanar} \Leftrightarrow |\, \mathbf{X}_1\ \mathbf{X}_2\, \mathbf{X}_3\, \mathbf{X}_4 \,| = 0 \tag{3.57}$$

Taking the dual equations from (3.56) and (3.57) one obtains for 2D lines $\mathbf{l}_i$ and planes $\mathbf{A}_i$:

$$\mathbf{l}_1, \mathbf{l}_2 \text{ and } \mathbf{l}_3 \text{ intersect in one point} \Leftrightarrow |\, \mathbf{l}_1\ \mathbf{l}_2\, \mathbf{l}_3 \,| = 0 \tag{3.58}$$

$$\mathbf{A}_1, \mathbf{A}_2, \mathbf{A}_3 \text{ and } \mathbf{A}_4 \text{ intersect in one point} \Leftrightarrow |\, \mathbf{A}_1\ \mathbf{A}_2\, \mathbf{A}_3\, \mathbf{A}_4 \,| = 0 \tag{3.59}$$

see also figure 2.8 on page 41.

To check collinearity, coplanarity or intersection, we need at least 3 entities. For the remainder of this work we concentrate on checking pairwise relations, for example the incidence of a point and a line in 3D. Checking collinearity between three points can then be accomplished by first constructing a line from two points and check the incidence of the third point with the constructed line.

3.3.1.2 Incidence of Dual Entities Equations (3.56)– (3.59) show how to check some geometric relations using the determinant form. As in equation (2.27) on page 42 we may reinterpret the entities involved in the determinant: by constructing a plane $\mathbf{A} = \mathbf{X}_1 \wedge \mathbf{X}_2 \wedge \mathbf{X}_3$ from the first three points and check whether the fourth point $\mathbf{X}_4$ is incident to the plane:

$$\begin{aligned}
\mathbf{X}_1, \mathbf{X}_2, \mathbf{X}_3 \text{ and } \mathbf{X}_4 \text{ are coplanar} &\Leftrightarrow |\, \mathbf{X}_1\ \mathbf{X}_2\, \mathbf{X}_3\, \mathbf{X}_4 \,| &= 0 \\
&\Leftrightarrow \mathbf{A}^{\mathsf{T}}\mathbf{X} &= 0 \\
&\Leftrightarrow < \mathbf{A}, \mathbf{X} > &= 0 \\
&\Leftrightarrow \mathbf{X}_4 \in \mathbf{A}
\end{aligned}$$

with $\mathbf{A} = (\mathbf{X}_1 \wedge \mathbf{X}_2 \wedge \mathbf{X}_3)$. For lines, we can derive a similar expression using the cap-product $< \mathbf{L}, \mathbf{M} >$ as already see in section 2.5.1:

$$\begin{aligned}
\mathbf{X}_1, \mathbf{X}_2, \mathbf{X}_3 \text{ and } \mathbf{X}_4 \text{ are coplanar} &\Leftrightarrow |\, \mathbf{X}_1\ \mathbf{X}_2\, \mathbf{X}_3\, \mathbf{X}_4 \,| &= 0 \\
&\Leftrightarrow \mathbf{L}^{\mathsf{T}}\overline{\mathbf{M}} &= 0 \\
&\Leftrightarrow < \mathbf{L}, \mathbf{M} > &= 0 \\
&\Leftrightarrow \mathbf{L} \cap \mathbf{M} \neq \emptyset
\end{aligned}$$

with $\mathbf{L} = \mathbf{X}_1 \wedge \mathbf{X}_2$ and $\mathbf{M} = \mathbf{X}_3 \wedge \mathbf{X}_4$, see also equations (2.28) and (2.29) on page 42.
Both incidence relations are checked by using the cap-product $< \cdot, \cdot >$: we can express the following incidence relations for dual entities using the cap-product: $\mathbf{x} \in \mathbf{l}$, $\mathbf{X} \in \mathbf{A}$ and $\mathbf{L} \cap \mathbf{M} \neq \emptyset$:

$$d = \langle \mathbf{x}, \mathbf{l} \rangle = \mathbf{x}^{\mathsf{T}} \mathbf{l} \stackrel{!}{=} 0 \qquad (3.60)$$

$$d = \langle \mathbf{X}, \mathbf{A} \rangle = \mathbf{X}^{\mathsf{T}} \mathbf{A} \stackrel{!}{=} 0 \qquad (3.61)$$

$$d = \langle \mathbf{L}, \mathbf{M} \rangle = \mathbf{L}^{\mathsf{T}} \overline{\mathbf{M}} \stackrel{!}{=} 0 \qquad (3.62)$$

To check whether the incidence relation holds, one has to check whether $d = 0$.
There exists no cap-product for the other projective relations, for example the incidence of a 3D line and a plane. For these relations, we have to find the corresponding algebraic expressions, which is the goal of the following paragraphs.

Remark For the incidence relation $\mathbf{L} \cap \mathbf{M} \neq \emptyset \;\Leftrightarrow\; \mathbf{L}^{\mathsf{T}} \overline{\mathbf{M}} = 0$ we can give another proof based in the algebraic expressions for constructions: assume two lines $\mathbf{L} = \mathbf{X} \wedge \mathbf{Y}$ and $\mathbf{M} = \mathbf{Z} \wedge \mathbf{W}$. The two lines are incident if and only if the four points are coplanar. Thus we construct a plane

$$\mathbf{A} = \mathbf{X} \wedge \mathbf{Z} \wedge \mathbf{W} = \mathbf{X} \wedge \mathbf{M} = \overline{\Pi}^{\mathsf{T}}(\mathbf{X})\mathbf{M} = \Pi^{\mathsf{T}}(\mathbf{X})\overline{\mathbf{M}}$$

We can now write:

$$\mathbf{Y} \in \mathbf{A} \Leftrightarrow \mathbf{Y}^{\mathsf{T}} \left(\Pi^{\mathsf{T}}(\mathbf{X})\overline{\mathbf{M}}\right) = 0 \Leftrightarrow (\Pi(\mathbf{X})\mathbf{Y})^{\mathsf{T}} \overline{\mathbf{M}} = 0 \Leftrightarrow \mathbf{L}^{\mathsf{T}} \overline{\mathbf{M}} = 0$$

3.3.1.3 Identity of 2D Points $\mathbf{x} \equiv \mathbf{y}$ and 2D Lines $\mathbf{l} \equiv \mathbf{m}$ Checking the identity of two homogeneous vectors from $\mathbb{P}^2$ is similar to checking the collinearity of two directional vectors $\boldsymbol{r}, \boldsymbol{v} \in \mathbb{R}^3$ in 3D: they are identical if and only if their cross product is zero, i.e. collinear$(\boldsymbol{r}, \boldsymbol{v}) \Leftrightarrow \boldsymbol{r} \times \boldsymbol{v} = \mathbf{0}$. The geometric interpretation is obvious as the cross-product yields the vector orthogonal to $\boldsymbol{r}$ and $\boldsymbol{v}$ and is undefined if they are collinear.
Thus the identity of points in 2D can be checked by

$$\mathbf{d} = \mathsf{S}(\mathbf{x})\mathbf{y} \stackrel{!}{=} \mathbf{0} \qquad (3.63)$$

Using the interpretation of $\mathsf{S}(\mathbf{x})$ from section 3.2, the equation $\mathsf{S}(\mathbf{x})\mathbf{y} = \mathbf{0}$ performs three incidence tests of the point $\mathbf{y}$ with the three lines $\mathbf{x} \wedge \mathbf{e}_i$, $i = 1, 2, 3$. In general only two incidence tests are sufficient to determine the validity of the relation, as the test has two degrees of freedom. Thus we can use the reduced matrix $\mathsf{S}^{[2]}(\mathbf{x})$, mentioned on page on page 63 and which contains two canonical lines. Thus the check for identity can be reformulated as

$$\mathbf{d}^{[2]} = \mathsf{S}^{[2]}(\mathbf{x})\mathbf{y} = -\mathsf{S}^{[2]}(\mathbf{y})\mathbf{x} \stackrel{!}{=} \mathbf{0} \tag{3.64}$$

The identity of 2D lines $\mathbf{l} \equiv \mathbf{m}$ is checked in exactly the same manner.

The reduction is computed as in algorithm 3.1. Note that if one does not apply the proposed coordinate transformation in step 1 of the algorithm, the reduction fails only for cases, where the two vectors contain zeroes and are exactly orthogonal, $\forall\, i: \; |x_i y_i| = 0$, see also equation (3.53). If one knows that the lines are close to being identical, thus very far from being orthogonal, the rotation step may be skipped.

3.3.1.4 Incidence $\mathbf{L} \in \mathbf{A}$ of a 3D Line $\mathbf{L}$ and a Plane $\mathbf{A}$ For the relation of a 3D line and a plane we can use a construction matrix as in the check for the identity of 2D points and lines: the line $\mathbf{L}$ has four canonical points, represented as columns of the Plücker matrix $\mathsf{\Gamma}^\mathsf{T}(\mathbf{L})$, see figure 3.7. In general it is sufficient to check whether two canonical points from $\mathsf{\Gamma}^\mathsf{T}(\mathbf{L})$ are incident to the plane $\mathbf{A}$, so the test has 2 degrees of freedom. The test may also

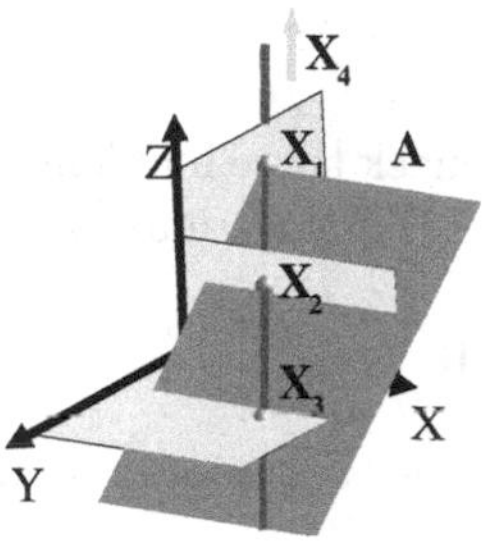

Fig. 3.7. Checking the incidence of a 3D line $\mathbf{L}$ with a plane $\mathbf{A}$ by checking the incidence of the canonical points of $\mathbf{L}$ with the plane $\mathbf{A}$.

be performed by choosing two columns of $\overline{\mathsf{\Pi}}(\mathbf{A})$ as they correspond to two canonical lines of the plane $\mathbf{A}$. The canonical lines may then be checked for incidence to the given line $\mathbf{L}$, by $\overline{\mathsf{\Pi}}(\mathbf{A})\overline{\mathbf{L}} = \mathsf{\Pi}(\mathbf{A})\mathbf{L}$, see also equation (3.62). So the algebraic expression of the test is bilinear:

$$\mathbf{D} = \mathsf{\Gamma}(\mathbf{L})\mathbf{A} = \mathsf{\Pi}^\mathsf{T}(\mathbf{A})\mathbf{L} \stackrel{!}{=} \mathbf{0} \tag{3.65}$$

We may use the reduced construction matrix $\mathsf{\Gamma}^{[2]}(\mathbf{L})$ cf. page 67 to test the incidence of the plane with only two out of four canonical points. Likewise, we choose two rows from $\mathsf{\Pi}^\mathsf{T}(\mathbf{A})$ yielding $\mathsf{\Pi}^{[2]\mathsf{T}}(\mathbf{A})$ according to the algorithm 3.1 on page 69.

$$\mathbf{D}^{[2]} = \mathsf{\Gamma}^{[2]}(\mathbf{L})\mathbf{A} = \mathsf{\Pi}^{[2]\mathsf{T}}(\mathbf{A})\mathbf{L} \stackrel{!}{=} \mathbf{0} \tag{3.66}$$

Again, the proposed transformation in step 1 of the reduction algorithm is only needed for special cases of lines and planes: $\forall\, i,j: \; |\overline{\mathsf{\Pi}}^\mathsf{T}_{i,j}(\mathbf{A})\, L_j| = 0$. Such cases happen only if the incidence relation is not fulfilled at all: for example, if $\mathbf{A}$ is the infinite plane and the line $\mathbf{L}$ is incident to the origin. In general, this may only happen if $\mathbf{A} \in \overline{\mathbf{L}}$.

3.3.1.5 Incidence $\mathbf{X} \in \mathbf{L}$ of a 3D Point $\mathbf{X}$ and a Line $\mathbf{L}$ This is the dual case of the previous test and thus we obtain the bilinear algebraic expression

$$\mathbf{D} = \overline{\Gamma}(\mathbf{L})\mathbf{X} = \overline{\Pi}^{\mathsf{T}}(\mathbf{X})\mathbf{L} \stackrel{!}{=} \mathbf{0}$$

and

$$\mathbf{D}^{[2]} = \overline{\Gamma}^{[2]}(\mathbf{L})\mathbf{X} = \overline{\Pi}^{[2]\mathsf{T}}(\mathbf{X})\mathbf{L} \stackrel{!}{=} \mathbf{0}$$

If one skips the coordinate transformation step in algorithm 3.1, then a reduction may be not possible for special cases and if $\mathbf{X} \in \overline{\mathbf{L}}$ instead of $\mathbf{X} \in \mathbf{L}$, see equation (3.54) on page 69 for an exact condition.

3.3.1.6 Identity of 3D Points $\mathbf{X} \equiv \mathbf{Y}$ and Planes $\mathbf{A} \equiv \mathbf{B}$ Since there is no simple test for the collinearity of directional vectors in $\mathbb{R}^4$, we adopt the above method of using the construction matrices: if the point $\mathbf{Y}$ is incident to the six canonical planes contained in the matrix $\Pi(\mathbf{X})$, then the points $\mathbf{X}$ and $\mathbf{Y}$ are identical:

$$\mathbf{D} = \Pi(\mathbf{X})\mathbf{Y} = -\Pi(\mathbf{Y})\mathbf{X} \stackrel{!}{=} \mathbf{0}$$

The test has 3 degrees of freedom, thus it is sufficient to check three incidences as long as the canonical planes are defined, i.e. non-zero. Thus we may write

$$\mathbf{D}^{[3]} = \Pi^{[3]}(\mathbf{X})\mathbf{Y} = -\Pi^{[3]}(\mathbf{Y})\mathbf{X} \stackrel{!}{=} \mathbf{0}$$

The reduction of the construction matrices is valid as long as the points are not at infinity.

The identity of 3D planes $\mathbf{A} \equiv \mathbf{B}$ is the dual case to the identity of points and thus

$$\mathbf{D} = \overline{\Pi}(\mathbf{A})\mathbf{B} = -\overline{\Pi}(\mathbf{B})\mathbf{A} \stackrel{!}{=} \mathbf{0}$$

and

$$\mathbf{D}^{[3]} = \overline{\Pi}^{[3]}(\mathbf{A})\mathbf{B} = -\overline{\Pi}^{[3]}(\mathbf{B})\mathbf{A} \stackrel{!}{=} \mathbf{0}$$

For the reduction of the matrices, we apply the algorithm 3.1 again. As in the case of identity of two lines, the proposed coordinate transformation in step 1 of the algorithm is only needed for special cases, where two planes are exactly orthogonal and only one coordinate of each plane is non-zero, cf. equation (3.53) on page 69

3.3.1.7 Identity of 3D Lines $\mathbf{L} \equiv \mathbf{M}$ The expression $\langle \mathbf{L}, \mathbf{M} \rangle = 0$ only checks if two 3D lines share one common point. To check the identity of all points on the lines, we use the dual Plücker-matrix $\overline{\Gamma}(\mathbf{L})$ for line $\mathbf{L}$ *and* the Plücker-matrix $\Gamma(\mathbf{M})$ for line $\mathbf{M}$: the matrix product

$$\mathsf{D} = \overline{\Gamma}(\mathbf{L})\Gamma(\mathbf{M}) \stackrel{!}{=} 0 \tag{3.67}$$

contains 16 cap-products of the canonical planes of $\mathbf{L}$ with the canonical points of $\mathbf{M}$. In general 4 incidences are sufficient for this test, as long as the corresponding columns and rows are non-zero. Applying a reduction to both matrices yield

$$\mathsf{D}^{[2,2]} = \overline{\Gamma}^{[2]}(\mathbf{L})\Gamma^{[2]}(\mathbf{M})$$

The matrix $\mathsf{D}^{[2,2]}$ is a 2×2 matrix containing 4 different incidence tests of canonical points of $\mathbf{M}$ with canonical planes of $\mathbf{L}$, thus we have the desired number of tests. We may choose those rows of $\overline{\Gamma}(\mathbf{L})$ resp. columns of $\overline{\Gamma}(\mathbf{M})$, which have the largest norm.

The difference of the 3D-line-identity relation in contrast to the previous ones is the fact that the distance test in equation (3.67) is a matrix-matrix product. To turn it into a matrix-vector product of the bilinear form $\mathbf{D} = \mathrm{vec}(\mathsf{D}) = \mathsf{U}(\mathbf{L})\mathbf{M} = \mathsf{V}(\mathbf{M})\mathbf{L}$, we can take advantage of the Kronecker product: using

$$\mathrm{vec}(\overline{\Gamma}(\mathbf{L})) = \mathsf{J}_\Gamma C\mathbf{L} \quad \text{with } \mathsf{J}_\Gamma{}^\mathsf{T} = \begin{pmatrix} 0 & 0 & 0 & 0 & 0 & 0 & -1 & 0 & 0 & 1 & 0 & 0 & 0 & 0 & 0 & 0 \\ 0 & 0 & 1 & 0 & 0 & 0 & 0 & 0 & -1 & 0 & 0 & 0 & 0 & 0 & 0 & 0 \\ 0 & -1 & 0 & 0 & 1 & 0 & 0 & 0 & 0 & 0 & 0 & 0 & 0 & 0 & 0 & 0 \\ 0 & 0 & 0 & 1 & 0 & 0 & 0 & 0 & 0 & 0 & 0 & 0 & -1 & 0 & 0 & 0 \\ 0 & 0 & 0 & 0 & 0 & 0 & 0 & 1 & 0 & 0 & 0 & 0 & 0 & -1 & 0 & 0 \\ 0 & 0 & 0 & 0 & 0 & 0 & 0 & 0 & 0 & 0 & 0 & 1 & 0 & 0 & -1 & 0 \end{pmatrix} \tag{3.68}$$

the bilinear forms for $\mathrm{vec}(\mathsf{D})$ depending on $\mathbf{L}$ and $\mathbf{M}$ are

$$\mathbf{D} = \mathrm{vec}(\mathsf{D}) = \mathrm{vec}(\overline{\Gamma}(\mathbf{L})\Gamma(\mathbf{M})) = (\Gamma^\mathsf{T}(\mathbf{M}) \otimes I_4)\,\mathrm{vec}(\overline{\Gamma}(\mathbf{L})) \tag{3.69}$$

$$= \left[(\Gamma^\mathsf{T}(\mathbf{M}) \otimes I_4)\mathsf{J}_\Gamma\, C\right]\mathbf{L} \;= \mathsf{V}(\mathbf{M})\,\mathbf{L} \tag{3.70}$$

$$= (I_4 \otimes \overline{\Gamma}(\mathbf{L}))\,\mathrm{vec}(\Gamma(\mathbf{M})) \tag{3.71}$$

$$= \left[(I_4 \otimes \overline{\Gamma}(\mathbf{L}))\mathsf{J}_\Gamma\right]\mathbf{M} \qquad = \mathsf{U}(\mathbf{L})\,\mathbf{M} \tag{3.72}$$

The matrices $\mathsf{U}(\mathbf{L})$ and $\mathsf{V}(\mathbf{M})$ are of dimension 16×6 and must be at least of rank 4.

The rows of $\mathsf{U}(\mathbf{L})$ resp. $\mathsf{V}(\mathbf{L})$ can be regarded as dual representations of special 3D lines: using the interpretation of the rows and columns of the Plücker matrices (table 3.3), the (i,j)-th element of the matrix $\mathsf{D} = \overline{\Gamma}(\mathbf{L})\Gamma(\mathbf{M})$ is

$$\mathsf{D}_{ij} = \langle \mathbf{L} \wedge \mathbf{E}_i,\; -(\mathbf{M} \cap \mathbf{E}_i)\rangle\rangle \tag{3.73}$$

$$= \left(\overline{\Pi}(\mathbf{E}_i)\mathbf{L}\right)^\mathsf{T} \Pi(\mathbf{E}_j)\mathbf{M} \tag{3.74}$$

$$= \mathbf{L}^\mathsf{T}\left(\overline{\Pi}^\mathsf{T}(\mathbf{E}_i)\Pi(\mathbf{E}_j)\right)\mathbf{M} \tag{3.75}$$

$$= \mathbf{M}^\mathsf{T}\left(\Pi(\mathbf{E}_j)\overline{\Pi}^\mathsf{T}(\mathbf{E}_i)\right)\mathbf{L} \tag{3.76}$$

As $\mathbf{L}_{ij} := \overline{\Pi}(\mathbf{E}_i)\Pi(\mathbf{E}_j)^\mathsf{T}\mathbf{L}$ is a valid 3D line, the matrix element D_{ij} can be interpreted as a cap product $\langle \overline{\mathbf{L}}_{ij}, \mathbf{M}\rangle$ between two 3D lines, and likewise $\langle \overline{\mathbf{M}}_{ij}, \mathbf{L}\rangle$.

Unfortunately, it is not easy to see how to simultaneously choose $r = 4$ rows of $\mathsf{U}(\mathbf{L})$ and $\mathsf{V}(\mathbf{M})$. In the following section we will describe an easier way to obtain $\mathsf{U}^{[r]}(\mathbf{L})$ and $\mathsf{V}^{[r]}(\mathbf{M})$.

3.3.1.8 Yet Another Algebraic Expression for Identity Relations

We finally describe a different way to check the equivalence of homogeneous vectors. We will see that it effectively leads to the same bilinear forms as proposed above, but can be more efficient to compute. We use the equivalence $\mathbf{L} \equiv \mathbf{M}$ of lines as an example, the same can be done for points and hyperplanes in 2D and 3D.

The method is based on the proportionality of the homogeneous coordinates: given two lines $\mathbf{L}$ and $\mathbf{M}$, we choose an index $i = 1, \cdots 6$ such that $|L_i M_i| > 0$. We then scale both vectors such that for both the i-th coordinate is $L_i M_i$. The difference $\mathbf{D}_{\Delta,i}$ of the scaled vectors must be zero:

$$\mathbf{D}_{\Delta,i} = M_i \mathbf{L} - L_i \mathbf{M} \stackrel{!}{=} \mathbf{0} \tag{3.77}$$

We can write this in a bilinear form

$$\mathbf{D}_{\Delta,i} = \Delta_i(\mathbf{L})\mathbf{M} = -\Delta_i(\mathbf{M})\mathbf{L} \quad \text{with } \Delta_i(\mathbf{L}) := \mathbf{L}\mathbf{E}_i^{\mathsf{T}} - L_i I_6 \tag{3.78}$$

Especially compared to equation (3.73), we gain computational efficiency at the expense of choosing an index i. One advantage here is though, that the matrices $\Delta_i(\mathbf{L})$ and $\Delta_i(\mathbf{M})$ share four linearly independent rows: e.g. consider $\Delta_i(\mathbf{L})$ for $i = 4$,

$$\Delta_4(\mathbf{L}) = \begin{pmatrix} -L_4 & 0 & 0 & L_1 & 0 & 0 \\ \hline 0 & -L_4 & 0 & L_2 & 0 & 0 \\ 0 & 0 & -L_4 & L_3 & 0 & 0 \\ \hline 0 & 0 & 0 & 0 & 0 & 0 \\ \hline 0 & 0 & 0 & L_5 & -L_4 & 0 \\ 0 & 0 & 0 & L_6 & 0 & -L_4 \end{pmatrix} \begin{matrix} \leftarrow \text{ not a valid 3D line} \\ \\ \\ \leftarrow \text{ not a valid 3D line} \\ \\ \\ \end{matrix}$$

In general, only the first and fourth row do not correspond to a valid 3D line - the first one due to the Plücker condition. Skipping both of these rows leaves a reduced 4×6 matrix $\Delta_4^{[r]}(\mathbf{L})$ with rank $r = 4$. The resulting rows of $\Delta_4^{[r]}(\mathbf{L})$ are contained in the 16×6 matrices $\mathsf{U}(\mathbf{L})$ and $\mathsf{V}(\mathbf{M})$: the 6×6 matrices $\Phi_{ij} = \overline{\Pi}^{\mathsf{T}}(\mathbf{E}_i)\Pi(\mathbf{E}_j),\ \ i \neq j$ are antisymmetric matrices with only two non-zero entries, +1 at position (s,t) and -1 at position (t,s) or vice versa. Only the pairs $(s,t) = (1,4)$, $(2,5)$ and $(3,5)$ do not occur due to the Plücker condition.

Thus the matrices $\Delta_i^{[r]}(\mathbf{L})$ and $\Delta_i^{[r]}(\mathbf{U})$ can be regarded as a (not necessarily order and sign-preserving) reduction of $\mathsf{U}(\mathbf{L})$ and $\mathsf{V}(\mathbf{M})$.

Note that the reduction only holds if there exists an i such that $|L_i M_i| > 0$. But if $L_i M_i = 0$ for all i, the reduction may fail. Using a coordinate transformation step as in algorithm 3.1, one can ensure a reduction for all

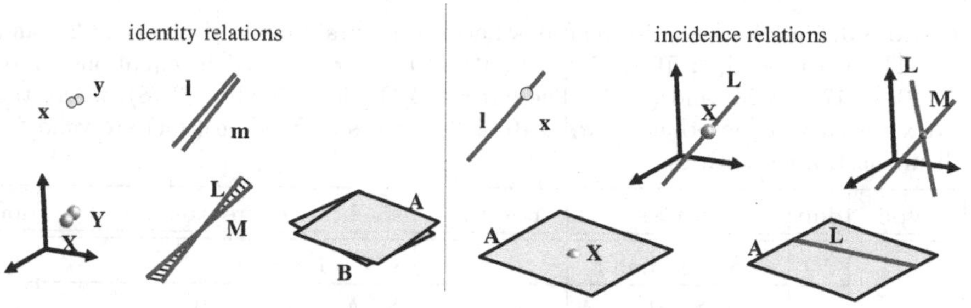

Fig. 3.8. Visualization of incidence (*right*) and identity relations (*left*) for 2D and 3D entities. The algebraic expressions for checking these relations can be found in table 3.5.

possible lines **L** and **M**. In this case, only a rotation is needed, e.g. a random, non-identical rotation would do.

To summarize the results of this section, all projective relations together with their algebraic expressions are listed in table 3.5 on the next page, see also figure 3.8.

3.3.2 Affine and Similarity Relations

The *parallelity relation* between lines and/or planes is an affine relation, i.e. it is invariant under affine transformations, but not under projective transformations. This observation is easy to prove since the intersection of parallel lines and/or planes is at infinity; the 2D line at infinity $\mathbf{e}_3$ respectively the 3D plane at infinity $\mathbf{E}_4$ remains invariant under affine transformation, cf. equation (2.21) on page 35.

With the proposed representation for lines and planes the check for parallelity is simple: The homogeneous parts $\boldsymbol{l}_h$ of a 2D line, $\boldsymbol{L}_h$ of a 3D line and $\boldsymbol{A}_h$ of a plane contain the directional information, either as a normal for hyperplanes or as the direction vector for the 3D line, thus collinearity of two lines or two planes in 3D can be checked using the above mentioned collinearity relation of directional vectors $\boldsymbol{r}$, $\boldsymbol{v}$:

$$\text{collinear}(\boldsymbol{r}, \boldsymbol{v}) \quad \Leftrightarrow \quad \boldsymbol{r} \times \boldsymbol{v} = \mathbf{0} \quad \Leftrightarrow \quad \boldsymbol{d} = \mathsf{S}(\boldsymbol{r})\boldsymbol{v} = \mathbf{0}$$

A plane and a line are parallel if the normal of the plane is orthogonal to the direction of the line, thus a simple dot-product $d = \boldsymbol{L}_h^\mathsf{T} \boldsymbol{A}_h \stackrel{!}{=} 0$ is sufficient. For two 2D lines $\mathbf{l}$ and $\mathbf{m}$, we compute the orthogonal direction $\boldsymbol{l}_h^\perp = (b, -a)^\mathsf{T}$ of a homogeneous part $\boldsymbol{l}_h = (a, b)^\mathsf{T}$ and check if it is perpendicular to $\boldsymbol{m}_h$ by $(\boldsymbol{l}_h^\perp)^\mathsf{T} \boldsymbol{m}_h \stackrel{!}{=} 0$.

Orthogonality in general is not an affine transformation but is invariant under similarity transformations. The tests for lines and planes in 2D and 3D use the same methods as above and are listed in table 3.5, see also figure 3.9.

Table 3.5. Possible pairwise relations between points, lines and planes in 2D and 3D. The matrices $\mathsf{S}(\cdot)$, $\mathsf{\Pi}(\cdot)$, $\overline{\mathsf{\Pi}}(\cdot)$, $\mathsf{\Gamma}(\cdot)$ and $\overline{\mathsf{\Gamma}}(\cdot)$ are defined in equations (3.5), (3.12), (3.17), (3.23) and (3.28). The matrix $\Delta_i(\mathbf{L})$ is defined in (3.78), where the index i is chosen such that $|L_i M_i| \gg 0$. The degrees of freedom (dof) are valid for all unconstrained entities.

type		dim	entities	notation	algebraic expression	dof
projective	Incidence	2D	point $\mathbf{x}$, line $\mathbf{l}$	$\mathbf{x} \in \mathbf{l}$	$\mathbf{x}^\mathsf{T}\mathbf{l} = \mathbf{l}^\mathsf{T}\mathbf{x} = 0$	1
		3D	point $\mathbf{X}$, plane $\mathbf{A}$	$\mathbf{X} \in \mathbf{A}$	$\mathbf{X}^\mathsf{T}\mathbf{A} = \mathbf{A}^\mathsf{T}\mathbf{X} = 0$	1
			point $\mathbf{X}$, line $\mathbf{L}$	$\mathbf{X} \in \mathbf{L}$	$\overline{\mathsf{\Pi}}^\mathsf{T}(\mathbf{X})\mathbf{L} = \overline{\mathsf{\Gamma}}^\mathsf{T}(\mathbf{L})\mathbf{X} = 0$	2
			line $\mathbf{L}$, line $\mathbf{M}$	$\mathbf{L} \cap \mathbf{M} \neq \emptyset$	$\overline{\mathbf{L}}^\mathsf{T}\mathbf{M} = \overline{\mathbf{M}}^\mathsf{T}\mathbf{L} = \mathbf{0}$	1
			line $\mathbf{L}$, plane $\mathbf{A}$	$\mathbf{L} \in \mathbf{A}$	$\mathsf{\Pi}(\mathbf{A})^\mathsf{T}\mathbf{L} = \mathsf{\Gamma}^\mathsf{T}(\mathbf{L})\mathbf{A} = \mathbf{0}$	2
	Identity	2D	points $\mathbf{x}, \mathbf{y}$	$\mathbf{x} \equiv \mathbf{y}$	$\mathsf{S}(\mathbf{x})\mathbf{y} = -\mathsf{S}(\mathbf{y})\mathbf{x} = \mathbf{0}$	2
			line $\mathbf{l}, \mathbf{m}$	$\mathbf{l} \equiv \mathbf{m}$	$\mathsf{S}(\mathbf{l})\mathbf{m} = -\mathsf{S}(\mathbf{l})\mathbf{m} = \mathbf{0}$	2
		3D	points $\mathbf{X}, \mathbf{Y}$	$\mathbf{X} \equiv \mathbf{Y}$	$\mathsf{\Pi}(\mathbf{X})\mathbf{Y} = -\mathsf{\Pi}(\mathbf{Y})\mathbf{X} = \mathbf{0}$	3
			lines $\mathbf{L}, \mathbf{M}$	$\mathbf{L} \equiv \mathbf{M}$	$\overline{\mathsf{\Gamma}}(\mathbf{L})\mathsf{\Gamma}(\mathbf{M}) = \mathsf{\Gamma}(\mathbf{M})\overline{\mathsf{\Gamma}}(\mathbf{L}) = 0$ $\Delta_i(\mathbf{L})\mathbf{M} = \Delta_i(\mathbf{M})\mathbf{L} = \mathbf{0}$	4
			planes $\mathbf{A}, \mathbf{B}$	$\mathbf{A} \equiv \mathbf{B}$	$\overline{\mathsf{\Pi}}(\mathbf{A})\mathbf{B} = -\overline{\mathsf{\Pi}}(\mathbf{B})\mathbf{A} = 0$	3
affine	Parallelity	2D	lines $\mathbf{l}, \mathbf{m}$	$\mathbf{l} \parallel \mathbf{m}$	$(\boldsymbol{l}_h^\perp)^\mathsf{T}\boldsymbol{m}_h = (\boldsymbol{m}_h^\perp)^\mathsf{T}\boldsymbol{l}_h = 0$	1
		3D	lines $\mathbf{L}, \mathbf{M}$	$\mathbf{L} \parallel \mathbf{M}$	$S(\boldsymbol{L}_h)\boldsymbol{M}_h = -S(\boldsymbol{M}_h)\boldsymbol{L}_h = \mathbf{0}$	2
			planes $\mathbf{A}, \mathbf{B}$	$\mathbf{A} \parallel \mathbf{B}$	$S(\boldsymbol{A}_h)\boldsymbol{B}_h = -S(\boldsymbol{B}_h)\boldsymbol{A}_h = \mathbf{0}$	2
			line $\mathbf{L}$, plane $\mathbf{A}$	$\mathbf{L} \parallel \mathbf{A}$	$\boldsymbol{L}_h^\mathsf{T}\boldsymbol{A}_h = \boldsymbol{A}_h^\mathsf{T}\boldsymbol{L}_h = 0$	1
similar	Ortho-gonality	2D	lines $\mathbf{l}, \mathbf{m}$	$\mathbf{l} \perp \mathbf{y}$	$\boldsymbol{l}_h^\mathsf{T}\boldsymbol{m}_h = \boldsymbol{m}_h^\mathsf{T}\boldsymbol{l}_h = 0$	1
		3D	lines $\mathbf{L}, \mathbf{M}$	$\mathbf{L} \perp \mathbf{M}$	$\boldsymbol{L}_h^\mathsf{T}\boldsymbol{M}_h = \boldsymbol{M}_h^\mathsf{T}\boldsymbol{L}_h = 0$	1
			planes $\mathbf{A}, \mathbf{B}$	$\mathbf{A} \perp \mathbf{B}$	$\boldsymbol{A}_h^\mathsf{T}\boldsymbol{B}_h = \boldsymbol{B}_h^\mathsf{T}\boldsymbol{A}_h = 0$	1
			line $\mathbf{L}$, plane $\mathbf{A}$	$\mathbf{L} \perp \mathbf{A}$	$S(\boldsymbol{L}_h)\boldsymbol{A}_h = -S(\boldsymbol{A}_h)\boldsymbol{L}_h = \mathbf{0}$	2
Euclidean (metric)	Distance	2D	points $\mathbf{x}, \mathbf{y}$	$d(\mathbf{x}, \mathbf{y}) = s$	$\dfrac{\lvert x_h \boldsymbol{y}_O - y_h \boldsymbol{x}_O \rvert}{\lvert x_h y_h \rvert} = s$	1
			point $\mathbf{x}$, line $\mathbf{l}$	$d(\mathbf{x}, \mathbf{l}) = s$	$\dfrac{\lvert \mathbf{x}^\mathsf{T}\mathbf{l} \rvert}{\lvert x_h \mathbf{l}_h \rvert} = s$	1
		3D	points $\mathbf{X}, \mathbf{Y}$	$d(\mathbf{X}, \mathbf{Y}) = s$	$\dfrac{\lvert X_h \boldsymbol{Y}_O - Y_h \boldsymbol{X}_O \rvert}{\lvert X_h Y_h \rvert} = s$	1
			lines $\mathbf{L}, \mathbf{M}$	$d(\mathbf{L}, \mathbf{M}) = s$	$\dfrac{\lvert \overline{\mathbf{L}}^\mathsf{T}\mathbf{M} \rvert}{\lvert S(\boldsymbol{L}_h)\boldsymbol{M}_h \rvert} = s$	1
			point $\mathbf{X}$ line $\mathbf{L}$	$d(\mathbf{X}, \mathbf{L}) = s$	$\dfrac{\lvert S(\boldsymbol{L}_h)\boldsymbol{X}_O + X_h \boldsymbol{L}_O \rvert}{\lvert X_h \boldsymbol{L}_h \rvert} = s$	1
			point $\mathbf{X}$, plane $\mathbf{A}$	$d(\mathbf{X}, \mathbf{A}) = s$	$\dfrac{\lvert \mathbf{X}^\mathsf{T}\mathbf{A} \rvert}{\lvert X_h \boldsymbol{A}_h \rvert} = s$	1

3.3.3 Distance Relations

The last type of relation we describe in this work are the Euclidean distance relations between two entities. For example the distance between a point $\mathbf{X}$ and a line $\mathbf{L}$ in 3D is assumed to be s, thus we check whether $d(\mathbf{X}, \mathbf{L}) = s$. A list of all possible distance relations can be found in table 3.5, we will skip a detailed proof in this work and only sketch the main ideas.

For all relations including a point $\mathbf{X}$, one can apply a translation T, such that the point $\mathsf{T}\mathbf{x}$ is equal to the origin. This transformation is also applied to the other entity, say $\mathbf{L}' = \mathsf{T}_L\mathbf{L}$ for lines, where T_L is from equation (3.42). Then $d(\mathbf{X}, \mathbf{L}) = \frac{|\boldsymbol{L}'_O|}{|\boldsymbol{L}'_h|}$. The formula for the distance between two lines $\mathbf{L}, \mathbf{M}$ can be followed from the proof that two lines are incident, see page 43 and BRAND 1966, p. 54.

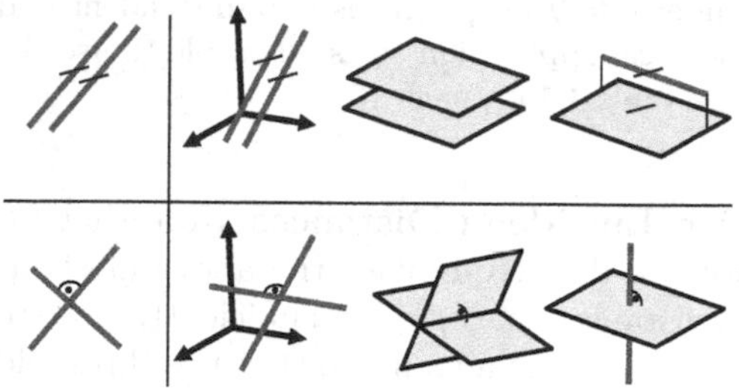

Fig. 3.9. Visualization of parallelity (*top*) and orthogonality relations (*bottom*) for 2D and 3D entities. The algebraic expressions for checking these relations can be found in table 3.5.

3.3.4 Checking Geometric Relations

Given the relations in table 3.5, we can formulate a generic algorithm for checking projective, affine and similarity relations R using the bilinear form

$$R(\mathbf{x}, \mathbf{y}) \Leftrightarrow \mathbf{d} = \mathsf{U}(\mathbf{x})\mathbf{y} = \mathsf{V}(\mathbf{y})\mathbf{x} = \mathbf{0} \qquad (3.79)$$

It is sufficient to check the r largest absolute values of the distance vector $\mathbf{d}$, where r is the degree of freedom of the relation $R(\mathbf{x}, \mathbf{y})$.

We may also use reduced matrices $\mathsf{U}(\mathbf{x})^{[r]}$ and $\mathsf{V}(\mathbf{y})^{[r]}$ obtained by the algorithm 3.1, yielding

$$R(\mathbf{x}, \mathbf{y}) \Leftrightarrow \mathbf{d}^{[r]} = \mathsf{U}^{[r]}(\mathbf{x})\mathbf{y} = \mathsf{V}^{[r]}(\mathbf{y})\mathbf{x} = \mathbf{0} \qquad (3.80)$$

Note that the coordinate transformation step of the reduction algorithm is only needed for special cases, cf. equations (3.53), (3.54) and (3.55) on page 69, where the relation $R(\mathbf{x}, \mathbf{y})$ is not fulfilled at all. As we have seen in section 3.3.1, these cases do usually not occur in practice. The reduced distance $\mathbf{d}^{[r]}$ will become handy when doing statistical tests, see chapter 4.

The simple method for check relations in 2D and 3D is summarized in algorithm 3.2.

Checking geometric relations

Objective: Given two geometric entities $\mathbf{x}$ and $\mathbf{y}$ (possibly in 2D and 3D) and a relation $R \in \{equal, incident, orthogonal, parallel\}$. Check whether the relation $R(\mathbf{x}, \mathbf{y})$ holds, given some tolerance ϵ.

1. Obtain Jacobian $\mathsf{U}(\mathbf{x})$ (or alternatively) $\mathsf{V}(\mathbf{y})$ according to table 3.5, such that $\mathbf{d} = \mathsf{U}(\mathbf{x})\mathbf{y}$ (or $\mathsf{V}(\mathbf{y})\mathbf{x}$).
2. Compute $\mathbf{d} = \mathsf{U}(\mathbf{x})\mathbf{y}$ and check whether $d_i < \epsilon$ for all coordinate d_i of $\mathbf{d}$. It is sufficient to check the first r largest d_i, where r is equal to the degrees of freedom of the relation, see last column of table 3.5.

2'. Alternative: one can compute the reduced distance vector $\mathbf{d}^{[r]} = \mathsf{U}^{[r]}(\mathbf{x})\mathbf{y}$ using the reduction algorithm 3.1 on page 69.

Algorithm 3.2: A non-statistical algorithm for checking geometric relations. The alternative of the second step is computationally more involving, but is useful for chapter 4. The choice of a small number $\epsilon \approx 0$ in general is data-dependent and thus can not be globally fixed. For practical applications it is advisable to use the statistical version of this algorithm, see table 4.4 on page 136.

3.3.4.1 Checking Relations with Euclidean Distances Note that for some of the projective relations we can directly replace the above methods by checking whether the Euclidean distance $d(\mathbf{x}, \mathbf{y}) < \epsilon$ is sufficiently close to zero. Checking for Euclidean distances has the advantage that the threshold ϵ can be interpreted with respect to the application. This is available only for 6 out of the 10 projective relations, see section 3.3.3 on the page before

Another option is to take advantage of the fact that for projective relations the distance vector $\mathbf{d} = \mathsf{U}(\mathbf{x})\mathbf{y}$ only contains incidence relations between dual entities. Thus we interpret each row of the construction matrix $\mathsf{U}(\mathbf{x})$ as an entity $\mathbf{z}_i^\mathsf{T}$ – the canonical entity – and compute the Euclidean distances $d(\mathbf{z}_i, \mathbf{y})$ between $\mathbf{y}$ and $\mathbf{z}_i$ as above. The Euclidean distances can be interpreted and checked against a given Euclidean distance threshold ϵ. Note that these distances depend on the choice of the coordinate system as the canonical entities of a construction matrix are defined by the coordinate system. We don't describe this approach in detail as we don't recommend the algorithm 3.2 or any variants that directly compute Euclidean distances: for practical applications one should use the statistical algorithm introduced in section 4.5.

3.4 General Construction of Entities

The construction methods in section 3.1 always result in a unique geometric entity, which may lie at infinity. But Computer Vision and Photogrammetry tasks often involve non-unique construction of entities: a simple example is the classical forward intersection of image points given two cameras, see figure 3.10(a): the entity point $\mathbf{X}$ is supposed to be fixed by two viewing rays $\mathbf{L}', \mathbf{L}''$, but in general these rays do not intersect exactly, due to noise and model errors. Thus we have to find a compromise, that fits best to the viewing rays. A more complex example is the construction of a 3D line given point and image observations, see figure 3.10(b).

There are lots of other examples of constructing geometric entities, including but not limited to the following tasks:

- *corner construction*: construct a junction point from a set of intersecting image line segments
- *grouping*: construct a line from a set of collinear image points and image line segments
- *3D-grouping*: construct a polyhedral line given incident, parallel or orthogonal lines and planes

In this section we describe a *direct* method to construct the unknown entity for the above and many other constructions. Note that we do not take explicitly the uncertainty of the involved entities into account. Therefore the solutions are sub-optimal, for a treatment with uncertainties see chapter 4.

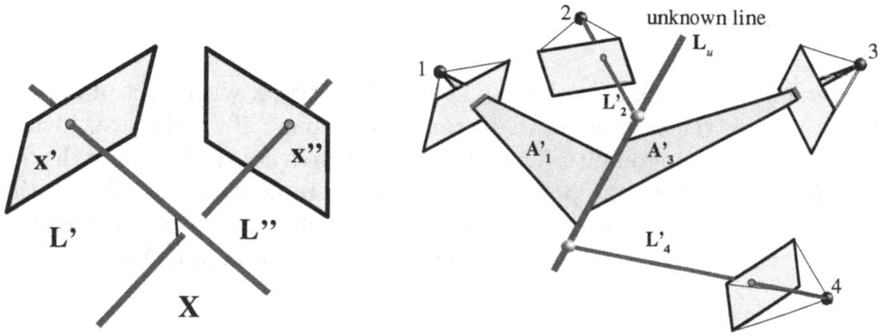

Fig. 3.10. Examples of non-unique construction of geometric entities: forward intersection with two viewing rays resulting a point $\mathbf{X}$ (a) and forward with two viewing rays and two viewing planes, resulting in a 3D line $\mathbf{L}_u$ (b).

3.4.1 Relations as Constraints

In the example of figure 3.10(a), the ideal situation is that the two rays intersect, $\mathbf{L}' \cap \mathbf{L}'' \neq \emptyset$, so that the point $\mathbf{X}$ is well-defined. If this is the case, the point $\mathbf{X}$ is incident to both rays $\mathbf{X} \in \mathbf{L}', \mathbf{L}''$. If this is not the case, we require that the point should be as close as possible to the rays, such that it is "almost" incident.

We can now reformulate the construction task as a *minimization problem*, where we want to minimize the contradiction $\mathbf{w}(\cdot)$, that the variable point $\mathbf{X}$ is incident to both rays:

$$\mathbf{w}\left(\{\mathbf{X} \in \mathbf{L}', \mathbf{X} \in \mathbf{L}''\}\right) \rightarrow \min \tag{3.81}$$

A point $\mathbf{X}$ which minimizes the contradiction $\mathbf{w}(\{\mathbf{X} \in \mathbf{L}', \mathbf{X} \in \mathbf{L}''\})$ is defined to be the best fit to the construction problem. We can now take advantage of the previous sections and directly give an algebraic expression for the above contradiction:

$$\mathbf{w}\left(\{\mathbf{X} \in \mathbf{L}', \mathbf{X} \in \mathbf{L}''\}\right) = \begin{pmatrix} \overline{\Gamma}^{\mathsf{T}}(\mathbf{L}') \\ \overline{\Gamma}^{\mathsf{T}}(\mathbf{L}'') \end{pmatrix} \mathbf{X} \tag{3.82}$$

since $\mathbf{X} \in \mathbf{L} \Leftrightarrow \overline{\Gamma}^{\mathsf{T}}(\mathbf{L})\mathbf{X} = 0$.

For the construction in figure 3.10(b), the fitted line $\mathbf{L}_u$ minimizes the contradiction

$$\mathbf{w}\left(\{\mathbf{L}_u \in \mathbf{A}'_1,\quad \mathbf{L}_u \cap \mathbf{L}'_2 \neq \emptyset,\quad \mathbf{L}_u \in \mathbf{A}'_3,\quad \mathbf{L}_u \cap \mathbf{L}'_4 \neq \emptyset\}\right) = \begin{pmatrix} \Pi(\mathbf{A}'_1) \\ (\overline{\mathbf{L}}'_2)^{\mathsf{T}} \\ \Pi(\mathbf{A}'_3) \\ (\overline{\mathbf{L}}'_4)^{\mathsf{T}} \end{pmatrix} \mathbf{L}_u \tag{3.83}$$

In general we may link an unknown geometric entity $\mathbf{x}$ with a set of observed entities by any relation given in section 3.3 on page 70: incidence, identity, parallelity and orthogonality. This always yields to a contradiction of the form $\mathsf{M}\mathbf{x} = 0$. In table 3.6, the possible relations for unknowns and observations are listed. A new observation can easily plugged into the set of contradictions by adding the appropriate construction matrix to the matrix M.

Table 3.6. Possible relations between a list of observed entities and unknown entities. In 2D, we may observed points $\mathbf{y}_i$ and lines $\mathbf{m}_i$ and want to estimate an unknown point $\mathbf{x}$ or an unknown line $\mathbf{l}$. In 3D, we may observe space points $\mathbf{Y}_i$, lines $\mathbf{M}_i$, and planes $\mathbf{B}_i$ and want to estimate an unknown space point $\mathbf{X}$, line $\mathbf{L}$ or plane $\mathbf{A}$.

2D			
obs.→ ↓ unk.		$\mathbf{y}_i$	$\mathbf{m}_i$
β	$\mathbf{x}$	$\equiv$	$\in$
	$\mathbf{l}$	$\in$	$\equiv$, $\parallel$, $\perp$

3D				
obs.→ ↓ unk.		$\mathbf{Y}_i$	$\mathbf{M}_i$	$\mathbf{B}_i$
β	$\mathbf{X}$	$\equiv$	$\in$	$\in$
	$\mathbf{L}$	$\in$	$\equiv$, $\in$, $\parallel$, $\perp$	$\in$, $\parallel$, $\perp$
	$\mathbf{A}$	$\in$	$\in$, $\parallel$, $\perp$	$\equiv$, $\parallel$, $\perp$

In general, we have n relations $R_i(\mathbf{y}_j, \boldsymbol{\beta})$, where $\mathbf{y}_j$ are w_i observations that are related to the unknown $\boldsymbol{\beta}$. In 3D, the maximum number n of different relations is 3 for unknown points, 8 for unknown lines and 7 for unknown planes, cf. the rows in table 3.6.

We obtain the following algebraic expression between the unknown and the observations:

$$\mathbf{w}\left(\{R_1(\mathbf{y}_1,\boldsymbol{\beta}),\cdots,R_1(\mathbf{y}_{w_1},\boldsymbol{\beta}),\cdots R_n(\mathbf{y}_{w_n},\boldsymbol{\beta})\}\right) = \begin{pmatrix} \mathsf{U}_1(\mathbf{y}_1) \\ \vdots \\ \mathsf{U}_n(\mathbf{y}_n) \end{pmatrix} \boldsymbol{\beta} = \mathsf{A}\boldsymbol{\beta} \stackrel{!}{=} \mathbf{0} \tag{3.84}$$

Alternatively, if we assume that the entities are not at infinity, we may also use the reduced construction matrices $\mathsf{U}^{[r_i]}(\mathbf{y}_j)$, where r_i is the degree of freedom of relation $R(\mathbf{y}_j, \boldsymbol{\beta})$, cf. last column in table 3.5. Note that we assume that the relation $R(\mathbf{y}_j, \boldsymbol{\beta})$ approximately holds, thus we do not have outliers such as an observed line that is supposed to be identical to the unknown, but actually is orthogonal. Therefore we can safely skip the coordinate transformation step in the reduction algorithm 3.1.

The modified minimization problem is now

$$\mathbf{w}'\left(\{R_1(\mathbf{y}_1,\boldsymbol{\beta}),\cdots,R_1(\mathbf{y}_{w_1},\boldsymbol{\beta}),\cdots R_n(\mathbf{y}_{w_n},\boldsymbol{\beta})\}\right) = \begin{pmatrix} \mathsf{U}_1^{[r_1]}(\mathbf{y}_1) \\ \vdots \\ \mathsf{U}_n^{[r_n]}(\mathbf{y}_n) \end{pmatrix} \boldsymbol{\beta} = \mathsf{A}'\boldsymbol{\beta} \stackrel{!}{=} \mathbf{0} \tag{3.85}$$

Assuming the reduced matrix A' is an $n \times u$ matrix, the problem is generally solvable, if $n \geq \mathrm{dof}(\boldsymbol{\beta})$, where $\mathrm{dof}(\boldsymbol{\beta})$ denotes the degree of freedoms of the unknown entity $\boldsymbol{\beta}$. Hence the redundancy is given by $red = n - \mathrm{dof}(\boldsymbol{\beta})$. To compute the redundancy for a specific 3D problem, check table 3.7.

The proposed formulation of the geometric estimation problem contains a large variety of combinations between an unknown entity and related observed entities.

3.4.2 Minimizing Algebraic Distance

To derive a solution to the minimization problem, we first reformulate it: instead of stacking the construction matrices U_i in a large matrix A as in the examples of the previous section, we may also minimize the squared sum

$$\Omega = \sum_i^N \mathbf{w}_i^{\mathsf{T}} \mathbf{w}_i = \boldsymbol{\beta}^{\mathsf{T}} \left(\sum_i^N \mathsf{U}_i^{\mathsf{T}} \mathsf{U}_i \right) \boldsymbol{\beta} \rightarrow \min \tag{3.86}$$

Since the unknown vector $\boldsymbol{\beta}$ is a homogeneous vector, we need an additional constraint to obtain a unique solution to the minimization problem, for example by fixing its length using $\boldsymbol{\beta}^{\mathsf{T}}\boldsymbol{\beta} = 1$.

Table 3.7. Calculating redundancy for geometric estimation problems: the cells of this table show the number of effective observations when estimating an unknown 3D entity using equation (3.84) resp. (3.85). For example, consider the forward intersection in figure 3.10(b): the unknown line **L** has $n_M = 2$ incident lines and $n_B = 2$ incident planes. Then the redundancy is given as $red = n_M + 2n_B - \text{dof}(\mathbf{L}) = 6 - 4 = 2$.

unknown entity	observations			
	$\in$	$\equiv$	$\parallel$	$\perp$
	n_Y points n_M lines n_B planes	m_Y points m_M lines m_B planes	p_M lines p_B planes	q_M lines q_B planes
point **X** (3 dof)	$2n_M + n_B$	$3m_Y$	-	-
line **L** (4 dof)	$2n_Y + n_M + 2n_B$	$4m_M$	$2p_M + p_B$	$q_M + 2q_B$
plane **A** (3 dof)	$n_Y + 2n_M$	$3m_B$	$p_M + 2p_B$	$2q_M + q_B$

The solution $\widehat{\beta}$ to the problem $\Omega \rightarrow \min$ with $\beta^{\mathsf{T}}\beta = 1$ is given by the eigenvector to the smallest eigenvalue of $\mathsf{U} = \sum_i \mathsf{U}_i \mathsf{U}_i^{\mathsf{T}}$ The solution is known to be the total least squares estimate cf. Duda and Hart 1973, p. 332f., 377f.

Practically, it is advisable to compute the total least squares solution using a singular value decomposition, Press *et al.* 1992: with U from above, compute

$$\mathsf{U} = \mathsf{E}\mathsf{D}\mathsf{G}^{\mathsf{T}} \tag{3.87}$$

where E and G are orthogonal matrices and D is a diagonal matrix with non-negative entries, that are sorted in descending order. Then the estimate $\widehat{\beta}$ is the last column of G. Alternatively one can use A' from (3.85) instead of U.

3.4.3 Enforcing Plücker Constraint

When computing the linear solution $\widehat{\beta}$ from the last section, only the homogeneous constraint was taken into account. In case of an unknown 3D line there is the additional Plücker constraint (2.20), which may not be fulfilled by the estimated $\widehat{\mathbf{L}}$. This happens, when the observed entities are subject to small noise. In this case we need to enforce the additional constraint to update the estimation to a new $\widehat{\widehat{\mathbf{L}}}$.

As suggested by Kanatani 1996 and Matei and Meer 2000 it is possible to use an orthogonal projection onto the manifold of all valid 3D lines, which is defined as the set of all homogeneous 6-vectors, that fulfill the Plücker constraint $p(\mathbf{L}) = \overline{\mathbf{L}}^{\mathsf{T}}\mathbf{L} = 0$. We can approximate the manifold locally by $\partial p(\mathbf{L})/\partial\mathbf{L} = \frac{1}{2}\overline{\mathbf{L}}^{\mathsf{T}}$

$$\widehat{\widehat{\mathbf{L}}} = \widehat{\mathbf{L}} - \frac{1}{2}\frac{\widehat{\overline{\mathbf{L}}}^{\mathsf{T}}\widehat{\mathbf{L}}}{|\widehat{\overline{\mathbf{L}}}|^2}\widehat{\overline{\mathbf{L}}} \tag{3.88}$$

Since this is only a local approximation and the constraint is non-linear, one has to iterate a couple of times until the Plücker condition is met.

To avoid the iteration process, one can enforce the constraint for the Euclidean and homogeneous part separately: one can either force the homogeneous part to be orthogonal to the Euclidean part:

$$\widehat{\widehat{\mathbf{L}}}_h = \widehat{\mathbf{L}}_h - \frac{\widehat{\mathbf{L}}_O^{\mathsf{T}}\widehat{\mathbf{L}}_O}{|\widehat{\mathbf{L}}_O|^2}\widehat{\mathbf{L}}_h \quad \text{such that} \quad \widehat{\widehat{\mathbf{L}}} = (\widehat{\widehat{\mathbf{L}}}_h, \widehat{\mathbf{L}}_O)^{\mathsf{T}}$$

or vice versa

$$\widehat{\widehat{\mathbf{L}}}_O = \widehat{\mathbf{L}}_O - \frac{\widehat{\mathbf{L}}_h^{\mathsf{T}}\widehat{\mathbf{L}}_h}{|\widehat{\mathbf{L}}_h|^2}\widehat{\mathbf{L}}_O \quad \text{such that} \quad \widehat{\widehat{\mathbf{L}}} = (\widehat{\mathbf{L}}_h, \widehat{\widehat{\mathbf{L}}}_O)^{\mathsf{T}}$$

It is advisable to weight both parts equally by using the bisector of $\widehat{\widehat{\mathbf{L}}}_h$ and $\widehat{\mathbf{L}}_h$ respectively $\widehat{\widehat{\mathbf{L}}}_O$ and $\widehat{\mathbf{L}}_O$. In equation (3.4.3) this is achieved by the factor $\frac{1}{2}$.

3.5 Estimating Projective Transformations

So far we have looked at ways to estimate geometric entities such as points, lines and planes from observations that were related to the unknown in some way. We now want to focus on estimating transformations, a common technique using point observations is the so-called DLT algorithm, see below. We will interpret a transformation of entities as a trilinear relation between the original entity, its image entity and the transformation itself: then it is easy to derive several kinds of DLT algorithms.

3.5.1 Collinearity Equations

The collinearity equations are the classical equations in Photogrammetry and are derived from a pinhole-camera model with a given inner orientation, let's say only the focal length c, a rotation matrix $\mathsf{R} = (r_{ij})$ and a projection center $(X_0, Y_0, Z_0)^{\mathsf{T}}$:

$$\begin{aligned} x' &= c\frac{r_{11}(X - X_0) + r_{12}(Y - Y_0) + r_{13}(Z - Z_0)}{r_{31}(X - X_0) + r_{32}(Y - Y_0) + r_{33}(Z - Z_0)} \\ y' &= c\frac{r_{21}(X - X_0) + r_{22}(Y - Y_0) + r_{23}(Z - Z_0)}{r_{31}(X - X_0) + r_{32}(Y - Y_0) + r_{33}(Z - Z_0)} \end{aligned} \tag{3.89}$$

We can include more parameters for the inner orientation to this basic equation, specifically skew c_s, scale difference m between the axis and the principle point $(x_H, y_H)^\mathsf{T}$. Then we obtain the Euclidean interpretation of a projective camera matrix P as in equation (2.22), which can be written as

$$\begin{aligned} x' &= \frac{p_{11}X + p_{12}Y + p_{13}Z + p_{14}}{p_{31}X + p_{32}Y + p_{33}Z + p_{34}} \\ y' &= \frac{p_{21}X + p_{22}Y + p_{23}Z + p_{24}}{p_{31}X + p_{32}Y + p_{33}Z + p_{34}} \end{aligned} \tag{3.90}$$

The abstract formulation of the collinearity equations have been described by DAS 1949 and ABDEL-AZIZ AND KARARA 1971.

In case a set of corresponding or homologous pairs of image points $\boldsymbol{x}'_i$ and $\boldsymbol{X}_i$ are known and the camera parameters p_{ij} are unknown, we can obtain p_{ij} by multiplying both equations in (3.90) with their common denominator and transforming it into a set of homogeneous equation:

$$\begin{pmatrix} -X_i & -Y_i & -Z_i & 1 & 0 & 0 & 0 & 0 & x'_iX_i & x'_iY_i & x'_iZ_i & x'_i \\ 0 & 0 & 0 & 0 & -X_i & -Y_i & -Z_i & 1 & y'_iX_i & y'_iY_i & y'_iZ_i & y'_i \end{pmatrix} \begin{pmatrix} p_{11} \\ p_{12} \\ \vdots \\ p_{34} \end{pmatrix} = \mathbf{0} \tag{3.91}$$

The equation is linear in the unknown parameter vector $\mathbf{p}$. Since it is of a form $A_i\mathbf{p} = \mathbf{0}$, it can be solved if at least 6 homologous point pairs are given by minimizing

$$\boldsymbol{\beta}^\mathsf{T} \left(\sum_i A_i^\mathsf{T} A_i \right) \boldsymbol{\beta} \ \rightarrow \ \min$$

with the constraint such as $\boldsymbol{\beta}^\mathsf{T}\boldsymbol{\beta} = 1$. One can use eigenvector solution as above, equation (3.86), advisably implemented using a singular value decomposition

This method for computing projection matrices is known as the *direct linear transformation* algorithm (DLT) since the equation (3.91) is linear in the unknown vector. It actually works not only for the projective camera matrix, but for any general projective transformation, including homographies and even affine transformations. A reference for the DLT in the context of computer graphics and vision is SUTHERLAND 1963, though DAS 1949 already describes the technique. It was named DLT by ABDEL-AZIZ AND KARARA 1971.

3.5.1.1 Collinearity Equations Interpreted as Geometric Relations

One can geometrically interpret equations (3.89) by the fact that the image point $\boldsymbol{x}'_i$ on the camera plane, the projective center and the object point $\boldsymbol{X}_i$

should be collinear, hence the name of the equations. Using this interpretation, the collinearity equation are nothing but an algebraic expression for a geometric relation involving the image point, the object point and the camera. With the algebraic expressions for constructions and relations introduced in this chapter, we can reformulate this collinearity relationship; using the homogeneous vectors $\mathbf{x}'_i$ and $\mathbf{X}_i$ the reformulation can be done in the image plane (2D) as well as in object space (3D):

(2D) The object point $\mathbf{X}_i$ is collinear to the image point $\mathbf{x}'_i$ on the camera plane, if the projected point $\mathsf{P}\mathbf{X}_\mathrm{i}$ is identical to $\mathbf{x}'_i$, thus if the 2D relation $\mathbf{x}_i \equiv \mathsf{P}\mathbf{X}_i$ holds. Algebraically, we obtain the bilinear forms (table 3.5):

$$\mathsf{S}(\mathbf{x}'_i)\mathsf{P}\mathbf{X}_i = \mathbf{0} \tag{3.92}$$

$$-\mathsf{S}(\mathsf{P}\mathbf{X}_i)\mathbf{x}'_i = \mathbf{0} \tag{3.93}$$

The above expressions are linear in $\mathbf{X}_i$ and $\mathbf{x}'_i$. Using proposition B.5 on page 185, we can reformulate equation (3.92) such that the same condition is linear in the parameter vector $\mathbf{p} = \mathrm{vec}(\mathsf{P}^\mathsf{T})$ of the projective camera:

$$\mathsf{S}(\mathbf{x}'_i)(I_3 \otimes \mathbf{X}_i^\mathsf{T})\mathbf{p} = \mathbf{0} \tag{3.94}$$

Note that the tests for identity of two 2D points has $r = 2$ degrees of freedom. Thus one can use the reduced matrix $\mathsf{S}^{[2]}(\mathbf{x}_i)$ instead of $\mathsf{S}(\mathbf{x}_i)$ in the last equation (3.94). If the first two rows are kept in $\mathsf{S}^{[2]}(\mathbf{x}_i)$, the resulting equation is equivalent to (3.91). Thus we have easily derived the formula for a DLT solution by using the algebraic expression for $\mathbf{x}'_i \equiv \mathsf{P}\mathbf{X}_i$, cf. Hartley and Zisserman 2000, p. 71f.

We obtained a *trilinear expression* for the collinearity equation, thus we sometimes write the collinearity equation as a relation with three entities, $\equiv (\mathbf{x}'_i; \mathbf{p}, \mathbf{X}_i)$, where the vector $\mathbf{p} = \mathrm{vec}(\mathsf{P}^\mathsf{T})$ represents the projective camera.

(3D) Another option to express the collinearity relation is to check whether the viewing ray $\mathbf{L}'(\mathbf{x}'_i)$ of $\mathbf{x}'_i$ goes through the entity point $\mathbf{X}_i$; in other words: the collinearity equation holds if the 3D relation $\mathbf{X}'_i \in \mathbf{L}'(\mathbf{x}_i)$ is true, which is equal to

$$\overline{\Gamma}^\mathsf{T}(\overline{\mathsf{Q}}^\mathsf{T}\mathbf{x}'_i)\mathbf{X}_i = \mathbf{0} \quad \Leftrightarrow \Gamma^\mathsf{T}(\mathsf{Q}^\mathsf{T}\mathbf{x}'_i)\mathbf{X}_i = \mathbf{0} \tag{3.95}$$

$$\overline{\Pi}^\mathsf{T}(\mathbf{X}_i)\overline{\mathsf{Q}}^\mathsf{T}\mathbf{x}'_i = \mathbf{0} \quad \Leftrightarrow \Pi^\mathsf{T}(\mathbf{X}_i)\mathsf{Q}^\mathsf{T}\mathbf{x}'_i = \mathbf{0} \tag{3.96}$$

with $\mathbf{L}'(\mathbf{x}'_i) = \overline{\mathsf{Q}}^\mathsf{T}\mathbf{x}'_i = C\mathsf{Q}^\mathsf{T}\mathbf{x}'_i$. cf. equation (3.48). Again, we can make the expression linear in the camera parameters, this time for the line projective camera $\mathbf{q} = \mathrm{vec}(\mathsf{Q}^\mathsf{T})$:

$$\Pi^\mathsf{T}(\mathbf{X}_i)(\mathbf{x}_\mathrm{i}'^\mathsf{T} \otimes I_6)\mathbf{q} = \mathbf{0} \tag{3.97}$$

Observe that this time, the expression is linear in the parameters of the line projection matrix Q. The check for incidence of a line and a point in 3D has $r = 2$ degrees of freedom, thus in 3.97 we can use the reduced matrix $\Pi^{[2]\mathsf{T}}(\mathbf{X}_i)$ to obtain two linearly independent equations, as expected. Again we denote the relationship with the triple $\in (\mathbf{X}_i; \mathbf{p}, \mathbf{x}'_i)$, where the vector $\mathbf{p}$ represents the projective camera and $\mathbf{q} = \mathsf{J}_{\mathsf{qp}}(\mathbf{p})\mathbf{p}$, cf. equation (3.51) on page 59.

3.5.2 Coplanarity Equations

One can not only estimate a projective camera matrix by homologous point pairs $(\mathbf{x}'_i, \mathbf{X}_i)$, but also by homologous line pairs $(\mathbf{l}'_i, \mathbf{L}_i)$. For this case we can formulate a coplanarity relation[3] for homologous line pairs: the projection center, the image line $\mathbf{l}'_i$ on the camera plane and the 3D line $\mathbf{L}_i$ must be coplanar. As in the last section, we can express this relationship in two ways:

(2D) For homologous line pairs, the following identity relation in 2D must hold: $\mathbf{l}'_i \equiv \mathsf{Q}\mathbf{L}$, which is equivalent to the trilinear form:

$$\mathsf{S}(\mathbf{l}'_i)\mathsf{Q}\,\mathbf{L}_i = \mathbf{0} \tag{3.98}$$

$$-\mathsf{S}(\mathsf{Q}\mathbf{L}_i)\,\mathbf{l}'_i = \mathbf{0} \tag{3.99}$$

$$\mathsf{S}(\mathbf{l}'_i)(I_3 \otimes \mathbf{L}_i^{\mathsf{T}})\,\mathbf{q} = \mathbf{0} \tag{3.100}$$

cf. table 3.5 and proposition B.5. We can also use the reduced matrix $\mathsf{S}^{[r]}(\mathbf{l}')$ with $r = 2$ to obtain a set of linearly independent equations for the relation $\equiv (\mathbf{l}'_i; \mathbf{p}, \mathbf{L})$

(3D) In 3D, the coplanarity condition for homologous line pairs uses the viewing plane of the image line: $\mathbf{L}_i \in \mathbf{A}'(\mathbf{l}'_i)$, which is equivalent to the trilinear form:

$$\Gamma^{\mathsf{T}}(\mathbf{L}_i)\mathsf{P}^{\mathsf{T}}\,\mathbf{l}'_i = \mathbf{0} \tag{3.101}$$

$$\Pi^{\mathsf{T}}(\mathsf{P}^{\mathsf{T}}\mathbf{l}_i)\,\mathbf{L}_i = \mathbf{0} \tag{3.102}$$

$$\Gamma^{\mathsf{T}}(\mathbf{L}_i)(\mathbf{l'}_i^{\mathsf{T}} \otimes I_4)\,\mathbf{p} = \mathbf{0} \tag{3.103}$$

The incidence of a line and a plane has $r = 2$ degrees of freedom, thus we can use $\Gamma^{[2]\mathsf{T}}(\mathbf{L}_i)$ in the last equation. This establishes a trilinear equation for $\in (\mathbf{L}_i; \mathbf{p}, \mathbf{x}'_i)$.

As for the collinearity relations, we obtain two trilinear forms, one expressed in 2D and one in 3D. But this time, the vector $\mathbf{p}$ from the point projection matrix P appears linear in the 3D case, while the vector $\mathbf{q}$ of the line projection

[3] Note that the coplanarity relation is not related to the coplanarity condition that was formulated in the context of *two* images: the condition of coplanarity included the projection centers of the cameras and an object point.

matrix appears linear in the 2D case. Thus a DLT algorithm which wants to estimate the projective camera matrix P should use equation (3.103), which is the analogy to the equation (3.94)

Table 3.8. Trilinear relations between the projective camera P and points and lines. Note that there are two possible expressions for point- resp. line-projection: the first expression constrains the entities in the image plane, thus testing if the projected point resp. line is identical to the observed one $\equiv(\mathbf{x}';\mathsf{P},\mathbf{X})$ resp. $\equiv(\mathbf{l}';\mathsf{Q},\mathbf{L})$. The other expression refers to the object space and tests whether the 3D entity is incident to the back-projected line resp. plane, $\in(\mathbf{X};\mathsf{P},\mathbf{x}')$ resp. $\equiv(\mathbf{L};\mathsf{Q},\mathbf{l}')$. Note that $\mathbf{q}=\operatorname{vec}(\mathsf{Q}^\mathsf{T})=\mathsf{J}_{\mathsf{qp}}\mathbf{p}$ with $\mathbf{p}=\operatorname{vec}(\mathsf{P}^\mathsf{T})$, cf. equation (3.51) on page 59.

transformation	relation	trilinear expressions
projection $\mathbf{x}'=\mathsf{P}\mathbf{X}$	$\equiv(\mathbf{x}';\mathbf{p},\mathbf{X})$	$\mathsf{S}(\mathbf{x}')\mathsf{P}\ \mathbf{X}=\mathbf{0}$
		$-\mathsf{S}(\mathsf{P}\mathbf{X})\ \mathbf{x}'=\mathbf{0}$
		$\mathsf{S}(\mathbf{x}')(\mathsf{I}_3\otimes\mathbf{X}^\mathsf{T})\ \mathbf{p}=\mathbf{0}$
	$\in(\mathbf{X};\mathbf{p},\mathbf{x}')$	$\Gamma^\mathsf{T}(\mathsf{Q}^\mathsf{T}\mathbf{x}')\ \mathbf{X}=\mathbf{0}$
		$\Pi^\mathsf{T}(\mathbf{X})\mathsf{Q}^\mathsf{T}\ \mathbf{x}'=\mathbf{0}$
		$\Pi^\mathsf{T}(\mathbf{X})(\mathbf{x}'^\mathsf{T}\otimes\mathsf{I}_6)\ \mathbf{q}=\mathbf{0}$
		$\Pi^\mathsf{T}(\mathbf{X})(\mathbf{x}'^\mathsf{T}\otimes\mathsf{I}_6)\mathsf{J}_{\mathsf{qp}}\ \mathbf{p}=\mathbf{0}$
projection $\mathbf{l}'=\mathsf{Q}\mathbf{L}$	$\equiv(\mathbf{l}';\mathbf{p},\mathbf{L})$	$\mathsf{S}(\mathbf{l}')\mathsf{Q}\ \mathbf{L}=\mathbf{0}$
		$-\mathsf{S}(\mathsf{Q}\mathbf{L})\ \mathbf{l}'=\mathbf{0}$
		$\mathsf{S}(\mathbf{l}')(\mathsf{I}_3\otimes\mathbf{L}^\mathsf{T})\ \mathbf{q}=\mathbf{0}$
		$\mathsf{S}(\mathbf{l}')(\mathsf{I}_3\otimes\mathbf{L}^\mathsf{T})\mathsf{J}_{\mathsf{qp}}\ \mathbf{p}=\mathbf{0}$
	$\in(\mathbf{L};\mathbf{p},\mathbf{l}')$	$\Gamma^\mathsf{T}(\mathbf{L})\mathsf{P}^\mathsf{T}\ \mathbf{l}'=\mathbf{0}$
		$\Pi^\mathsf{T}(\mathsf{P}^\mathsf{T}\mathbf{l})\ \mathbf{L}=\mathbf{0}$
		$\Gamma^\mathsf{T}(\mathbf{L})(\mathbf{l}'^\mathsf{T}\otimes\mathsf{I}_4)\ \mathbf{p}=\mathbf{0}$

3.5.3 Simultaneous DLT with Points and Lines

A DLT algorithm which *simultaneously* uses homologous points and homologous lines is now quite easy to describe using equations (3.103) and (3.94): then we have equations of the form $\mathsf{A}_i\mathbf{p}=\mathbf{0}$ and $\mathsf{B}_j\mathbf{p}=\mathbf{0}$ which can be solved by minimizing

$$\Omega=\boldsymbol{\beta}^\mathsf{T}\left(\sum_i\mathsf{A}_i^\mathsf{T}\mathsf{A}_i+\sum_j\mathsf{B}_j^\mathsf{T}\mathsf{B}_j\right)\boldsymbol{\beta}\ \rightarrow\ \min \tag{3.104}$$

with the constraint $\boldsymbol{\beta}^\mathsf{T}\boldsymbol{\beta} = 1$. In general, any mixture of point and line observations is allowed as long as there are 6 homologous pairs, as each pair adds 2 constraints to the 11 unknown projective camera parameters (12 unknown entries of matrix P minus 1 homogeneous constraint).

Note that one might also want to use the equations (3.100) and (3.97) using $\mathbf{q}$ to obtain an estimate of the projective camera, but as $\mathbf{q}$ is a 18-vector which represents 11 parameters, this is computationally not advisable. One may also use the Jacobian $\mathsf{J}_{\mathsf{qp}} = \partial\mathbf{q}/\partial\mathbf{p}$ to transform the equations (3.100) and (3.97) back to $\mathbf{p}$, but then the resulting equation is not linear in $\mathbf{p}$, cf. table 3.8.

Our approach of simultaneously using points and line correspondences is similar to the work of HARTLEY AND ZISSERMAN 2000 and LUXEN AND FÖRSTNER 2001: both represent an observed 3D line $\mathbf{L}_i$ indirectly by two incident points $\mathbf{X}_{i_1}, \mathbf{X}_{i_2}$, then the coplanarity equation can be written as

$$\mathbf{A}'(\mathbf{l}'_i)^\mathsf{T}\mathbf{X}_{i_j} = \mathbf{l}'^{\mathsf{T}}_{\mathbf{i}}\mathsf{P}\mathbf{X}_{i_j} = 0 \quad \text{with } j = 1, 2$$

In our approach though, a 3D line is directly incorporated in the DLT by their homogeneous 6-vector. This has the conceptual advantage that both points and lines are treated in the same way: the formulas can be directly used if the 3D lines are given by their homogeneous 6-vector or if they are constructed from other entities—one does not have to choose incident points representing the line.

3.5.4 Simultaneous DLT Algorithms for Homographies

We now develop a DLT algorithm for estimating homographies in 2D and 3D: given a set of observed entities $\mathbf{x}_i$ and the corresponding transformed entities $\mathbf{x}'_i$, which are supposed to be images of an unknown homography, thus ideally $\tilde{\mathbf{x}}_i{}' = \tilde{\mathsf{H}}\tilde{\mathbf{x}}_i$, where the tilde denotes true but unknown values.

3.5.4.1 2D Homography For the 2D case, we assume a set of homologous point pairs $(\mathbf{x}_i, \mathbf{x}'_i)$, $i = 1, \cdots, n_x$ and a set of homologous line pairs $(\mathbf{l}_j, \mathbf{l}'_j)$, $j = 1, \cdots, n_l$, for which the triple relations $\equiv (\mathbf{x}'_i; \mathsf{H}, \mathbf{x}_i)$ and $\equiv (\mathbf{l}'_j; \mathsf{H}, \mathbf{l}_j)$ are supposed to hold. Identity in 2D has 2 degrees of freedom, so each homologous pair adds 2 constraints on the unknown homography H. There are 8 unknown parameters for H (9 matrix entries less 1 homogeneous constraint), so we need at least $2n_x + 2n_l \geq 8$ equations or $n = n_x + n_l \geq 4$ homologous pairs to estimate the homography – in case of $n = n_x + n_l = 4$, the homography is uniquely determined. For the triple identity relations, we can easily write trilinear algebraic expressions using the $\mathsf{S}(\cdot)$ construction matrix, analogous to the 2D case of the collinearity and coplanarity relations. The resulting six equations for point and line pairs are listed in table 3.9, which can be used in a minimizing approach as in equation (3.104).

Table 3.9. Trilinear relations between 2D and 3D homographies and homologous geometric entities. The last row of each relation yields the DLT for the homography H.

transformation	relation	trilinear expressions
2D homography $\mathbf{x}' = \mathsf{H}\mathbf{x}$	$\equiv (\mathbf{x}'; \mathsf{H}, \mathbf{x})$	$\mathsf{S}(\mathbf{x}')\mathsf{H} \quad \mathbf{x} \quad = \mathbf{0}$
		$-\mathsf{S}(\mathsf{H}\mathbf{x}) \quad \mathbf{x}' \quad = \mathbf{0}$
		$\mathsf{S}(\mathbf{x}')(I_3 \otimes \mathbf{x}^\mathsf{T}) \operatorname{vec}(\mathsf{H}^\mathsf{T}) = \mathbf{0}$
	$\equiv (\mathbf{l}'; \mathsf{H}, \mathbf{l})$	$\mathsf{S}(\mathbf{l})\mathsf{H}^\mathsf{T} \quad \mathbf{l}' \quad = \mathbf{0}$
		$-\mathsf{S}(\mathsf{H}^\mathsf{T}\mathbf{l}') \quad \mathbf{l} \quad = \mathbf{0}$
		$\mathsf{S}(\mathbf{l})(\mathbf{l}'^\mathsf{T} \otimes I_3) \operatorname{vec}(\mathsf{H}^\mathsf{T}) = \mathbf{0}$
3D homography $\mathbf{X}' = \mathsf{H}\mathbf{X}$	$\equiv (\mathbf{X}'; \mathsf{H}, \mathbf{X})$	$\Pi(\mathbf{X}')\mathsf{H} \quad \mathbf{X} \quad = \mathbf{0}$
		$\Pi(\mathsf{H}\mathbf{X}) \quad \mathbf{X}' \quad = \mathbf{0}$
		$\Pi(\mathbf{X}')(I_4 \otimes \mathbf{X}^\mathsf{T}) \operatorname{vec}(\mathsf{H}^\mathsf{T}) = \mathbf{0}$
	$\equiv (\mathbf{L}'; \mathsf{H}, \mathbf{L})$	$(\overline{\Gamma}^\mathsf{T}(\mathbf{L}')\mathsf{H} \otimes I_4)\mathsf{J}_\Gamma \quad \mathbf{L} \quad = \mathbf{0}$
		$(I_4 \otimes \Gamma(\mathbf{L})\mathsf{H}^\mathsf{T})\mathsf{J}_\Gamma C \quad \mathbf{L}' \quad = \mathbf{0}$
		$(\overline{\Gamma}^\mathsf{T}(\mathbf{L}') \otimes \Gamma(\mathbf{L})) \operatorname{vec}(\mathsf{H}^\mathsf{T}) = \mathbf{0}$
	$\equiv (\mathbf{A}'; \mathsf{H}, \mathbf{A})$	$\overline{\Pi}(\mathbf{A})\mathsf{H}^\mathsf{T} \quad \mathbf{A}' \quad = \mathbf{0}$
		$\overline{\Pi}(\mathsf{H}^\mathsf{T}\mathbf{A}') \quad \mathbf{A} \quad = \mathbf{0}$
		$\overline{\Pi}(\mathbf{A})(\mathbf{A}'^\mathsf{T} \otimes I_4) \operatorname{vec}(\mathsf{H}^\mathsf{T}) = \mathbf{0}$

3.5.4.2 3D Homography To estimate a 3D homography, we can have three different homologous sets: homologous point pairs $(\mathbf{X}_i, \mathbf{X}'_i)$, $i = 1, \cdots, n_X$, homologous plane pairs $(\mathbf{A}_j, \mathbf{A}'_j)$, $j = 1, \cdots, n_A$ and homologous line pairs $(\mathbf{L}_k, \mathbf{L}'_k)$, $j = 1, \cdots, n_L$, such that the triple identity relations $\equiv (\mathbf{X}'_i; \mathsf{H}, \mathbf{X}_i)$, $\equiv (\mathbf{A}'_j; \mathsf{H}, \mathbf{A}_j)$ and $\equiv (\mathbf{L}'_k; \mathsf{H}, \mathbf{L}_k)$ hold. A 3D homography H contains 16 matrix elements, thus 15 free parameters. Therefore we need $3n_X + 4n_L + 3n_A \geq 15$ equations, for example at least 3 lines, 1 point and 1 plane in general position, which in this case uniquely defines the full homography.

For the identity relations of points, the trilinear algebraic expressions are obtained using the bilinear form $\Pi(\mathbf{X}'_i)\mathsf{H}\mathbf{X}_i = -\Pi(\mathsf{H}\mathbf{X}_i)\mathbf{X}'_i = \mathbf{0}$ and proposition B.5 on page 185. The algebraic expressions are listed in table 3.9.

For planes, one uses $\mathbf{A}'_j = \mathsf{H}^{-\mathsf{T}}\mathbf{A}_j \Leftrightarrow \mathsf{H}^\mathsf{T}\mathbf{A}'_j = \mathbf{A}_j$ to obtain the trilinear forms, again listed in table 3.9.

The algebraic expression for the identity of homologous 3D lines $(\mathbf{L}_k, \mathbf{L}'_k)$ is a bit more involving, as the identity expression is based on a product of Plücker matrices, $\mathsf{D} = \overline{\Gamma}(\mathbf{L}')\Gamma(\mathbf{L}) = 0$, cf. equation (3.67) on page 74. The goal is to find an expression for $\operatorname{vec}(\mathsf{D})$, that is linear in $\mathbf{L}_k$, $\mathbf{L}'_k$ and $\operatorname{vec}(\mathsf{H}^\mathsf{T})$.

Unfortunately, this is impossible: the 3D line homography H_L, defined in equation (3.41) on page 57 is quadratic with respect to the parameters of H. Thus an algebraic expression for the identity of $\mathbf{L}'_k = \mathsf{H}_L \mathbf{L}_k$ can not be written in a linear form $\text{vec}(\mathsf{D}) = \mathsf{U}(\mathbf{L}'_k, \mathbf{L})\text{vec}(\mathsf{H}^\mathsf{T}) = \mathbf{0}$.
The solution to this problem is to choose a different distance matrix D', which is based on the interpretation of the rows resp. columns of the Plücker matrix $\Gamma(\mathbf{L})$: As described in section 3.2, especially table 3.3, the columns can be interpreted as four canonical points $\mathbf{Y}_s = \mathbf{L}_k \cap \mathbf{E}_s, s = 1, \cdots 4$ that are incident to the line $\mathbf{L}$. We can transform these points by the point-homography H and obtain four points $\mathbf{Y}'_s$, compactly written as $\mathsf{H}\Gamma(\mathbf{L}_k)$. Assuming $\mathbf{L}'_k \equiv \mathsf{H}_L \mathbf{L}_k$ the transformed canonical points are incident to the $\mathbf{L}'_k$.
Note that although $\mathsf{H}\Gamma(\mathbf{L}_k)$ is a 4×4 matrix, in general this is *not* a valid Plücker-matrix – but nevertheless the transformed canonical points can be tested to be incident to the line $\mathbf{L}'_k$ by $\overline{\Gamma}(\mathbf{L}'_k)\mathbf{Y}'_s = \mathbf{0}$. Thus we obtain an algebraic expression for $\equiv (\mathbf{L}'_k; \mathsf{H}, \mathbf{L}_k)$, which is not quadratic in the parameters of H:

$$\equiv (\mathbf{L}'_k; \mathsf{H}, \mathbf{L}_k) \quad \Leftrightarrow \quad \mathsf{D}' = \overline{\Gamma}(\mathbf{L}'_k)\mathsf{H}\Gamma(\mathbf{L}_k) = 0 \tag{3.105}$$

$$\Leftrightarrow \quad \mathsf{D}'' = \Gamma(\mathbf{L}_k)\mathsf{H}^\mathsf{T}\overline{\Gamma}(\mathbf{L}'_k) = 0 \tag{3.106}$$

The second expression for D'' in (3.106) uses the same reasoning as described above, but this time by transforming the canonical *planes* of the image lines in $\overline{\Gamma}(\mathbf{L}'_k)$ instead of the canonical points in $\Gamma(\mathbf{L}'_k)$. Alternatively equation (3.106) can be simply obtained by transposing (3.105), thus $\mathsf{D}'' = \mathsf{D}'^\mathsf{T}$.
Using D'' and the matrix J_Γ with $\text{vec}(\Gamma(\mathbf{L})) = \mathsf{J}_\Gamma \mathbf{L}$ defined in equation (3.68) on page 75, it is now easy to derive the trilinear expressions for homologous line pairs, see table 3.3.
Summarizing the previous results, we have equations of the form $\mathsf{A}_i \mathbf{h} = \mathbf{0}$, $\mathsf{B}_j \mathbf{h} = \mathbf{0}$ resp. $\mathsf{G}_k \mathbf{h} = \mathbf{0}$ depending on homologous point, plane resp. line pairs that are linear in the parameters of the 3D homography vector $\mathbf{h} = \text{vec}(\mathsf{H})$. Thus minimizing

$$\Omega = \boldsymbol{\beta}^\mathsf{T} \left(\sum_i \mathsf{A}_i^\mathsf{T} \mathsf{A}_i + \sum_j \mathsf{B}_j^\mathsf{T} \mathsf{B}_j + \sum_k \mathsf{G}_k^\mathsf{T} \mathsf{G}_k \right) \boldsymbol{\beta} \rightarrow \min \tag{3.107}$$

under the constraint $\boldsymbol{\beta}^\mathsf{T}\boldsymbol{\beta} = 1$ yields an estimate for the unknown homography.

3.5.5 Estimating Constrained Transformations

Up to now we have constructed projective transformations without any constraints on the parameters other than the homogeneous constraint. We want to make a short comment on estimating special homographies, such as affinity or similarity but also the affine camera P^a without providing an explicit algorithm.

In 2D, an affine transformation H_a is defined by the upper two rows $\mathsf{H}_a^{(1,1),(2,3)}$ of a 3×3 matrix, the third row being $(0,0,1)^\mathsf{T}$, cf. equation (2.21) on page 35. This is a linear constraint on the unknown parameter vector. If we assume a similarity transformation, we even have two quadratic constraints, cf. section 2.3.1.

Non-linear constraints can not be as easily incorporated as previously described when minimizing the algebraic distance. But one can use a two-step procedure similar to the estimation of Plücker coordinates, cf. section 3.4.3: first compute a homography $\widehat{\mathsf{H}}$ as in the previous section. In a second step the additional constraint has to be enforced by choosing the "closest" affine transformation $\widehat{\mathsf{H}}_a$ to $\widehat{\mathsf{H}}$. The closeness term is not uniquely defined and may underlie different criteria. In the case of affinity one could just skip estimating the first two elements of the third row; the resulting 7 parameters can again be written in a homogeneous vector. For similarity and other special transformations, one need more sophisticated methods, which will not be discussed further in this work.

In the field of Computer Vision, the two-step method is well-known, especially within the task of estimating the fundamental matrix from point-correspondences, cf. HARTLEY 1995B. In this work the 8-point algorithm is discussed: first derive an equation of the form $(\mathbf{x}'_i \otimes \mathbf{x}''_i)\mathbf{f}$, with $\mathbf{f} = \text{vec}(\mathsf{F}^\mathsf{T})$, which can be regarded as a DLT for the fundamental matrix. The equation is solved by algebraic minimization and in a second step the singularity constraint is enforced by minimizing the Frobenius norm, cf. TSAI AND HUANG 1984. An overview of methods for estimating the fundamental matrix can be found in ZHANG 1998 and TORR AND MURRAY 1997.

3.5.6 Conditioning of Entities for Minimization

As mentioned in the last section, HARTLEY 1995B used the estimation of a fundamental matrix as an example for a DLT and investigated the effect of (homogeneous) transformations of the 2D image points to the precision of the solution. Indeed, given two related sets of homologous, but slightly inconsistent image points $(\mathbf{x}'_i, \mathbf{x}''_i)$ and $(\mathsf{T}_1\mathbf{x}'_i, \mathsf{T}_2\mathbf{x}''_i)$, then the fundamental matrix F for the first set is numerically not equal to $\mathsf{T}_1^{-\mathsf{T}}\mathsf{F}\mathsf{T}_2$, as different algebraic errors are minimized.

It turns out that it is very much advisable to "normalize" the points such that their centroid is at the origin and the points are scaled so that the average distance from the origin is equal to $\sqrt{2}$. Because of the improved condition of the design matrix, the stability of the linear equation drastically improves.

As we assume that the coordinate are approximately close to the centroid, we concentrate here on the scaling operation: it has already been introduced in section 2.4 and was called conditioning of geometric entities. Thus contrary to HARTLEY 1995B, we will call the coordinate transformation *conditioning* of the geometric scene to avoid confusion with the normalization operation of homogeneous vectors introduced in section 2.1.1 on page 21.

The same positive effect of conditioning occurs not only when estimating the fundamental matrix, but also when we construct unknown entities based on known related entities: as an example, assume four observed 2D lines $\mathbf{l}_i$, where each of them is either parallel or orthogonal and should be incident to an unknown point $\mathbf{x} = (t, t, 1)^\mathsf{T}$, cf. figure 3.11. Each line has an error-distance e from the unknown point. If $e = 0$, we get an exact solution by minimizing (3.86) with $\mathsf{U}_i = \mathbf{l}_i^\mathsf{T}$. Numerically though, the matrix $\mathsf{U} = \sum_i \mathbf{l}_i \mathbf{l}_i^\mathsf{T}$ becomes ill-conditioned when the point $\mathbf{x}$ moves away from the origin, using the ratio of the two largest singular values as the condition number (note that ideally, the matrix U has rank 2). This is getting worse when $e > 0$. By translating the centroid of the lines, $\mathbf{x}' = (0, 0, 1)^\mathsf{T}$ and scaling the entities such that $e' < 1$, the matrix U' remains well-conditioned.

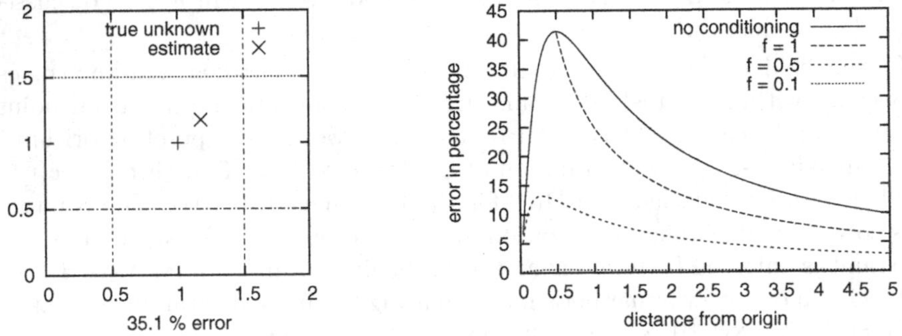

Fig. 3.11. *Left:* this is an example of an estimation error when not conditioning a geometric scene: the goal is to construct an unknown point from four supposedly incident lines. *Right:* The estimation error is plotted in percentage depending on the distance of the unknown point from the origin. The error is small if the point is close to the origin or far away. The error is plotted for various conditioning factor f. The error is small for $f = 0.1$ independent of the distance of the origin.

In section 2.4 an algorithm for conditioning homogeneous vectors of geometric entities was proposed, which is similar to the one mentioned in HARTLEY AND ZISSERMAN 2000, p. 113. For the purpose of estimation of geometric entities, a conditioning factor f_{min} of the order of 0.1 was empirically found to be useful when applying the algorithm 3.3, which is introduced in the next section.

Algebraic optimization

Objective: find an estimation for an unknown geometric entity $\tilde{\boldsymbol{\beta}}$ using a set of observed entities $\mathbf{y}_i$, that are supposed to be related to the unknown entity by the geometric relations $R_i(\mathbf{y}_i, \tilde{\boldsymbol{\beta}})$.

If the unknown $\tilde{\boldsymbol{\beta}}$ is a transformation, then the observed entities consist of corresponding pairs of entities $(\mathbf{y}_i, \mathbf{y}'_i)$, such that $\mathbf{y}'_i = \mathsf{M}(\tilde{\boldsymbol{\beta}})\mathbf{y}_i$ with $\mathsf{M}(\tilde{\boldsymbol{\beta}})$ being the matrix representation of the unknown transformation.

1. If necessary, condition the observations $\mathbf{y}_i$ with a conditioning factor f determined as in algorithm 2.1 with $f_{min} = 0.1$. For transformations, obtain two different factors f, f', see equation (2.25) on page 39.
2. For each $\mathbf{y}_i$, obtain the construction matrix U_i, depending on the relation R_i and the type of the unknown entity $\tilde{\boldsymbol{\beta}}$, cf. table 3.5. For unknown transformations, check the tables 3.8 and 3.9.
3. For $\mathsf{U} = \sum_i \mathsf{U}_i^\mathsf{T} \mathsf{U}_i$, compute a singular value decomposition $\mathsf{U} = \mathsf{EDG}^\mathsf{T}$. The estimate $\widehat{\boldsymbol{\beta}}$ is equal to the last row of G.
4. If necessary, recondition the estimated $\widehat{\boldsymbol{\beta}}$ by $1/f$; for a transformation $\widehat{\boldsymbol{\beta}}$ use equation (2.25) with f, f'.

Algorithm 3.3: Algorithm for directly computing unknown geometric entities and transformations defined by a set of observations, that are related to the unknown.

3.5.7 Generic Construction Algorithm

To summarize the section on construction and estimation we introduce a generic construction algorithm, which can be used for estimating geometric entities in 2D and 3D, but also for projective transformations. Note that this algorithm can even be taken for computing unique constructions, thus estimations with zero redundancy, which is a bit more involving than the unique construction algorithm.

In the next chapter we describe a statistical version of this algorithm, for which examples will be shown on artificial and real data.

4 Statistical Geometric Reasoning

In the last chapter we have developed simple algorithms for geometric reasoning on points, lines and planes in 2D and 3D. They were designed with a perfect world in mind: we have not explicitly represented possible errors in the coordinates of the geometric entities. Only for optimal estimations of objects we have allowed the entities to have small errors, but have not given any error measures for observations and unknowns. The weakness of having no error representation becomes apparent when testing relations between entities, see section 3.3: to check a geometric relation, a distance measure has to be equal to zero, but in practice, the computed distances are only "almost equal to zero".

In part this is due to the finite precision of machine floating operations, but the real reason is that we usually deal with imperfect data: geometric entities are initially obtained by some *measurement process*. It is obvious that, the result of a measurement is generally corrupted by the imperfection of the measuring device, influences from the environment or the experience of the human operator. Additionally, the data model that is used is always imperfect and does not reflect the reality appropriately. All these influences add a certain amount of *noise* or *uncertainty* to the true, ideal values and thus we only observe a distorted view of the ideal data. For practical applications it is of most importance to take the uncertainty of the observed data into account. The main question that is investigated in this chapter is how to represent the uncertainty such that the noisy nature of the data is considered appropriately and at the same time ensure that the complexity of the reasoning algorithms from the last chapter remains tractable.

The outline of this chapter is as follows: we first develop an uncertainty representation of geometric entities and analyze the necessary steps for geometric reasoning with respect to the uncertainty. As the entities are parametrized by homogeneous vectors, the uncertainty representation includes entities at infinity. To keep the transformations during the reasoning steps sufficiently simple, it is practical to use an *approximate* representation of the uncertainty based on the second moments. We explore the errors of the approximate representation and the conditions under which the approximation is valid. The last three sections of this chapter use the results of the error analysis and introduce relatively simple and generic algorithms for constructing new un-

S. Heuel: Uncertain Projective Geometry, LNCS 3008, pp. 97-148, 2004.

certain entities and test relations between them. We have implemented these algorithms in a software library called SUGR (Statistically Uncertain Geometric Reasoning) and demonstrate the validity with various Monte Carlo tests. Note that we do not explicitly cover the uncertainty of line segments and planar patches. It is possible to represent segments and patches by their geometric entity – either an infinite line or a plane – and e.g. their bordering elements, endpoints and border lines, see chapter 2.

4.1 Representation of Uncertain Geometric Entities

In the last chapter we have derived ways to do geometric reasoning with homogeneous vectors that represent points, lines and planes. This was achieved by doing both the constructions and the tests for relations completely in the homogeneous vector space. Only when interpreting the homogeneous coordinates in Euclidean space a vector was normalized by the norm of its homogeneous part, cf. section 2.2 on page 24. The transfer of the Euclidean parameters to the homogeneous vectors can be done by a simple transformation.

In this context, we now introduce the notion of uncertainty and develop an approximate representation for uncertain geometric entities within the framework of projective geometry.

4.1.1 General Representation

The desired uncertainty representation has to accomplish the following tasks:

- transfer the Euclidean uncertainty to the homogeneous uncertainty as all imprecise measurements originate in Euclidean space
- derive a new homogeneous uncertainty after constructions
- normalize the homogeneous uncertainty back to the Euclidean space, since only the Euclidean interpretation of uncertainties are meaningful for Euclidean reconstruction.

To model the uncertainty of homogeneous vectors, we assume each homogeneous coordinate to be a continuous random variable, $\underline{\mathbf{x}} = (\underline{x}_1, \cdots, \underline{x}_n)^\mathsf{T}$, such that we can define a probability density function (pdf) for homogeneous vectors:

Definition 7 (Probability density function) *The probability density function $p_\mathbf{x}$ of a homogeneous vector $\underline{\mathbf{x}} \in \mathbb{R}^{n+1}$ is a non-negative valued function $p : \mathbb{R}^{n+1} \rightarrow \mathbb{R}^+$ such that*

$$\int_{\mathbf{y} \in \mathbb{R}^{n+1}} p_\mathbf{x}(\mathbf{y}) d\mathbf{y} = 1$$

The probability $P(\mathbf{x} \in \mathcal{S})$ *that a point* $\mathbf{x}$ *is within a region* $\mathcal{S} \subseteq \mathbb{R}^{n+1}$ *is*

$$P(\mathbf{x} \in \mathcal{S}) = \int_{\mathbf{y} \in \mathcal{S}} p_{\mathrm{x}}(\mathbf{y}) d\mathbf{y}$$

Note that the pdf of $\underline{\mathbf{x}}$ is written as p_{x}, where the index identifies the random variable, cf. PAPOULIS AND PILLAI 2002; sometimes we write $\underline{\mathbf{x}} \sim p_{\mathrm{x}}(\mathbf{x})$.
A statistical projective point of dimension n may now be defined using its homogeneous vector representation $\underline{\mathbf{x}} \in \mathbb{R}^{n+1}$ with its distribution $p_{\mathrm{x}}(\mathbf{x})$. Examples of pdfs for projective points are depicted in figure 4.1 for $\mathbb{P}^1$: the pdf $p_{\mathrm{x}}(\mathbf{x})$ has an arbitrary form, whereas the pdf $p_{\mathrm{y}}(\mathbf{x})$ is only non-zero along the Euclidean line parallel to u-axis. One can argue that both pdfs represent the same uncertain projective points: only the directional variation of their homogeneous vectors refer to Euclidean uncertainty, thus we may project the pdfs onto the unit sphere, then $p_{x_s}(\mathbf{x}) = p_{y_s}(\mathbf{x})$.

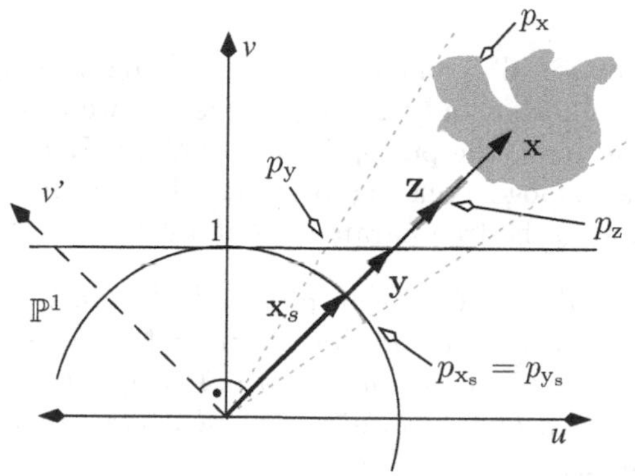

Fig. 4.1. Examples of uncertain homogeneous vectors in $\mathbb{P}^1$ with different pdfs. The two pdfs p_{x} and p_{y} represent the same uncertain Euclidean entity, since their normalized pdfs are $p_{\mathrm{x_s}}$ and $p_{\mathrm{y_s}}$ are identical. If the pdf is only non-zero along the direction of its vector as for p_{z}, the underlying geometric entity has no errors and p_{z} is not a proper pdf, see definition 8.

Thus analogous to the the equivalence relation (2.2) on page 22 , we can resolve the ambiguity caused by the homogeneity by defining

$$\underline{\mathbf{x}} \cong \underline{\mathbf{y}} \quad :\Leftrightarrow \quad p_{\mathrm{x_s}} = p_{\mathrm{y_s}} \text{ with } \underline{\mathbf{x}}_s := \mathsf{N}_s(\underline{\mathbf{x}}) = \frac{\underline{\mathbf{x}}}{|\underline{\mathbf{x}}|} \text{ and } \underline{\mathbf{y}}_s := \mathsf{N}_s(\underline{\mathbf{y}}) \qquad (4.1)$$

Thus two stochastical homogeneous vectors are equivalent if their corresponding pdfs of the *normalized* vectors are identical, where the new pdfs $p_{\mathrm{x_s}}$ and $p_{\mathrm{y_s}}$ are obtained by error propagation.
Considering a homogeneous vector is $\underline{\mathbf{x}} = (\underline{\lambda}\, \underline{\boldsymbol{x}}^{\mathsf{T}}, \underline{\lambda})^{\mathsf{T}}$, the homogeneous factor λ is a non-stochastic value, but we may still regard it as a random variable $\underline{\lambda}$ with a *Dirac delta function* as pdf, $p_{\underline{\lambda}} = \delta(\lambda - x)$. Thus we allow generalized functions as pdfs. This has the effect that the equivalence relation of (4.1) might yield generalized functions as well. Therefore we reformulate equation (4.1) by using the cumulative pdf (cpdf) $P_{\mathrm{x}}(\mathbf{y}) = \int_{\mathbf{x} \geq \mathbf{y}} p_{\mathrm{x}}(\mathbf{x}) d\mathbf{x}$

$$\underline{\mathbf{x}} \cong \underline{\mathbf{y}} \quad \Leftrightarrow \quad P_{\mathrm{x_s}} = P_{\mathrm{y_s}} \qquad \text{with } \underline{\mathbf{x}}_s := \mathsf{N}_s(\underline{\mathbf{x}}) = \frac{\underline{\mathbf{x}}}{|\underline{\mathbf{x}}|} \text{ and } \underline{\mathbf{y}}_s := \mathsf{N}_s(\underline{\mathbf{y}}) \tag{4.2}$$

Proper Density Functions As we have pointed out earlier, we assume that all free parameters of our underlying geometric entities are uncertain, thus the marginal distributions of the free parameters are not generalized functions. For an uncertain homogeneous vector $\underline{\mathbf{x}}$, this means that orthogonal to the direction of an observed $\mathbf{x}$ all marginal distributions must not be a Delta function. Technically, this means that the following property should hold:

Definition 8 (Proper pdf for geometric entities)
Given an observation $\mathbf{x} = (x_1, \cdots, x_n)$ *from a pdf* p_{x}*, an orthogonal transformation* $\mathbf{x}' = \mathsf{M}\mathbf{x}$ *exists such that* $\mathbf{x}$ *maps onto* $(1, 0, \cdots, 0)^{\mathsf{T}}$*. If for all* $i = 2, \cdots, n$ *the marginal distribution for* x'_i *is not a generalized function, then we call the* p_{x} *a proper pdf for* $\mathbf{x}$.

Looking at figure 4.1, definition 8 states that the pdf for v' must not be a Delta function, $p_{v'}(x) \neq \delta(x)$.
The above property is not sufficient for all geometric entities, though: as a 3D line is represented by a 6-vector, but has only 4 free parameters, we have to take the Plücker condition into account; the pdf $p_{\overline{\mathrm{L}}}$ for the dual line $\overline{\mathbf{L}}$ also has to fulfill property 4.1 as it is a homogeneous vector. Thus the length of the vector is $\overline{\mathbf{L}} = C\mathbf{L}$ is not relevant to the free parameters of a 3D line.

Definition 9 (Proper pdf for 3D lines) *Given an observed 3D line* $\mathbf{L} = (L_1, \cdots, L_n)$ *from a pdf* p_{L}*, an orthogonal transformation* $\mathbf{L}' = \mathbf{ML}$ *exists such that* $\mathbf{L}$ *maps onto* $(1, 0, \cdots, 0)^{\mathsf{T}}$ *and* $\overline{\mathbf{L}}$ *maps onto* $(0, 0, 0, 1, 0, 0)^{\mathsf{T}}$*. If for all* $i = 2, 3, 5, 6$ *the marginal distribution for* L'_i *is not a generalized function, then we call the* p_{L} *a proper pdf for* $\mathbf{L}$.

We conclude that a proper uncertainty representation of an observed geometric entity is a pair $(\mathbf{x},\ p_{\mathrm{x}})$, consisting of the observation $\mathbf{x}$ as an instance of the stochastical variable $\underline{\mathbf{x}}$ and a probability density function p_x for $\underline{\mathbf{x}}$. The probability density function has to satisfy definition 8 resp. 9 for 3D lines.

Transfer, Construction, Normalizing We now explore the previously described steps for doing geometric reasoning, namely *transfer* of Euclidean entities, *construction* of new entities and *normalizing* back from homogeneous to Euclidean entities, see also figure 4.2.

- *transfer*
  We assume that a pdf p_{x} is given for the Euclidean parameters contained in the uncertain vector $\underline{\boldsymbol{x}}$. The corresponding homogeneous vector is $\underline{\mathbf{x}} = (\underline{\lambda}\,\underline{\boldsymbol{x}}^{\mathsf{T}}, \underline{\lambda})^{\mathsf{T}}$, where $\underline{\lambda}$ is the homogeneous factor which also has a pdf p_λ. Thus it is necessary to compute $p_{(\mathrm{x},\lambda)}$.

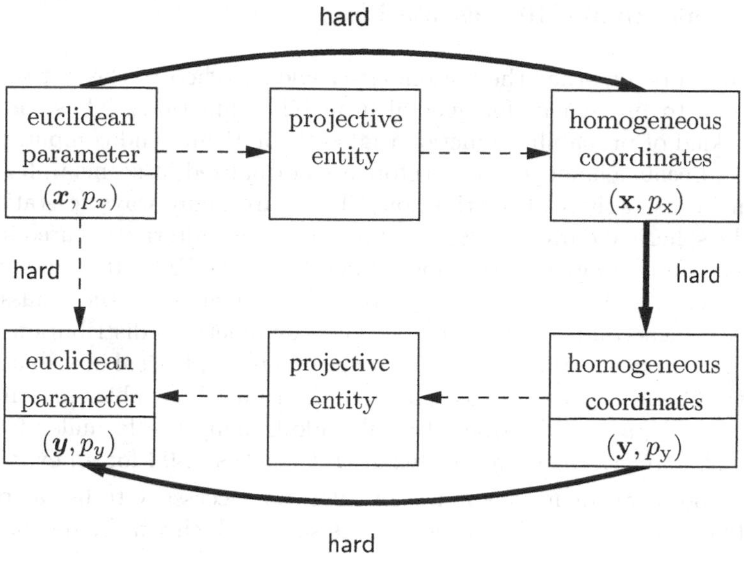

Fig. 4.2. General Statistical Geometric Reasoning using Homogeneous coordinates. As opposed to the diagram 3.1 in chapter 2, the observed entities are now represented by its vector $\boldsymbol{x}$ and its pdf p_x. The task of reasoning now includes not only the transformation of the vector $\boldsymbol{x}$ including the propagation its pdf. With general pdfs, this is quite hard to accomplish.

- *construction*
  As we have seen in the previous chapter, constructions of new entities given two homogeneous vectors are bilinear (or trilinear) functions $\underline{\mathbf{z}} = \underline{\mathbf{z}}(\underline{\mathbf{x}}, \underline{\mathbf{y}})$. Thus it is necessary to compute the pdf p_z.
- *normalizing*
  Normalizing back to Euclidean coordinates requires to compute the pdf p_{x^O} with $\mathbf{x}^O = \mathsf{N}_O(\underline{\mathbf{x}}) = \frac{\underline{\mathbf{x}}}{|\underline{\mathbf{x}}_h|}$.

An additional remark is related to the term *statistically well-conditioned*, coined by DURRANT-WHYTE 1989 and which is unrelated to the term "conditioning" introduced in section 3.5.6. In few words, a parameter representation is statistically well-conditioned, if the entity–representation relation is one-to-one and that the parameter vector of the representation must change "smoothly" with physical motion. Obviously homogeneous vectors do not satisfy the one-to-one condition. The proposed normalization of pdfs though ensures that for Euclidean or spherical entities the one-to-one relation can be established except for the sign ambiguity which is not important for our application. Therefore the comparability of two entities is still unique and applicable.

4.1.2 Approximate Representation

It is fairly obvious that the reasoning scheme sketched in the last section is very hard to implement for general probability functions. Thus we need a special kind of probability function that is theoretically and computationally feasible. Because homogeneous vectors are normalized, it seems to be advantageous to use spherical distributions. There are many ways to statistically describe spherical data, cf. FISHER *et al.* 1987. One alternative used in Computer Vision is Bingham's distribution, cf. COLLINS 1993. It is an antipodal distribution for directions on a sphere and is – similar to the Gaussian for the plane – a second-order approximation of symmetrical distributions on the sphere. This has the theoretical advantage of a real spherical distribution. On the other hand, operations on Bingham distributed variables in general do not yield a Bingham distributed result, additionally the formulas for doing reasoning may still be quite involving, cf. COLLINS 1993 for an example.

We propose that for most applications it is not necessary to use a spherical distribution as long as the uncertainty is small, which will be demonstrated in section 4.3.

4.1.2.1 First and Second Moments and Gaussian Distribution As an approximation of general pdfs, only some characterizing parts will be used, specifically the first and second (central) moments of p_{x}:

$$\begin{aligned} E(\underline{\mathbf{x}}) = \boldsymbol{\mu}_{\mathrm{x}} &= \int_{\mathbf{y}\in\mathbb{R}^{n+1}} \mathbf{y} p_{\mathrm{x}}(\mathbf{y}) d\mathbf{y} \\ C(\underline{x}_i, \underline{x}_j) = \sigma_{ij} &= E[(\underline{x}_i - \mu_i)(\underline{x}_j - \mu_j)] \\ &= \int_{-\infty}^{\infty} \cdots \int_{-\infty}^{\infty} (\underline{x}_i - \mu_i)(\underline{x}_j - \mu_j) p_{\mathrm{x}}(\mathbf{y}) d\mathbf{y} \end{aligned}$$

The first moment $\boldsymbol{\mu}_{\mathrm{x}}$ is the mean value of $\underline{\mathbf{x}}$; if we have only one observed instance $\mathbf{x}$, then it is reasonable to assume $\boldsymbol{\mu}_{\mathrm{x}} = \mathbf{x}$. The second central moment σ_{ij} is the covariance of the coordinates x_i and x_j of $\mathbf{x}$. The matrix $\boldsymbol{\Sigma}_{\mathrm{xx}} = (\sigma_{ij})$ is the covariance matrix of the uncertain homogeneous vector $\underline{\mathbf{x}}$. For the variances on the diagonal of $\boldsymbol{\Sigma}_{\mathrm{xx}}$, we use the notation $\sigma_i^2 = \sigma_{ii}$.

Usually we have only one observation $\mathbf{x}$ available of a geometric entity. Then we postualte $\mathbf{x}$ to be the best estimators for the mean $\boldsymbol{\mu}_{\mathrm{x}}$. The proposed uncertainty representation for a geometric entity is now a pair $(\mathbf{x}, \boldsymbol{\Sigma}_{\mathrm{xx}})$ consisting of the actual observed entity, represented as a homogeneous vector $\mathbf{x}$ and the covariance matrix $\boldsymbol{\Sigma}_{\mathrm{xx}}$ of the homogeneous vector.

As we disregard all moments of higher order, we only approximate the ideal pdf. But if we assume a Gaussian distribution function

$$g_{\mu,\Sigma}(\underline{\mathbf{x}}) = \frac{1}{\sqrt{(2\pi)^n |\boldsymbol{\Sigma}|}} e^{-\frac{1}{2}(\mathbf{x}-\boldsymbol{\mu})^{\mathsf{T}} \boldsymbol{\Sigma}^{-1} (\mathbf{x}-\boldsymbol{\mu})} \tag{4.3}$$

then the pdf $g_{\mu,\Sigma}(\underline{\mathbf{x}})$ is uniquely determined by the mean $\boldsymbol{\mu}$ and the covariance $\boldsymbol{\Sigma}$. If $\underline{\mathbf{x}}$ is Gaussian or normal distributed, we write $\underline{\mathbf{x}} \sim N(\boldsymbol{\mu}, \boldsymbol{\Sigma})$.

4.1.2.2 Singular Covariance Matrices In the following sections we will be confronted with cases where an uncertain vector $\underline{\mathbf{x}}$ is assumed to be normally distributed $\underline{\mathbf{x}} \sim N(\boldsymbol{\mu}, \boldsymbol{\Sigma})$ but $\boldsymbol{\Sigma}$ is singular, hence $|\boldsymbol{\Sigma}| = 0$; the covariance matrix $\boldsymbol{\Sigma}$ is positive semi-definite instead of positive definite for $|\boldsymbol{\Sigma}| \neq 0$. A singular covariance matrix indicates that at least one marginal distribution is not a proper function. As an example, assume $\underline{\mathbf{y}} = (\underline{y}_1, \underline{y}_2)^\mathsf{T}$ with $\boldsymbol{\Sigma}_{\mathrm{yy}} = \left(\begin{smallmatrix}1&0\\0&0\end{smallmatrix}\right)$, then $\underline{y}_1 \sim N(\mu_{y_1}, 1)$ and $\underline{y}_2 \sim N(\mu_{y_2}, 0)$, cf. figure 4.3.

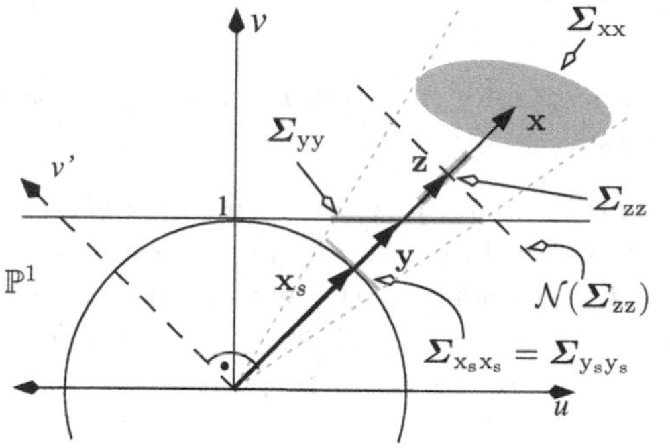

Fig. 4.3. Examples of uncertain homogeneous vectors in $\mathbb{P}^1$ with different Gaussian pdfs, see also figure 4.1. The two covariances $\boldsymbol{\Sigma}_{\mathrm{xx}}$ and $\boldsymbol{\Sigma}_{\mathrm{yy}}$ represent the same Euclidean uncertainty, since their normalized covariance matrices are identical. If the covariance matrix has a nullspace which is orthogonal to an observed vector $\mathbf{x}$, then it does not represent a proper covariance matrix.

Although the covariance matrix $\boldsymbol{\Sigma}_{\mathrm{xx}}$ may be singular, it has to fulfill the property that it refers to a full-rank covariance matrix of the parameters of the underlying geometric entity. We can reformulate definition 8 for general pdfs to a property of the covariance matrix:

Proposition 2 (Proper covariance matrix for entities) *Given an observation $\mathbf{x} = (x_1, \cdots, x_n)$ being an homogeneous vector and a covariance matrix $\boldsymbol{\Sigma}_{\mathrm{xx}}$. If the nullspace $\mathcal{N}(\boldsymbol{\Sigma}_{\mathrm{xx}})$ of $\boldsymbol{\Sigma}_{\mathrm{xx}}$ does not intersect with the subspace $\{\mathbf{x}\}^\perp$ orthogonal to the vector $\mathbf{x}$:*

$$\mathcal{N}(\boldsymbol{\Sigma}_{\mathrm{xx}}) \cap \{\mathbf{x}\}^\perp = \{\mathbf{0}\} \tag{4.4}$$

then the Gaussian with $\boldsymbol{\mu}_{\mathrm{x}} = \mathbf{x}$ and $\boldsymbol{\Sigma}_{\mathrm{xx}}$ is a proper pdf for $\mathbf{x}$. We call $\boldsymbol{\Sigma}_{\mathrm{xx}}$ a proper covariance matrix *for $\mathbf{x}$.*

Proof: Referring to definition 8, we look at the orthogonal transformation $\mathbf{x}' = \mathsf{M}\mathbf{x}$, such that $\mathbf{x}$ is mapped to $\mathbf{e}_i = (1, 0, \cdots, 0)^\mathsf{T}$. Using first-order error propagation, see page 104, the transformed covariance matrix is $\boldsymbol{\Sigma}_{\mathrm{x'x'}} = \mathsf{M}\boldsymbol{\Sigma}_{\mathrm{xx}}\mathsf{M}^\mathsf{T}$. For all coordinates x'_i the variances $\sigma^2_{x'_i}$ must not be zero, only $\sigma^2_{x'_1}$ for the first coordinate may be zero. In other words, the nullspace $\mathcal{N}(\boldsymbol{\Sigma}_{\mathrm{x'x'}})$ of $\boldsymbol{\Sigma}_{\mathrm{x'x'}}$ may be empty or include $\mathbf{e}_i$. Transforming back using M^{-1} yields equation (4.4). □

Thus a covariance matrix for homogeneous vectors has a rank of $r \geq n$, with $\mathbf{x} \in \mathbb{R}^{n+1}$. Again, for the special case of 3D lines, definition 9 has to be reformulated as only 4 free parameters are contained in the six Plücker coordinates:

Proposition 3 (Proper covariance matrix for 3D lines) *Given an observed 3D line* $\mathbf{L} = (L_1, \cdots, L_6)$ *and a covariance matrix* $\boldsymbol{\Sigma}_{\mathrm{LL}}$. *If the nullspace*$\mathcal{N}(\boldsymbol{\Sigma}_{\mathrm{LL}})$ *of* $\boldsymbol{\Sigma}_{\mathrm{LL}}$ *does not intersect with the subspace* $\{\mathbf{L}, \overline{\mathbf{L}}\}^{\perp}$ *orthogonal to the vectors* $\mathbf{L}$ *and* $\overline{\mathbf{L}}$*:*

$$\mathcal{N}(\boldsymbol{\Sigma}_{\mathrm{LL}}) \cap \{\mathbf{L}, \overline{\mathbf{L}}\}^{\perp} = \{\mathbf{0}\}$$

then the Gaussian with $\boldsymbol{\mu}_{\mathrm{L}} = \mathbf{L}$ *and* $\boldsymbol{\Sigma}_{\mathrm{xx}}$ *is a proper pdf for* $\mathbf{L}$. *We call* $\boldsymbol{\Sigma}_{\mathrm{LL}}$ *a proper covariance matrix for* $\mathbf{L}$.

We are now prepared to define the uncertainty representation of a geometric entity based on homogeneous vectors:

Definition 10 (Uncertain geometric entity) *A point, line or a plane in 2D or 3D has been observed and represented as a homogeneous vector* $\mathbf{x}$, *cf. section 2 on page 19. We represent the* geometric entity with its uncertainty *or* uncertain homogeneous vector *as a pair* $(\mathbf{x}, \boldsymbol{\Sigma}_{\mathrm{xx}})$ *where the covariance matrix* $\boldsymbol{\Sigma}_{\mathrm{xx}}$ *is a proper covariance matrix according to propositions 2 resp. 3 for 3D lines.*

Note that this definition handles entities at infinity in the same way as entities, that have a finite distance to the origin.

4.2 Transformation of Uncertain Homogeneous Vectors

It is important to explore the effects of transformations to uncertain homogeneous vectors. We will see that for our applications there are mostly linear or bilinear transformations, only normalizations are completely nonlinear. This section will describe the Euclidean representations of geometric entities and how to transform them to homogeneous representations and vice versa using first order error propagation.
We will concentrate on geometric entities such as points, lines and planes. We only briefly comment on geometric transformations that are represented by homogeneous vectors.

4.2.1 First Order Error Propagation

One advantage of using first and second moments for characterizing the pdf of a stochastic variable is the simplicity for first order error propagation, cf. Koch 1999, p. 100:

Proposition 4 (Error propagation) *Let* $\underline{\mathbf{x}} \in \mathbb{R}^{n+1}$ *be an uncertain homogeneous vector with the covariance matrix* $\boldsymbol{\Sigma}_{\mathrm{xx}}$ *and let* A *be a fixed* $(m+1) \times (n+1)$ *matrix and* $\mathbf{b}$ *a fixed* $(m+1)$ *vector , then the covariance matrix* $\boldsymbol{\Sigma}_{\mathrm{yy}}$ *of the linear transformation* $\underline{\mathbf{y}} = \mathsf{A}\underline{\mathbf{x}} + \mathbf{b}$ *is given as*

$$\boldsymbol{\Sigma}_{\mathrm{yy}} = \mathsf{A}\,\boldsymbol{\Sigma}_{\mathrm{xx}}\mathsf{A}^{\mathsf{T}} \tag{4.5}$$

Note that we have not assumed any pdf for $\underline{\mathbf{x}}$, thus the error propagation is true even if $\underline{\mathbf{x}}$ is not Gaussian distributed. But an important observation can be made if $\underline{\mathbf{y}}$ is known to be Gaussian:

Proposition 5 (Error propagation of a Gaussian) *If* $\underline{\mathbf{x}}$ *is normally distributed,* $\underline{\mathbf{x}} \sim N(\boldsymbol{\mu}_x, \boldsymbol{\Sigma}_{\mathrm{xx}})$, *the linear transformed* $\underline{\mathbf{y}}$ *is again normally distributed:*

$$\underline{\mathbf{y}} \sim N(\,\mathsf{A}\boldsymbol{\mu}_x + \mathbf{b},\ \mathsf{A}\boldsymbol{\Sigma}_{\mathrm{xx}}\mathsf{A}^{\mathsf{T}}\,)$$

As we have seen, some of our mappings are bilinear thus are not strict linear transformations as products of different coordinates appear in the formulas, cf. section 3.1 on page 48. In general we assume a real differentiable vector function $\mathbf{f}(\mathbf{x})$ of a homogeneous vector $\mathbf{x} \in \mathbb{R}^{n+1}$. To approximate a function value $\mathbf{y} = \mathbf{f}(\mathbf{x} + \Delta\mathbf{n})$ we can apply the Taylor series

$$\mathbf{f}(\mathbf{x} + \Delta\mathbf{n}) = \mathbf{f}(\mathbf{x}) + \left(\frac{\partial \mathbf{f}(\mathbf{z})}{\partial \mathbf{z}}\right)_{\mathbf{z}=\mathbf{x}} \Delta\mathbf{n} + \mathcal{O}(\Delta\mathbf{n}^2) \tag{4.6}$$

If the terms of second degree and higher, $\mathcal{O}(\Delta\mathbf{n}^2)$, are sufficiently small enough, we can approximate the covariance matrix $\boldsymbol{\Sigma}_{\mathrm{yy}}$ of $\mathbf{y}$ by

$$\boldsymbol{\Sigma}_{\mathrm{yy}} = \mathsf{J}\boldsymbol{\Sigma}_{\mathrm{yy}}\mathsf{J}^{\mathsf{T}} \quad \text{with} \quad \mathsf{J} = \left(\frac{\partial \mathbf{f}(\mathbf{z})}{\partial \mathbf{z}}\right)_{\mathbf{z}=\mathbf{x}} \tag{4.7}$$

With the first order error propagation, it is now possible to investigate in detail the three steps for doing statistical geometric reasoning with the proposed approximate representation: *transfer* of uncertain Euclidean entities to Homogeneous coordinates, uncertainty propagation for *construction* of new uncertain homogeneous entities and *normalization* back to uncertain Euclidean entities. In section 4.3 on page 113 we will discuss the validity of the approximation in equation (4.7) with respect to these operations.

4.2.2 Transfer to Homogeneous Coordinates

In this subsection we explore how to transform the uncertainty of a Euclidean representation in the proposed homogeneous representation.

4.2.2.1 Points Without loss of generality, we only deal with 2D points, as it is basically the same as the 3D case. Given a Euclidean 2D point $\boldsymbol{x} = (x_1, x_2)^\mathsf{T} \in \mathbb{R}^2$ we have seen that a simple mapping yields a corresponding homogeneous vector $\mathbf{x} = (x_1, x_2, 1)^\mathsf{T}$, see equation (2.3) on page 22. We can reformulate this mapping to

$$\mathbf{x} = \begin{pmatrix} 1 & 0 \\ 0 & 1 \\ 0 & 0 \end{pmatrix} \boldsymbol{x} + \begin{pmatrix} 0 \\ 0 \\ 1 \end{pmatrix} \tag{4.8}$$

If $\boldsymbol{\Sigma}_{xx}$ is the covariance matrix for $\boldsymbol{x}$, then the covariance matrix $\boldsymbol{\Sigma}_{\mathrm{xx}}$ for $\mathbf{x}$ is according to theorem 4:

$$\boldsymbol{\Sigma}_{\mathrm{xx}} = \begin{pmatrix} \sigma_{x_1}^2 & \sigma_{x_1 x_2} & 0 \\ \sigma_{x_1 x2} & \sigma_{x_2}^2 & 0 \\ 0 & 0 & 0 \end{pmatrix} \tag{4.9}$$

In general, a vector $\mathbf{x}_\lambda = (\lambda x_1, \lambda x_2, \lambda)^\mathsf{T}$ yields

$$\boldsymbol{\Sigma}_{\mathrm{x}_\lambda \mathrm{x}_\lambda} = \lambda^2 \begin{pmatrix} \sigma_{x_1}^2 & \sigma_{x_1 x_2} & 0 \\ \sigma_{x_1 x2} & \sigma_{x_2}^2 & 0 \\ 0 & 0 & 0 \end{pmatrix} \tag{4.10}$$

The transfer of uncertain Euclidean points to homogeneous vectors is rigorous as the transformation is linear, no loss of information is involved. This is important as in most applications a reasoning process starts with points as observations, either in 2D or in 3D.

We assume that only non-infinite or Euclidean points can be *directly* observed. Infinite points, such as points on the horizon in an image, do exist, but are usually constructed using Euclidean entities, such as using a projective camera and a 2D image point. Thus the covariance matrix of an infinite point is determined by the covariance matrices of its constructing entities using error propagation.

4.2.2.2 Hyperplanes Hyperplanes are 2D lines and 3D planes. Again we concentrate on the 2D case: directly observed 3D planes are not common in the computer vision context, but 2D lines are frequently observed in images. Given a normal vector $\underline{\boldsymbol{n}} = (\underline{u}, \underline{v})^\mathsf{T}$ with $\boldsymbol{\Sigma}_n = \begin{pmatrix} \sigma_u^2 & \sigma_{uv} \\ \sigma_{uv} & \sigma_v^2 \end{pmatrix}$ and a distance $\underline{d}$ with σ_d and covariances σ_{ud} and σ_{vd} the covariance matrix for $\mathbf{l}(\underline{\boldsymbol{n}}, \underline{d})$ is obviously

$$\boldsymbol{\Sigma}_{\mathrm{ll}} = \begin{pmatrix} \sigma_u^2 & \sigma_{uv} & \sigma_{ud} \\ \sigma_{uv} & \sigma_v^2 & \sigma_{vd} \\ \sigma_{ud} & \sigma_{vd} & \sigma_d^2 \end{pmatrix}$$

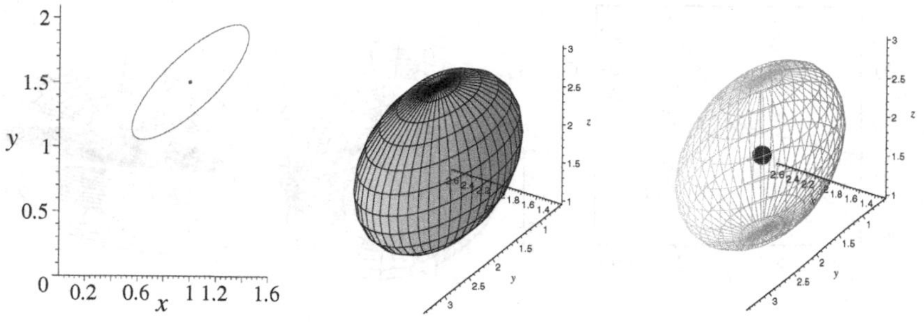

(a) point in 2D (b) point in 3D

Fig. 4.4. Visualization of confidence region of an uncertain 3D point

As observed in section 2.2.1 on page 25, there are lots of Euclidean representations for a 2D line and not all transformations to the homogeneous representation can be covered in this work. We particularly use one representation of an uncertain 2D line, which is based on the point-angle form and consists of a 5-tuple

$$\boldsymbol{l}_e = (x_m, y_m, \varphi, \sigma_d, \sigma_\varphi) \tag{4.11}$$

as proposed by FÖRSTNER 1992 and used by FUCHS 1998. A detailed derivation of this representation is found in the appendix C.1.1 on page 187. The point $\boldsymbol{x}_m = (x_m, y_m)^\mathsf{T}$ is the *center point* or *point of gravity* of the line, where the displacement error σ_v orthogonal to the line is minimal and equal to σ_d. The further away we move from $\boldsymbol{x}_m$ along the line by u, the larger is the displacement error, namely $\sigma_v = \sqrt{\sigma_\varphi^2 u^2 + \sigma_d^2}$. The resulting hyperbola in figure 4.5(a) can be interpreted as the union of all 1D-confidence regions of points on the line, that are uncertain only orthogonal to the given line, see appendix C.1.1. Note that σ_φ and σ_d are uncorrelated in representation (4.11), see section C.1.1 for details.

The transformation from (4.11) to the proposed representation is

$$(\mathbf{l}, \boldsymbol{\Sigma}_{ll}) = \left(\begin{pmatrix} \cos(\varphi) \\ \sin(\varphi) \\ -d \end{pmatrix}, \mathsf{M} \begin{pmatrix} \sigma_\varphi^2 & 0 & 0 \\ 0 & 0 & 0 \\ 0 & 0 & \sigma_d^2 \end{pmatrix} \mathsf{M}^\mathsf{T} \right)$$

$$\text{with } \mathsf{M} = \begin{pmatrix} 1 & 0 & 1 \\ 0 & 1 & 1 \\ -x_m & -y_m & 1 \end{pmatrix} \begin{pmatrix} \cos(\varphi) & \sin(\varphi) & 0 \\ -\sin(\varphi) & \cos(\varphi) & 0 \\ 0 & 0 & 1 \end{pmatrix} \tag{4.12}$$

with $d = x_m \cos(\varphi) + y_m \sin(\varphi)$.

For a 3D plane, the confidence volume is a hyperboloid as shown in figure 4.5(b). Similarly to the 2D case, there is a center point $\boldsymbol{X}_m$, where the displacement error orthogonal to the plane is minimal. If the plane was con-

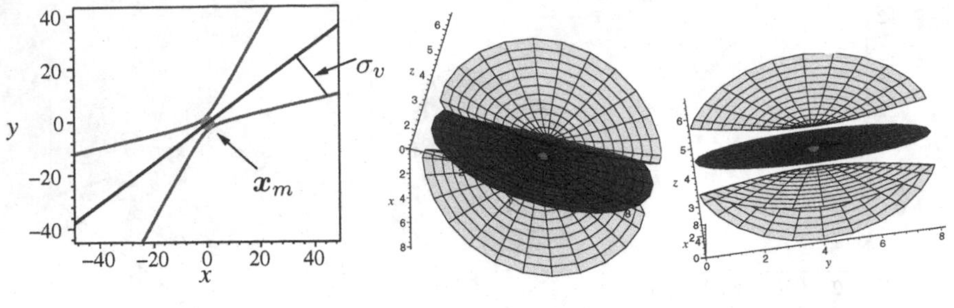

(a) line in 2D (b) plane in 3D

Fig. 4.5. Visualization of the confidence region of a 2D line and a 3D plane. In (a) the confidence region is a hyperbola, with the center point $\boldsymbol{x}_m$ being the point on the line with a minimal displacement error σ_v orthogonal to the line. The hyperboloid in (b) is the confidence volume for a plane again with a center point $\boldsymbol{X}_m$ at the minimal displacement error.

structed by three uncertain points, the center point is the weighted point of gravity of the three points.

4.2.2.3 3D Lines For 3D lines there are even more ways to represent the Euclidean parameters than for hyperplanes, but none of them are used frequently enough to serve as a reference. This is due to the fact that 3D lines are hard to observe directly; in computer vision applications they are usually observed indirectly by two 3D points or by two image lines. For both cases we have provided simple algebraic formulas which can be used for error propagation, see section 4.2.6 on page 111.

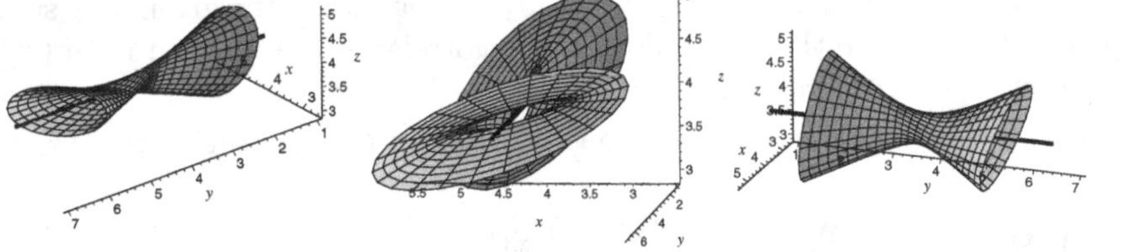

Fig. 4.6. Visualization of the confidence region of a 3D line. Its shape is a twisted cone, which is bent at two locations.

A visualization of the confidence volume of a 3D line is shown in figure 4.6, see also FÖRSTNER 1992. The shape of the volume is a twisted cone, which is bent at two locations. The shape can be explained as follows: a 3D line may be given by two 3D points, where the first point has an error ellipsoid elongated along the x-axis and the second point has an error ellipsoid elongated along the y- axis, thus the two ellipsoids are twisted against each other.

4.2.2.4 Geometric Transformations We represent the uncertainty of projective point transformations H by a covariance matrix $\boldsymbol{\Sigma}_{\mathrm{hh}}$ of the vector $\mathbf{h} = \mathrm{vec}(\mathsf{H}^\mathsf{T})$. Here, we assume that the covariance $\boldsymbol{\Sigma}_{\mathrm{hh}}$ is a proper, cf. proposition 2. In case the transformation and its uncertainty is given in Euclidean parameters, the covariance can be built using error propagation. The functional relationship between the given Euclidean representation and the homogeneous matrix is application dependent: it is not easy to describe a general method for establishing such a functional relationship: as an example, there are seemingly countless ways to describe the inner and outer orientation of a camera.

Assuming that the covariance of a point transformation H is given. To obtain the covariances for hyperplane transformations $\mathsf{H}_{\mathrm{A}} = \mathsf{H}^{-\mathsf{T}}$, one can compute (FÖRSTNER 2002)

$$\boldsymbol{\Sigma}_{\mathrm{H_A H_A}} = (\mathsf{H}^{-\mathsf{T}} \otimes \mathsf{H}^{-1})\boldsymbol{\Sigma}_{\mathrm{hh}}(\mathsf{H}^{-1} \otimes \mathsf{H}^{-\mathsf{T}}) \tag{4.13}$$

The covariance matrix for a 3D line transformation H_L is also computed by error propagation using the Jacobians mentioned in (3.41) on page 57; for the projective camera matrix Q for lines, see (3.51) on page 59.

4.2.3 Normalization to Euclidean Coordinates

Transferring a homogeneous vectors $\mathbf{x}$ back to Euclidean entities was done using the normalization operation $\mathsf{N}_O(\mathbf{x}) = \mathbf{x}/|\boldsymbol{x}_h|$, where the norm was skipped for point entities. We now want to extend the normalization operation to uncertain geometric entities $(\mathbf{x}, \boldsymbol{\Sigma}_{\mathrm{xx}})$. Again we apply first order error propagation and as the normalization is a non-linear function, we need to derive the Jacobians $\mathsf{J} = \frac{\partial \mathsf{N}_O(\mathbf{x})}{\partial \mathbf{x}}$. The normalization operation $\mathsf{N}_O((\mathbf{x}, \boldsymbol{\Sigma}_{\mathrm{xx}}))$ is then given by

$$\mathsf{N}_O((\mathbf{x}, \boldsymbol{\Sigma}_{\mathrm{xx}})) = \left(\frac{\mathbf{x}}{|\boldsymbol{x}_h|}, \mathsf{J}(\mathbf{x})\boldsymbol{\Sigma}_{\mathrm{xx}}\mathsf{J}^\mathsf{T}(\mathbf{x})\right) = (\mathbf{x}', \boldsymbol{\Sigma}_{\mathrm{x'x'}}) \tag{4.14}$$

Note that the ambiguity in the sign of a normalization $\mathsf{N}_O(\cdot)$ does not affect the error propagation. The Jacobians for 2D points and lines and for 3D points, lines and planes are listed in table 4.1. The nullspace of the normalized covariance matrix $\boldsymbol{\Sigma}_{\mathrm{x'x'}}$ is the nullspace of the Jacobian J^T, see table 4.1. Note that the nullspace is equal to the embedded homogeneous part of the vector; for example if $\mathbf{l}$ is a 2D line, the nullspace of $\boldsymbol{\Sigma}_{\mathrm{l'l'}}$ is $(\boldsymbol{l}_h, 0)^\mathsf{T}$.

It is possible to use the acquired pair $\mathsf{N}_O((\mathbf{x}, \boldsymbol{\Sigma}_{\mathrm{xx}}))$ to transform to any other Euclidean representation. For example, the transform for a 2D line in representation (4.11), see on page 107 is described in the appendix C.1.

Table 4.1. Jacobians for Euclidean normalizations $\mathsf{N}_O(\cdot)$ for points, lines and planes in 2D and 3D.

dim	entity	normalization	Jacobian J for $\mathsf{N}_O(\cdot)$	nullspace of J^T
2D	point $\mathbf{x}$	$\mathsf{N}_O(\mathbf{x}) = \frac{\mathbf{x}}{x_h}$	$\frac{\partial \mathsf{N}_O(\mathbf{x})}{\partial \mathbf{x}} = \frac{1}{x_h}\begin{pmatrix} I & -\frac{1}{x_h}\boldsymbol{x}_O \\ \mathbf{0}^\mathsf{T} & 0 \end{pmatrix}$	$\begin{pmatrix} \mathbf{0} \\ x_h \end{pmatrix}$
	line $\mathbf{l}$	$\mathsf{N}_O(\mathbf{l}) = \frac{1}{\lvert \boldsymbol{l}_h \rvert}$	$\frac{\partial \mathsf{N}_O(\mathbf{l})}{\partial \mathbf{l}} = \frac{1}{\lvert \boldsymbol{l}_h \rvert}\begin{pmatrix} I - \frac{\boldsymbol{l}_h \boldsymbol{l}_h^\mathsf{T}}{\lvert \boldsymbol{l}_h \rvert^2} & \mathbf{0} \\ \frac{-l_0 \boldsymbol{l}_h^\mathsf{T}}{\lvert \boldsymbol{l}_h \rvert^2} & 1 \end{pmatrix}$	$\begin{pmatrix} \boldsymbol{l}_h \\ 0 \end{pmatrix}$
3D	point $\mathbf{X}$	$\mathsf{N}_O(\mathbf{X}) = \frac{\mathbf{X}}{X_h}$	$\frac{\partial \mathsf{N}_O(\mathbf{X})}{\partial \mathbf{X}} = \frac{1}{X_h}\begin{pmatrix} I & -\frac{1}{X_h}\boldsymbol{X}_O \\ \mathbf{0}^\mathsf{T} & 0 \end{pmatrix}$	$\begin{pmatrix} \mathbf{0} \\ X_h \end{pmatrix}$
	line $\mathbf{L}$	$\mathsf{N}_O(\mathbf{L}) = \frac{\mathbf{L}}{\lvert \boldsymbol{L}_h \rvert}$	$\frac{\partial \mathsf{N}_O(\mathbf{L})}{\partial \mathbf{L}} = \frac{1}{\lvert \boldsymbol{L}_h \rvert}\begin{pmatrix} I - \left(\frac{\boldsymbol{L}_h \boldsymbol{L}_h^\mathsf{T}}{\lvert \boldsymbol{L}_h \rvert^2}\right) & 0 \\ -\frac{\boldsymbol{L}_h \boldsymbol{L}_O^\mathsf{T}}{\lvert \boldsymbol{L}_h \rvert^2} & I \end{pmatrix}$	$\begin{pmatrix} \boldsymbol{L}_h \\ \mathbf{0} \end{pmatrix}$
	plane $\mathbf{A}$	$\mathsf{N}_O(\mathbf{A}) = \frac{\mathbf{A}}{\lvert \boldsymbol{A}_h \rvert}$	$\frac{\partial \mathsf{N}_O(\mathbf{A})}{\partial \mathbf{A}} = \frac{1}{\lvert \boldsymbol{A}_h \rvert}\begin{pmatrix} I - \frac{\boldsymbol{A}_h \boldsymbol{A}_h^\mathsf{T}}{\lvert \boldsymbol{A}_h \rvert^2} & \mathbf{0} \\ \frac{-A_0 \boldsymbol{A}_h^\mathsf{T}}{\lvert \boldsymbol{A}_h \rvert^2} & 1 \end{pmatrix}$	$\begin{pmatrix} \boldsymbol{A}_h \\ 0 \end{pmatrix}$

4.2.4 Normalization to Spherical Coordinates

We may also want to normalize a homogeneous vector to unit length, called spherical normalization, cf. page 22. Given any homogeneous vector $\mathbf{x} \in \mathbb{P}^n$, the Jacobian J_s to the spherical normalization $\mathsf{N}_s(\mathbf{x}) = \frac{\mathbf{x}}{\lvert \mathbf{x} \rvert}$ is

$$\mathsf{J}_s(\mathbf{x}) := \frac{\partial \mathsf{N}_s(\mathbf{x})}{\partial \mathbf{x}} = \frac{1}{\lvert \mathbf{x} \rvert}\left(I - \frac{\mathbf{x}\mathbf{x}^\mathsf{T}}{\lvert \mathbf{x} \rvert^2}\right) \tag{4.15}$$

Thus

$$\mathsf{N}_s((\mathbf{x}, \boldsymbol{\Sigma}_{\mathbf{xx}})) = \left(\frac{\mathbf{x}}{\lvert \mathbf{x} \rvert}, \mathsf{J}_s(\mathbf{x})\boldsymbol{\Sigma}_{\mathbf{xx}}\mathsf{J}_s^\mathsf{T}(\mathbf{x})\right) \tag{4.16}$$

This equation holds for all geometric entities and transformations. It is easy to see that the nullspace of J_s and thus of the transformed covariance matrix is equal to $\mathbf{x}$.

4.2.5 Changing Nullspaces Using Orthogonal Projections

It is easy to see that the Jacobian J_s for the spherical normalization is essentially an orthogonal projection Ψ onto the space $\{\mathbf{x}\}^\perp$ (Koch 1999) and the appendix B.2 on page 184:

$$\mathsf{J}_s = \frac{1}{\lvert \mathbf{x} \rvert}\Psi_{\mathbf{x}} \quad \text{with } \Psi_{\mathbf{x}} = I - \mathbf{x}(\mathbf{x}^\mathsf{T}\mathbf{x})^{-1}\mathbf{x}^\mathsf{T} \tag{4.17}$$

Thus we may also apply only the projection matrix Ψ to the covariance matrix. Then we only change the nullspace of the covariance matrix but don't change the metric information. This is valid as long as we don't impose a nullspace that is a subspace of $\{\mathbf{x}\}^\perp$, see definition 10 on page 104; for spherical normalization, the nullspace is $\mathbf{x}$. Thus we define a new operator $\mathcal{P}_\mathbf{y}$ for uncertain homogeneous vectors, which only changes the nullspace of the covariance matrix $\boldsymbol{\Sigma}_{\mathrm{xx}}$ to $\mathbf{y}$, but does not change the underlying vector:

$$\mathcal{P}_\mathbf{y}\left((\mathbf{x},\ \boldsymbol{\Sigma}_{\mathrm{xx}})\right) := \left(\mathbf{x},\ \Psi_\mathbf{y}\boldsymbol{\Sigma}_{\mathrm{xx}}\Psi_\mathbf{y}^\mathsf{T}\right) \quad \text{with } \Psi_\mathbf{y} \text{ being an orthogonal projection onto } \{\mathbf{y}\}^\perp \tag{4.18}$$

For an uncertain 3D line $\mathbf{L}$, we may project $\boldsymbol{\Sigma}_{\mathrm{LL}}$ onto $\{\mathbf{L}, \overline{\mathbf{L}}\}^\perp$:

$$\mathcal{P}_{(\mathbf{L},\overline{\mathbf{L}})}\left((\mathbf{L},\ \boldsymbol{\Sigma}_{\mathrm{LL}})\right) := \left(\mathbf{L},\ \Psi_{(\mathbf{L},\overline{\mathbf{L}})}\boldsymbol{\Sigma}_{\mathrm{LL}}\Psi^\mathsf{T}_{(\mathbf{L},\overline{\mathbf{L}})}\right) \quad \text{with } \Psi_{(\mathbf{L},\overline{\mathbf{L}})} \text{ being an orthogonal projection onto } \{\mathbf{L}, \overline{\mathbf{L}}\}^\perp \tag{4.19}$$

Using a general dot-product $\mathbf{x}^\mathsf{T}\mathsf{W}\mathbf{y}$ with W being a positive semidefinite matrix, one can also express the Euclidean normalization: with the general orthogonal projection $\Psi_{\mathsf{A},\mathsf{W}}$ from proposition B.4 on page 184, one can show that for any geometric entity $\mathbf{x}$,

$$\frac{\partial \mathsf{N}_O(\mathbf{x})}{\partial \mathbf{x}} = \frac{1}{|\boldsymbol{x}_h|}\Psi_{\mathbf{x},\mathsf{M}} \quad \text{with } \Psi_{\mathbf{x},\mathsf{M}} = I - \mathbf{x}(\mathbf{x}^\mathsf{T}\mathsf{M}\mathbf{x})^{-1}\mathbf{x}^\mathsf{T}\mathsf{M} \tag{4.20}$$

where M is the identity matrix with zero entries on the diagonal elements, which correspond to the Euclidean part of the vector.

4.2.6 Construction

In chapter 3, constructions were shown to be bilinear operation, so two entities $\mathbf{x}$ and $\mathbf{y}$ form a new entity $\mathbf{z} = \mathbf{f}(\mathbf{x}, \mathbf{y})$, which can be expressed as

$$\mathbf{z} = \mathsf{U}(\mathbf{x})\mathbf{y} = \mathsf{V}(\mathbf{y})\mathbf{x} \tag{4.21}$$

Assuming the entities to be uncertain, the pairs $(\mathbf{x}, \boldsymbol{\Sigma}_{\mathrm{xx}})$ and $(\mathbf{y}, \boldsymbol{\Sigma}_{\mathrm{yy}})$ and possibly the covariances between $\mathbf{x}$ and $\mathbf{y}$, namely $\boldsymbol{\Sigma}_{\mathrm{xy}}$ are necessary for doing error propagation:

$$(\mathbf{z},\ \boldsymbol{\Sigma}_{\mathrm{zz}}) = \left(\mathsf{U}(\mathbf{x})\mathbf{y},\ \ [\mathsf{V}(\mathbf{y}), \mathsf{U}(\mathbf{x})]\begin{pmatrix}\boldsymbol{\Sigma}_{\mathrm{xx}} & \boldsymbol{\Sigma}_{\mathrm{xy}} \\ \boldsymbol{\Sigma}_{\mathrm{xy}} & \boldsymbol{\Sigma}_{\mathrm{yy}}\end{pmatrix}\begin{bmatrix}\mathsf{V}^\mathsf{T}(\mathbf{y}) \\ \mathsf{U}^\mathsf{T}(\mathbf{x})\end{bmatrix}\right) \tag{4.22}$$

In case of independence between $\mathbf{x}$ and $\mathbf{y}$ one obtains

$$(\mathbf{z},\ \boldsymbol{\Sigma}_{\mathrm{zz}}) = \left(\mathsf{U}(\mathbf{x})\mathbf{y},\ \ \mathsf{U}(\mathbf{x})\boldsymbol{\Sigma}_{\mathrm{yy}}\mathsf{U}^\mathsf{T}(\mathbf{x}) + \mathsf{V}(\mathbf{y})\boldsymbol{\Sigma}_{\mathrm{xx}}\mathsf{V}^\mathsf{T}(\mathbf{y})\right) \tag{4.23}$$

As an example, the uncertain 2D line $(\mathbf{l}, \boldsymbol{\Sigma}_{\mathrm{ll}})$ joining two points $(\mathbf{x}, \boldsymbol{\Sigma}_{\mathrm{xx}})$, $(\mathbf{y}, \boldsymbol{\Sigma}_{\mathrm{yy}})$ is given by

$$(\mathbf{l}, \boldsymbol{\Sigma}_{\mathrm{ll}}) = \left(\mathbf{x} \times \mathbf{y},\ \mathsf{S}(\mathbf{x})\boldsymbol{\Sigma}_{\mathrm{yy}}\mathsf{S}^{\mathsf{T}}(\mathbf{x}) + \mathsf{S}(\mathbf{y})\boldsymbol{\Sigma}_{\mathrm{xx}}\mathsf{S}^{\mathsf{T}}(\mathbf{y})\right) \tag{4.24}$$

with the skew-symmetric matrix $\mathsf{S}(\mathbf{x})$. Note that the error propagation is also contained in equation (4.18) of KANATANI 1996.

Generally the covariance matrix $\boldsymbol{\Sigma}_{\mathrm{ll}}$ has rank 3 unless the two points are identical. This is an example that we cannot assume that homogeneous covariance matrices have a rank which is equal to the degree of freedom of the geometric entity.

In this work, we usually assume that the geometric entities are independent from each other, thus equation (4.23) will be used for most of the following text.

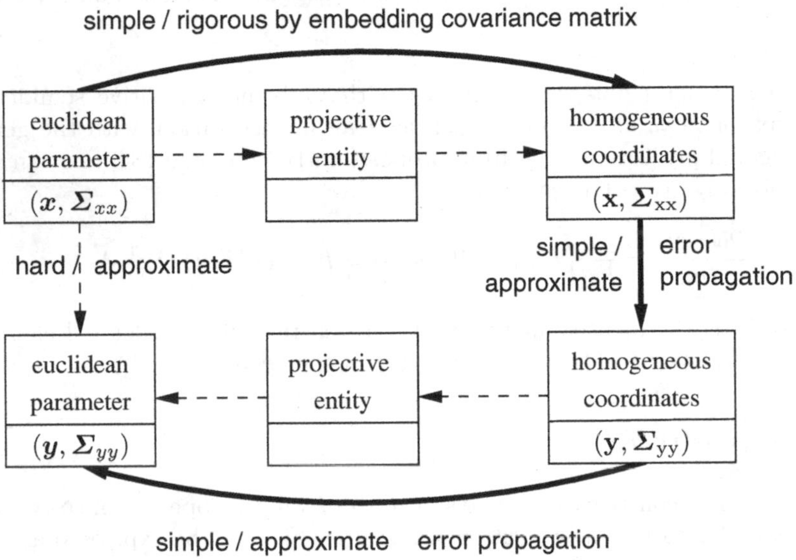

Fig. 4.7. Approximate Statistical Geometric Reasoning using Projective Geometry. Compared to figure 4.2, the pdfs are replaced by the covariance matrices of the vectors. The error propagation of covariance matrices is much simpler than for general pdfs.

Summing up section 4.2, we are now capable of doing statistical geometric reasoning with unique constructions, as shown in figure 4.7 using an approximate uncertainty representations. In the next section we explore the errors that occur when using the proposed approximate representation.

4.3 Errors in Approximated Uncertainty Representation

At certain stages of the reasoning process pictured in figure 4.7, we have mentioned approximations to the rigorous analytical method: an approximated representation and approximation in error propagation. This section describes the nature of these approximation in detail.

As spatial reasoning with homogeneous coordinates is based on bilinear forms (cf. table 3.1) and as normalization of homogeneous coordinates is a division, the analysis of the degree of approximation using homogeneous entities is much simpler than using only Euclidean entities, as has already been observed by CRIMINISI 2001 and CLARKE 1998.

We especially investigate the *bias* of the first-order error propagation, i.e. the difference between the true value and the approximated or computed value. It is important to remark that the bias has to be considered with respect to the measurement error that is to be expected: a bias that is 1% or even 10% of the true standard deviation is negligible as it has practically no consequences to the application. An example: assume that within a system that reconstructs buildings from images, a measurement error was estimated to be $\hat{\sigma} = 2$ cm. Then practically one needs not to care about the fact, that the true value is $\tilde{\sigma} = 2.2$ cm.

We will see that – assuming the original errors are less than 10% – the bias is usually small or can be controlled in a simple way to obtain valid approximations of the true uncertainty.

4.3.1 Second Moments and Gaussian Assumption

Using the first and second moments of a pdf means that we use the Laplacian approximation by cutting off the Taylor series after the linear term. Of course, this causes systematic errors (bias) in mean and variance. But if no other information than the first two moments are available, then in general the Gaussian distribution is the most likely distribution due to the maximum entropy principle COVER 1991. Additionally, the central limit theorem states that the average of a sufficiently large number of independent stochastical variables with identical standard deviations tend to form a Gaussian distribution.

Previous work on uncertain geometry in the Computer Vision (COLLINS 1993, HADDON AND FORSYTH 2001) state that the Gaussian distribution is not useful for geometric reasoning, especially when the uncertainty is large and the operations on the entities are highly nonlinear. We shortly discuss these arguments:

- *Large uncertainty*
  Consider the sphere as a unique representation space for homogeneous vectors, then it is true that the Gaussian does not approximate well for

large uncertainties: since only the tangential plane at the observed point on the sphere is considered, the bias becomes large when the uncertainty of the error becomes large. On the other hand, the tangential plane may very well approximate the actual distribution on the sphere as long as the error is small enough: assume an angle of $\varphi < 20°$ then $sin\varphi \approx \varphi$ with a bias of less than 1%. In our application we assume small Euclidean errors which result in small directional errors for homogeneous vectors[1].
In particular, we assume Gaussian distributions for Euclidean point coordinates and for Euclidean directional vectors, namely the homogeneous part of hyperplanes and both parts of the 3D line vector. While Gaussians are widely accepted for point coordinates, there exists many different distributions for directional vectors (for example *von Mises* distribution, *Bingham* distribution). Again, we argue that for small uncertainties the bias of a Gaussian distribution compared to a spherical distribution is small. In the following subsections, we will see that an error of less than 10% is small enough to assume Gaussian distributions.

- *Nonlinear operations*
  The error of only using second moments can be arbitrarily high when assuming highly nonlinear operation on random variables. Thus in general, no assessment can be made for the accuracy of the approximation. The operations that we use in this work are bilinear operations on homogeneous vectors. In the next two sections, we will investigate the bias caused by the particular bilinear function from chapter 3. Again, we will see that for small errors in the observed geometric entities the bias can be neglected.

4.3.2 Bias in Scalar Multiplication

The error propagation for constructions as in equation (4.22) and (4.23) on page 111 is not rigorous since generally products of stochastic variables occur, thus being non-linear.
The most simple bilinear operation is the product $\underline{z} = \underline{x}\,\underline{y}$ of two stochastic scalars $\underline{x}$ and $\underline{y}$. This is the only non-linear form appearing in any unique construction using join or intersection, thus in the following it is explored in detail. $\underline{x}$ and $\underline{y}$ are assumed to be normally distributed by σ_x resp. σ_y and having a covariance σ_{xy}.
The stochastical variables $\underline{x}, \underline{y}$ are modeled by adding a stochastical noise vector $\underline{e}_x$, $\underline{e}_y$ to the mean thus e.g. $\underline{x} = \mu(\underline{x}) + \underline{e}_x$, where $\underline{e}_x \sim N(0, \sigma_x^2)$ is normally distributed. For $\underline{z}$ we get

$$\underline{z} = \mu(\underline{z}) + \underline{e}_z = (\mu_x + \underline{e}_x)(\mu_y + \underline{e}_y) = \mu_x\mu_y + \mu_y\underline{e}_x + \mu_x\underline{e}_y + \underline{e}_x\underline{e}_y \tag{4.25}$$

[1] This is a similar argument as mentioned in the footnote 1 on page 19: it is commonly accepted to approximate the world as planar as long as we deal with man-made structures within an area of a few kilometers.

Note that equation (4.25) is equivalent to the Taylor series at the mean values. In this case the Taylor series is finite and we showed all terms.
Using (4.25), the mean value $E(\underline{z})$ is given by

$$E(\underline{z}) = E(\mu_x\mu_y + \mu_y\underline{e}_x + \mu_x\underline{e}_y + \underline{e}_x\underline{e}_y) = \mu_x\mu_y + \sigma_{xy} \tag{4.26}$$

The variance $V(\underline{z}) = E(\underline{z}^2) - E(\underline{z})^2$ can be computed by

$$E(\underline{z})^2 = \mu_x^2\mu_y^2 + 2\mu_x\mu_y\sigma_{xy} + \sigma_{xy}^2$$
$$E(\underline{z}^2) = \mu_x^2\mu_y^2 + \mu_x^2\sigma_x^2 + \mu_y^2\sigma_y^2 + \sigma_x^2\sigma_y^2 + 2\sigma_{xy}^2 + 4\mu_x\mu_y\sigma_{xy}$$

The last formula can be derived by expanding $\underline{z}^2$ and using the equality $E(\underline{e}_x^2\underline{e}_y^2) = \sigma_x^2\sigma_y^2 + \sigma_{xy}^2$ and thus

$$V(\underline{z}) = \sigma_z^2 = \mu_y^2\sigma_x^2 + \mu_x^2\sigma_y^2 + 2\mu_x\mu_y\sigma_{xy} + \sigma_x^2\sigma_y^2 + \sigma_{xy}^2 \tag{4.27}$$

$E(\underline{z})$ and $V(\underline{z})$ are analytically strict any distribution. Adapting equation (4.23) on page 111, thus using first order approximation for mean and variance yield:

$$E^{(1)}(\underline{z}) = \mu_x\mu_y \tag{4.28}$$
$$V^{(1)}(\underline{z}) = \mu_y^2\sigma_x^2 + \mu_x^2\sigma_y^2 + 2\mu_x\mu_y\sigma_{xy} \tag{4.29}$$

Comparing first order approximation with the strict solution, the relative bias r_E and r_V for variance and mean is given by

$$r_E = \frac{b_E}{\sigma_z} = \frac{E(\underline{z}) - E^{(1)}(\underline{z})}{\sigma_z} = \frac{\sigma_{xy}}{\sigma_z} = \frac{\rho_{xy}\sigma_x\sigma_y}{\sigma_z} \tag{4.30}$$
$$r_V = \frac{b_v}{\sigma_z^2} = \frac{V(\underline{z}) - V^{(1)}(\underline{z})}{\sigma_z^2} = \frac{\sigma_x^2\sigma_y^2 + \sigma_{xy}^2}{\sigma_z^2} = \frac{\sigma_x^2\sigma_y^2(1 + \rho_{xy}^2)}{\sigma_z^2} \tag{4.31}$$

We chose to normalize the bias by the standard deviation σ_z resp. variance σ_z^2 to obtain an objective measure relative to the expected error σ_z.
If the correlation is zero, there is no bias in the mean , thus $r_E = 0$. In the case of no correlation, the relative bias in variance reduces to $\sigma_x^2\sigma_y^2/\sigma_z^2$. In case of $\sigma = \sigma_x = \sigma_y$ and $\sigma_{xy} = 0$ the relative bias reduces to

$$r_V = \frac{\sigma^2}{\mu_x^2 + \mu_y^2 + \sigma^2} \tag{4.32}$$

Thus only if $\mu_x^2 + \mu_y^2 < 9\,\sigma^2$ (i.e. $\underline{x}$ and $\underline{y}$ are both expected to be very close to zero with respect to σ) the relative bias in variance is larger than 10%. It is important to note that the bias is always positive, thus the true standard deviation is always larger than the approximated one.
In case the correlation is non-zero, we consider the case of $\mu_x = \mu_y = 1$ and $\sigma_x = 0.05$ and $\sigma_y = 0.2$, therefore assuming an unusually large standard

deviation of 20% for $\underline{y}$. The relative bias of the mean is less than 4%, the relative bias of the variance is less than 1 %, cf. figures 4.8(a) and 4.8(b). The influence of the variation of μ_x and μ_y are depicted in figures 4.8(c) and 4.8(d): only if $\mu_x = \mu_y < \sigma$, the bias becomes larger than 10 % for the variance and larger than 20% for the mean.

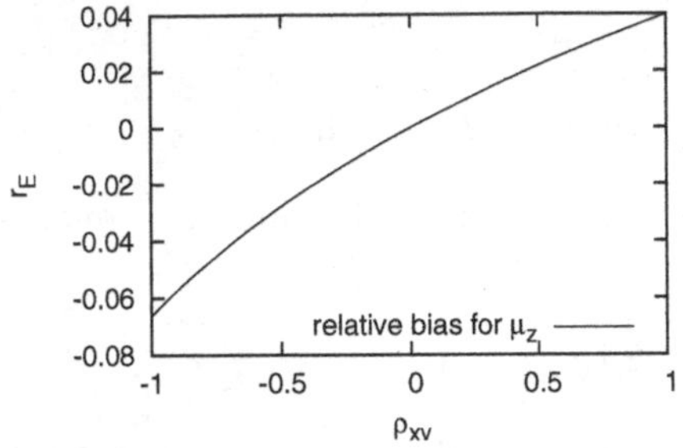

(a) relative bias r_E of mean with respect to correlation ρ_{xy}

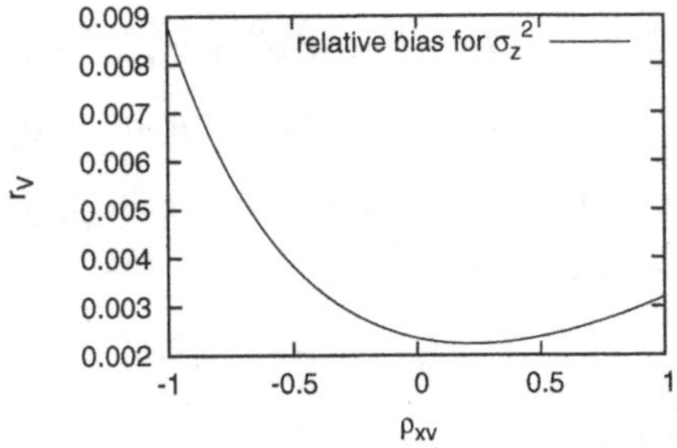

(b) relative bias r_V of variance with respect to correlation ρ_{xy}

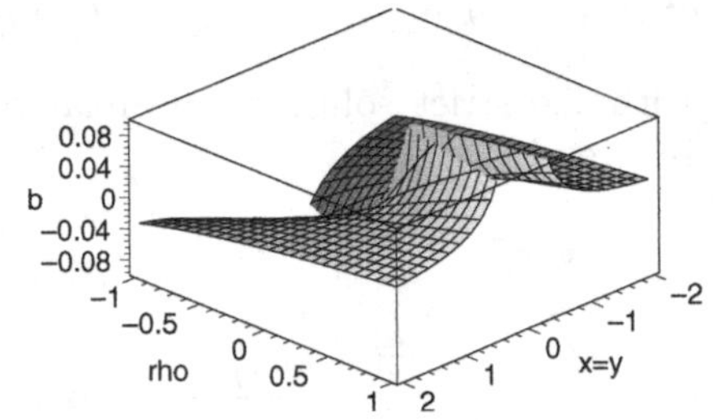

(c) relative bias r_E of mean with respect to correlationρ and $\mu_x = \mu_y$

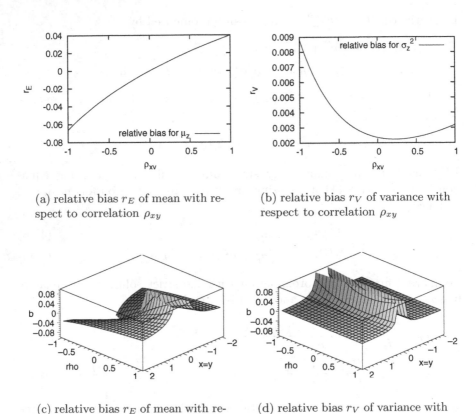

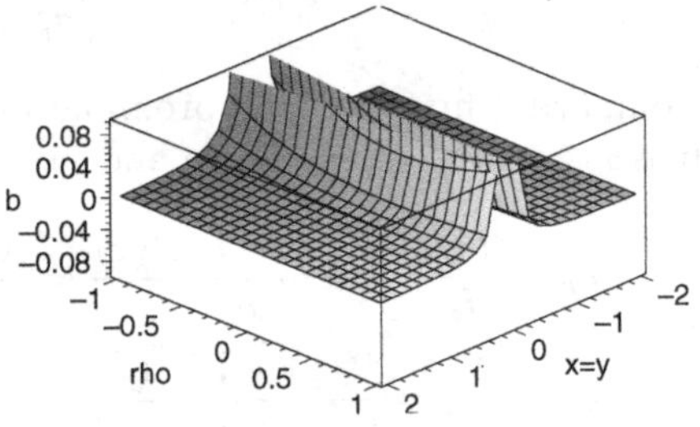

(d) relative bias r_V of variance with respect to correlationρ and $\mu_x = \mu_y$

Fig. 4.8. Relative bias r_E and r_V of bilinear operations with scalar multiplication as an example. For figures 4.8(a) and 4.8(b) the means of $\underline{x}$ and $\underline{y}$ were fixed to 1, $\mu_x = \mu_y = 1$ and the standard deviations to $\sigma_x = \sigma_y = 0.1$. The correlation factor ρ_{xy} varies between -1 and 1. In 4.8(c) and 4.8(d) the mean of $\underline{x}$ and $\underline{y}$ varies simultaneously between $\mu_x = \mu_y = [-2 \cdots 2]$ and the correlation varies again between -1 and 1. The bias is usually around 2% or less and becomes intractable only if $\mu_x = \mu_y \approx 0$, see text for details.

From this analysis we conclude that for the most simple bilinear form – namely scalar multiplication – the bias is significant if the standard deviation of both numbers is larger than their mean. This has been described for more complex bilinear forms by HADDON AND FORSYTH 2001. Usually in vision problems the relative accuracy is less than 10 %: Thus we do not expect significant problems caused by bilinear scalar multiplication.

4.3.3 Bias in Bilinear Constructions

Next we want to explore the bias for bilinear operations on homogeneous vectors instead of scalars. Again, we choose a simple example for a theoretical analysis.

Assume two stochastic homogeneous vectors $\underline{\mathbf{x}}, \underline{\mathbf{l}} \in \mathbb{R}^2$ on the projective axis, where $\mathbf{x} = (u,v)^\mathsf{T}$ represents a projective point and $\mathbf{l} = (a,b)^\mathsf{T}$ represents a projective hyperplane in $\mathbb{P}^1$, see figure 2.1 on page 21. Furthermore two covariance matrices $\boldsymbol{\Sigma}_{\mathrm{xx}}$ and $\boldsymbol{\Sigma}_{\mathrm{ll}}$ are given, which fulfill definition 10.

As a bilinear operation we choose the dot-product $\underline{d} = \underline{\mathbf{x}}^\mathsf{T}\mathbf{l} = \underline{\mathbf{l}}^\mathsf{T}\mathbf{x}$ and are especially interested in the mean value $E(d)$ and the variance σ_d of $V(d)$. Assuming Gaussian distribution, we can compute the true values by reformulating the dot-product into a quadratic form:

$$\underline{d} = \underline{\mathbf{z}}^\mathsf{T} W \underline{\mathbf{z}} = (u,v,a,b)^\mathsf{T} \frac{1}{2} \begin{pmatrix} 0 & 0 & 1 & 0 \\ 0 & 0 & 0 & 1 \\ 1 & 0 & 0 & 0 \\ 0 & 1 & 0 & 0 \end{pmatrix} \begin{pmatrix} u \\ v \\ a \\ b \end{pmatrix}$$

Mean and variance of $\underline{d}$ are now given by

$$E(d) = \operatorname{trace}(W\boldsymbol{\Sigma}_{zz}) + \boldsymbol{\mu}_z^\mathsf{T} W \boldsymbol{\mu}_z = au + bv \tag{4.33}$$

$$\begin{aligned} V(d) &= 2\operatorname{trace}(W\boldsymbol{\Sigma}_{zz}W\boldsymbol{\Sigma}_{zz}) + 4\boldsymbol{\mu}_x^\mathsf{T} W\boldsymbol{\Sigma}_{zz}W\boldsymbol{\mu}_x \\ &= \sigma_a^2\sigma_u^2 + 2\sigma_{ab}\sigma_{uv} + \sigma_b^2\sigma_v^2 + u^2\sigma_a^2 + 2uv\sigma_{ab} + v^2\sigma_b^2 \\ &\qquad + a^2\sigma_u^2 + 2ab\sigma_{uv} + b^2\sigma_v^2 \end{aligned} \tag{4.34}$$

cf. KOCH 1999. As we assume $\mathbf{x}$ and $\mathbf{l}$ to be uncorrelated, there is no bias between the first order approximation $E^{(1)}(d) = \boldsymbol{\mu}_x^\mathsf{T}\boldsymbol{\mu}_l$ and $E(d)$. The variance $V^{(1)}$ obtained by the first order error propagation and the bias b_V with respect to the ideal case is

$$V^{(1)} = u^2\sigma_a^2 + 2uv\sigma_{ab} + v^2\sigma_b^2 + a^2\sigma_u^2 + 2ab\sigma_{uv} + b^2\sigma_v^2 \tag{4.35}$$

$$b_V = V(d) - V^{(1)}(d) = \sigma_a^2\sigma_u^2 + \sigma_b^2\sigma_v^2 + 2\sigma_{ab}\sigma_{uv} \tag{4.36}$$

It is easy to see that the addends in the bias of the variance are a consequence of the discussion of the last section, cf. equation 4.31.

Projecting the Covariance Matrix Unfortunately there is an important difference to the last section: no statement can be made about the upper bound of the bias within this general setting. We may even obtain a zero-variance $V^{(1)}$ if the nullspaces of the covariance matrices are $\mathcal{N}(\boldsymbol{\Sigma}_{\mathsf{xx}}) = \mathbf{l}$ and $\mathcal{N}(\boldsymbol{\Sigma}_{\mathsf{ll}}) = \mathbf{x}$: then the first order error propagation yields $V^{(1)}(d) = \boldsymbol{\mu}_x^{\mathsf{T}}\boldsymbol{\Sigma}_{\mathsf{ll}}\boldsymbol{\mu}_x + \boldsymbol{\mu}_l^{\mathsf{T}}\boldsymbol{\Sigma}_{\mathsf{xx}}\boldsymbol{\mu}_l = 0$, independent from the ideal variance $V(d)$. The difference to the last section is due to the fact that the covariance matrices $\boldsymbol{\Sigma}_{\mathsf{xx}}$ and $\boldsymbol{\Sigma}_{\mathsf{ll}}$ are not unique, the values of the matrices can be arbitrarily distorted as long as the normalized covariances matrices are the identical, $\mathsf{N}_s(\boldsymbol{\Sigma}_{\mathsf{xx}}) = \mathsf{N}_s(\boldsymbol{\Sigma}_{\mathsf{ll}})$, see also figure 4.3. To avoid the ambiguity, we may normalize the covariance matrix prior to computing the bilinear operation. Since we do not want to change the vector $\mathbf{x}$, we apply an orthogonal projection to the covariance matrix $\boldsymbol{\Sigma}_{\mathsf{xx}} = \begin{pmatrix} \sigma_u^2 & \sigma_{uv} \\ \sigma_{uv} & \sigma_v^2 \end{pmatrix}$, cf. equation (4.18) and with some calculation one obtains:

$$\mathcal{P}_{\mathbf{x}}(\boldsymbol{\Sigma}_{\mathsf{xx}}) = \frac{(v^2\sigma_u^2 - 2uv\sigma_{uv} + u^2\sigma_v^2)}{(u^2+v^2)^2} \begin{pmatrix} v^2 & -uv \\ -uv & u^2 \end{pmatrix} \tag{4.37}$$

and in the same manner to $\boldsymbol{\Sigma}_{\mathsf{ll}}$. Note that the numerator of the factor in (4.37) is a quadratic form:

$$(v^2\sigma_u^2 - 2uv\sigma_{uv} + u^2\sigma_v^2) = (v,u)^{\mathsf{T}} \begin{pmatrix} \sigma_u^2 & -\sigma_{uv} \\ -\sigma_{uv} & \sigma_v^2 \end{pmatrix} \begin{pmatrix} u \\ v \end{pmatrix}$$

With the projected covariance matrices we obtain a new bias

$$b'_V = \frac{(v^2\sigma_u^2 - 2uv\sigma_{uv} + u^2\sigma_v^2)(b^2\sigma_{aa}^2 - 2ab\sigma_{ab} + a^2\sigma_{bb}^2)(v^2b^2 + 2uvab + u^2a^2)}{(u^2+v^2)^2(a^2+b^2)^2} \tag{4.38}$$

Although the new bias looks more complicated, we gained some new insights: first, the bias solely depends on the directional uncertainty of the homogeneous vector, thus it depends solely on the underlying Euclidean information and can not be arbitrarily large. Second the bias b'_V is non-negative, as the factors of the numerator can be written as a quadratic form of the covariance matrices, which are positive semidefinite by definition. Thus we obtain only optimistic approximations of the true variance. This is an important observation which holds for all constructions with projected covariance matrices, as we will see in practical tests, cf. section 4.4.2 on page 125 and section 4.5.4 on page 135.

Conditioning the Vectors Although the bias was found to be non-negative, it may still be quite large: consider that the homogeneous vectors for the point and the hyperplane are identical, $\mathbf{x} = \mathbf{l}$. Then $\mathcal{N}(\boldsymbol{\Sigma}_{\mathsf{xx}}) = \mathbf{x}$ and $\mathcal{N}(\boldsymbol{\Sigma}_{\mathsf{ll}}) = \mathbf{l}$ and $V^{(1)}(d) = \mathbf{x}^{\mathsf{T}}\boldsymbol{\Sigma}_{\mathsf{ll}}\mathbf{x} + \mathbf{l}^{\mathsf{T}}\boldsymbol{\Sigma}_{\mathsf{xx}}\mathbf{l} = 0$ and therefore the bias is maximal. On the other hand, assume the covariances were obtained by embedding the Euclidean covariance into a homogeneous vector, thus

$$\boldsymbol{\Sigma}^e_{\mathbf{xx}} = \begin{pmatrix} \sigma_x^2 & 0 \\ 0 & 0 \end{pmatrix} \quad ; \quad \boldsymbol{\Sigma}^e_{\mathbf{ll}} = \begin{pmatrix} 0 & 0 \\ 0 & \sigma_l^2 \end{pmatrix}$$

If we don't apply any projection on the matrices, the bias is zero, independent from the values σ_x and σ_l. This suggests to transform an uncertain homogeneous vector ($\mathbf{x}$, $\mathcal{P}_{\mathbf{x}}(\boldsymbol{\Sigma}_{\mathbf{xx}})$) with a proper $\boldsymbol{\Sigma}_{\mathbf{xx}}$, such that the variance for the homogeneous part approaches zero.

A simple way to achieve this goal is to condition the two geometric entities $\mathbf{x}$ and $\mathbf{y}$ according to the algorithm 2.1 on page 40: given a point $\mathbf{x} = (u, v)^{\mathsf{T}}$ with a projected covariance matrix $\mathcal{P}_{\mathbf{x}}(\boldsymbol{\Sigma}_{\mathbf{xx}})$, the variance σ_v of the homogeneous part v of the vector approaches zero when conditioned with a factor $f < 1$, cf. figure 4.9, since the conditioned and projected covariance matrix $\mathcal{P}_{\mathbf{x}^\circ}(\boldsymbol{\Sigma}_{\mathbf{x}^\circ\mathbf{x}^\circ})$ is close to $\boldsymbol{\Sigma}^e_{\mathbf{xx}}$. For a hyperplane $\mathbf{l}$, the same reasoning applies and the conditioned and projected covariance matrix $\mathcal{P}_{\mathbf{l}^\circ}(\boldsymbol{\Sigma}_{\mathbf{l}^\circ\mathbf{l}^\circ})$ is close to $\boldsymbol{\Sigma}^e_{\mathbf{ll}}$. Note that in practice, we first condition the two vectors and then project

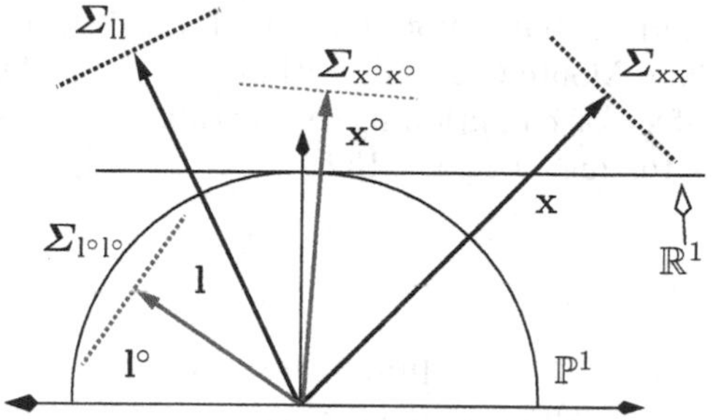

Fig. 4.9. Effect of conditioning to uncertain homogeneous vectors $\mathbf{x}, \mathbf{y}$. Note that the covariance matrix $\boldsymbol{\Sigma}_{\mathbf{x}^\circ\mathbf{x}^\circ}$ of the conditioned vector $\mathbf{x}^\circ$ is close to the Euclidean case if $\mathcal{N}(\boldsymbol{\Sigma}_{\mathbf{x}^\circ\mathbf{x}^\circ}) = \mathbf{x}^\circ$, see text for details.

the covariance matrix with $\mathcal{P}_{\mathbf{x}}$ to avoid two projecting steps. This is due to the fact that the conditioning with the matrix $\mathsf{W}_x(f)$ changes the nullspace in an undesirable manner and this has to be corrected.

The reduced bias is dependent on the conditioning factor f:

$$b''_v = \frac{f^4(au+bv)^2(v^2\sigma_u^2 - 2uv\sigma_{uv} + u^2\sigma_v^2)(b^2\sigma_a^2 - 2ab\sigma_{ab} + a^2\sigma_b^2)}{(a^2+f^2b^2)^2\,(f^2u^2+v^2)^2} \tag{4.39}$$

Note that b''_V is already reconditioned by $1/f^2$ to compare it with the previous bias. As an example assume $a = b = u = v = 1$, we then obtain

$$b''_V = \frac{4f^4}{(f^2+1)^4} b'_V = g(f) b'_V \tag{4.40}$$

such that $f < 1$ yield a smaller bias b''_V than b'_V, for $f = 0.1$, the reduction is 99.96%, see figure 4.10(left). The relative bias $r''_V = b''_V / V(d)$ is shown in figure 4.10(right) and is about 2% for $f = 0.1$ and about 0.5% for $f = 0.05$.

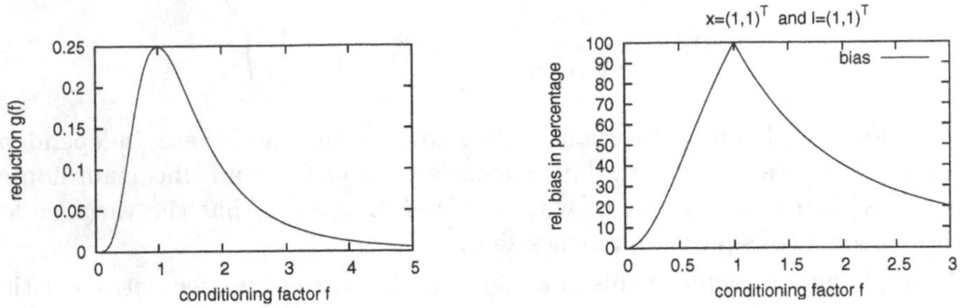

Fig. 4.10. *Left:* Sample curve for the reduction factor $g(f)$ between bias b''_V and b''_V: $b''_V = g(f)''b_V$, cf. equation (4.40). One can see that a conditioning factor $f < 1$ – thus scaling the Euclidean part down – greatly reduces the bias. A condition factor $f > 1$ scales the homogeneous part of a vector down and the bias gets smaller, too, but slower than for $f < 1$. *Right:* This curve shows the relative bias r''_V in percentage for σ_d^2 dependent on f with $d = \mathbf{x}^\mathsf{T}\mathbf{l}$ and $\mathbf{x} = \mathbf{l}$.

The theoretical analysis of the bias for uncertain homogeneous vectors coming from $\mathbb{P}^1$ can be generalized for $\mathbb{P}^2$ and $\mathbb{P}^3$, but is not elaborated in this work. We rather rely on the validation using Monte Carlo algorithms, see 4.4.2. An additional demonstration of the effect of conditioning/projecting for 2D is shown in the context of hypothesis testing, see page 132

4.3.4 Bias in Normalization

To investigate the bias of using first order error propagation in normalization, we only consider Euclidean normalization as defined in equation (2.14). The case of spherical normalization has already been informally discussed in section 4.3.1: as long as the uncertainty is small, a normalization to the tangent plane of a sphere is a valid approximation and the bias is expected to be small.

For the Euclidean normalization, we again consider the most simple case: assume a point $\mathbf{x} \in \mathbb{P}^1$ on the projective axis, thus in homogeneous coordinates $\mathbf{x} = (u, v)^\mathsf{T}$. Assume $\underline{u}$ and $\underline{v}$ being stochastical variables, with the covariance matrix

$$\boldsymbol{\Sigma}_{\mathrm{xx}} = \begin{pmatrix} \sigma_u^2 & \sigma_{uv} \\ \sigma_{uv} & \sigma_v^2 \end{pmatrix}$$

The normalization $x = \mathsf{N}_O(\mathbf{x}) = u/v$ is nonlinear and as in section 4.3.2 we consider the Taylor series:

$$\underline{x} = \mu_x + \underline{e}_x = \frac{\mu_u}{\mu_v} + \frac{\underline{e}_u}{\mu_v} - \frac{\mu_u \underline{e}_v}{\mu_v^2} - \frac{\underline{e}_v \underline{e}_u}{\mu_v^2} + \frac{\mu_u \underline{e}_v^2}{\mu_v^3} + \mathcal{O}(|\underline{e}_u + \underline{e}_v|^3) \qquad (4.41)$$

Contrary to the bilinear product the Taylor series is infinite and we consider it only up to second order to estimate the bias between first order and second order approximation. This approach reflects the dominant error caused by considering first order approximation.

The mean value and variance for the first and second order approximation are computed similarly to the last section and after some calculations one obtains

$$E^{(1)}(\underline{x}) = \frac{\mu_u}{\mu_v} \tag{4.42}$$

$$E^{(2)}(\underline{x}) = \frac{\mu_u}{\mu_v} - \frac{\sigma_{uv}}{\mu_v^2} + \frac{\mu_u \sigma_v^2}{\mu_v^3} \tag{4.43}$$

$$V^{(1)}(\underline{x}) = \frac{\sigma_u^2}{\mu_v^2} - 2\frac{\mu_u \sigma_{uv}}{\mu_v^3} + \frac{\mu_u^2 \sigma_v^2}{\mu_v^4} \tag{4.44}$$

$$V^{(2)}(\underline{x}) = \frac{\sigma_u^2}{\mu_v^2} - 2\frac{\mu_u \sigma_{uv}}{\mu_v^3} + \frac{\mu_u^2 \sigma_v^2}{\mu_v^4} + 3\frac{\sigma_u^2 \sigma_v^2}{\mu_v^4} + 2\frac{\sigma_{uv}^2}{\mu_v^4} + 2\frac{\sigma_{uv} \mu_u \sigma_v^2}{\mu_v^5} + 4\frac{\mu_u^2 \sigma_v^4}{\mu_v^6} \tag{4.45}$$

Thus we obtain the approximate bias $b_E^{(2)}$ and $b_V^{(2)}$ for mean and variance:

$$b_E^{(2)} = \frac{\mu_u \sigma_v^2}{\mu_v^3} - \frac{\sigma_{uv}}{\mu_v^2} \tag{4.46}$$

$$= \sigma_v \left(\frac{\mu_u \sigma_v - \rho_{uv} \sigma_u \mu_v}{\mu_v^3} \right) \tag{4.47}$$

$$b_V^{(2)} = 3\frac{\sigma_u^2 \sigma_v^2}{\mu_v^4} + 2\frac{\sigma_{uv}^2}{\mu_v^4} + 2\frac{\sigma_{uv} \mu_u \sigma_v^2}{\mu_v^5} + 4\frac{\mu_u^2 \sigma_v^4}{\mu_v^6} \tag{4.48}$$

$$= \sigma_v^2 \left(\frac{4\mu_u^2 \sigma_v^2 + 3\sigma_u^2 \mu_v^2 + 2\rho_{uv}^2 \sigma_u^2 \mu_v^2 + 2\rho_{uv} \sigma_u \sigma_v \mu_u \mu_v}{\mu_v^6} \right) \tag{4.49}$$

The formulas for the bias are more complicated than for the scalar multiplication, but because of the factor σ_v^2 one can see that the bias of both mean and variance is 0, if the homogeneous factor $\underline{v}$ is known to be certain. Thus if we normalize an embedded covariance matrix as in (4.10) on page 106 back to the Euclidean space, we obtain the same covariance matrix.

Now assume that the vector $\mathbf{x}$ is conditioned, such that $|\mu_u| \leq f_{min}|\mu_v|$ with $f_{min} < 1$ and the covariance matrix is equal to its orthogonal projection: $\boldsymbol{\Sigma}_{\mathrm{xx}} = \mathcal{P}_{\mathbf{x}}(\boldsymbol{\Sigma}_{\mathrm{xx}})$, cf. equation (4.37). Then σ_v is comparably very small compared to σ_u, see also figure 4.9; therefore $\boldsymbol{\Sigma}_{\mathrm{xx}}$ is close to a form as in (4.10) on page 106 and the bias must be very small.

As an example, assume $\mu_u = 10, \mu_v = 1, \sigma_u = 0.5, \sigma_v = 0.1$. A plot of the relative bias $rb_V^{(2)} = b_V^{(2)}/V^{(2)}$ depending on the conditioning factor f is shown in figure 4.11, where the pair $(\mathbf{x}, \boldsymbol{\Sigma}_{\mathrm{xx}})$ was conditioned to $(\mathbf{x}^\circ, \boldsymbol{\Sigma}_{\mathrm{x^\circ x^\circ}})$ and the nullspace was enforced to be equal to the conditioned vector $\mathbf{x}^\circ$.

Summary As an important result of the analysis of the error made by the approximated representation, we conclude that it is absolutely necessary to condition an uncertain homogeneous vector and to make sure that the nullspace

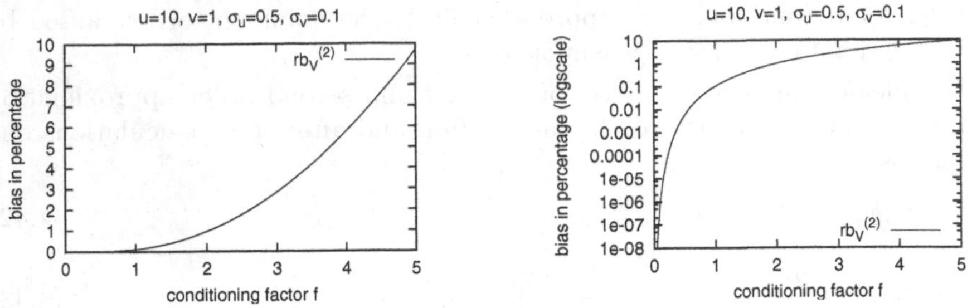

Fig. 4.11. Plot of relative bias of Euclidean normalization in normal scale (*left*) and logscale (*right*). To obtain this plot, a point $\mathbf{x} = (u, v)^{\mathsf{T}}$ was conditioned by factor f, obtaining $\mathbf{x}^{\circ}$ and the nullspace of the covariance matrix $\boldsymbol{\Sigma}_{\mathbf{x}^{\circ}\mathbf{x}^{\circ}}$ was forced to be $\mathbf{x}^{\circ}$.

of the accompanying covariance matrix contains only the homogeneous vector itself. If these two conditions are met, the bias in the approximation can be neglected for all operations of geometric reasoning, provided the uncertainty in the Euclidean parameters is sufficiently small. Additionally, the bias is non-negative yielding only too optimistic approximations of the true variance.

4.4 Construction of Entities

In the last section we have analyzed the errors that are involved in the approximate representation of uncertainty. We now present a method for the unique construction of geometric entities, for which the uncertainty is represented using definition 10. Furthermore we will shortly discuss constructions of entities by projective transformations.

4.4.1 A Statistical Approach to Join and Intersection

The algorithm in table 4.1 implements statistical versions of the unique constructions introduced in chapter 3. It takes advantage of the first-order error propagation and uses conditioning of the homogeneous vectors as well as projection of covariance matrices to reduce the bias.

To guarantee that the proposed algorithms yields proper uncertain geometric entities as defined in definition 10 on page 104, we have to prove the following proposition:

Proposition 6 (Uncertain constructions yield proper entities)
Assume two uncorrelated uncertain geometric entities $(\mathbf{x}, \boldsymbol{\Sigma}_{\mathbf{xx}})$ *and* $(\mathbf{y}, \boldsymbol{\Sigma}_{\mathbf{yy}})$. *The homogeneous part of the vectors is larger than the Euclidean*

Join and intersection with uncertainty
Objective: given two uncorrelated uncertain geometric entities $(\mathbf{x},\ \boldsymbol{\Sigma}_{xx})$ and $(\mathbf{y},\ \boldsymbol{\Sigma}_{yy})$, compute the uncertain join resp. intersection $(\mathbf{z}, \boldsymbol{\Sigma}_{zz}) = (\mathbf{x},\ \boldsymbol{\Sigma}_{xx}) \wedge (\mathbf{y},\ \boldsymbol{\Sigma}_{yy})$ resp. $(\mathbf{z}, \boldsymbol{\Sigma}_{zz}) = (\mathbf{x},\ \boldsymbol{\Sigma}_{xx}) \cap (\mathbf{y},\ \boldsymbol{\Sigma}_{yy})$.

1. If necessary, condition $\mathbf{x}$ and $\mathbf{y}$ with a factor f as determined in algorithm 2.1 using $f_{min} = 0.1$. Scale the Euclidean parts of the covariance matrices in the same way. We obtain $(\mathbf{x}^\circ, \boldsymbol{\Sigma}_{x^\circ x^\circ})$ and $(\mathbf{y}^\circ, \boldsymbol{\Sigma}_{y^\circ y^\circ})$.
2. Change the nullspaces of $\boldsymbol{\Sigma}_{x^\circ x^\circ}$ and $\boldsymbol{\Sigma}_{y^\circ y^\circ}$ to $\{\mathbf{x}\}$ resp. $\{\mathbf{y}\}$ using orthogonal projection $\mathcal{P}_\mathbf{x}$ and $\mathcal{P}_\mathbf{y}$, cf. (4.18) on page 111.
3. Compute the new conditioned uncertain geometric entity $(\mathbf{z}^\circ,\ \boldsymbol{\Sigma}_{z^\circ z^\circ}) = \big(\mathsf{U}(\mathbf{x}^\circ)\mathbf{y}^\circ,\ \mathsf{U}(\mathbf{x}^\circ)\boldsymbol{\Sigma}_{yy}\mathsf{U}^\mathsf{T}(\mathbf{x}^\circ) + \mathsf{V}(\mathbf{y}^\circ)\boldsymbol{\Sigma}_{xx}\mathsf{V}^\mathsf{T}(\mathbf{y}^\circ)\big)$ where the Jacobians $\mathsf{U}(\mathbf{x}^\circ)$ and $\mathsf{V}(\mathbf{y}^\circ)$ are chosen according to table 3.1.
4. If necessary, recondition $(\mathbf{z}^\circ,\ \boldsymbol{\Sigma}_{z^\circ z^\circ})$ by scaling its Euclidean part by $1/f$ to obtain the constructed uncertain geometric entity $(\mathbf{z},\ \boldsymbol{\Sigma}_{zz})$.

Algorithm 4.1: Algorithm for computing join and intersection of uncertain geometric entities. The conditioning of the coordinates and the change of the nullspace of the covariance matrices reduce the bias in the resulting uncertainty.

part and the nullspaces of the covariance matrices are $\mathcal{N}(\boldsymbol{\Sigma}_{xx}) = \mathbf{x}$ *and* $\mathcal{N}(\boldsymbol{\Sigma}_{yy}) = \mathbf{y}$. *Then for the join* $\cap$ *or the intersection* $\wedge$ *the pair*

$$(\mathbf{z},\ \boldsymbol{\Sigma}_{zz}) = \big(\mathsf{U}(\mathbf{x})\mathbf{y},\ \mathsf{U}(\mathbf{x})\boldsymbol{\Sigma}_{yy}\mathsf{U}^\mathsf{T}(\mathbf{x}) + \mathsf{V}(\mathbf{y})\boldsymbol{\Sigma}_{xx}\mathsf{V}^\mathsf{T}(\mathbf{y})\big) \tag{4.50}$$

with $\mathsf{U}(\mathbf{x})$ *and* $\mathsf{V}(\mathbf{y})$ *from table 3.1 is a proper uncertain geometric entity according to definition 10.*

Proof: We have to show that $\mathcal{N}(\boldsymbol{\Sigma}_{zz}) \cap \mathcal{E}(\mathbf{x} \circ \mathbf{y}) = \{\mathbf{0}\}$ with $\circ$ being the join or intersection operator and $\mathcal{E}(\mathbf{x}) = \{\mathbf{x}\}^\perp$ for points and hyperplanes in 2D and 3D resp. $\mathcal{E}(\mathbf{L}) = \{\mathbf{L}, \overline{\mathbf{L}}\}^\perp$ for 3D lines. From proposition 1 on the following page, we know the nullspaces of the addends of $\boldsymbol{\Sigma}_{zz} = \boldsymbol{\Sigma}^1_{zz} + \boldsymbol{\Sigma}^2_{zz}$. In the remainder of this proof, we will only cover the join-operations, the intersection operation is skipped due to duality reasons.

- Case 1: $\mathbf{l} = \mathbf{x} \wedge \mathbf{y} = \mathsf{S}(\mathbf{x})\mathbf{y} = -\mathsf{S}(\mathbf{y})\mathbf{x}$
  As $\mathcal{N}(\boldsymbol{\Sigma}^1_{ll}) \subseteq \{\mathbf{x}\}$ and $\mathcal{N}(\boldsymbol{\Sigma}^2_{ll}) \subseteq \{\mathbf{y}\}$, then because of proposition B.2 $\mathcal{N}(\boldsymbol{\Sigma}_{ll}) \subseteq \{\mathbf{l}\}^\mathsf{T}$ if $\mathbf{x} \neq \mathbf{y}$. If they are identical, the line $\mathbf{l}$ is undefined and $\mathrm{rk}(\boldsymbol{\Sigma}_{ll}) = 2$.
- Case 2: $\mathbf{L} = \mathbf{X} \wedge \mathbf{Y} = \Pi(\mathbf{X})\mathbf{X} = -\Pi(\mathbf{Y})\mathbf{X}$
  Because of propositions B.2 and 1, $\mathcal{N}(\boldsymbol{\Sigma}_{LL}) \subseteq \big(\mathcal{I}(\overline{\Pi}(\mathbf{X})) \cap \mathcal{I}(\overline{\Pi}(\mathbf{Y}))\big) \cup \{\mathbf{L}\}$. The image $\mathcal{I}(\overline{\Pi}(\mathbf{X})) = \mathcal{I}(C\Pi(\mathbf{X}))$ contains the duals of all lines incident to the point $\mathbf{X}$. As the same applies for $\mathbf{Y}$, the nullspace $\mathcal{N}(\boldsymbol{\Sigma}_{LL})$

contains the dual of all lines $\mathbf{L}$ incident to both $\mathbf{X}$ and $\mathbf{Y}$. If $\mathbf{X} \neq \mathbf{Y}$, then $\mathcal{N}(\boldsymbol{\Sigma}_{\mathrm{LL}}) = \{\mathbf{L}, \overline{\mathbf{L}}\}$ and thus the assertion $\mathcal{N}(\boldsymbol{\Sigma}_{\mathrm{LL}}) \cap \{\mathbf{L}, \overline{\mathbf{L}}\}^{\perp} = \{\mathbf{0}\}$ holds. If $\mathbf{X} = \mathbf{Y}$, the line is undefined and the nullspace of $\boldsymbol{\Sigma}_{\mathrm{LL}}$ is at least three-dimensional, thus $\mathrm{rk}(\boldsymbol{\Sigma}_{\mathrm{LL}}) \geq 3$.

- Case 3: $\mathbf{A} = \mathbf{X} \wedge \mathbf{L} = \overline{\mathsf{\Pi}}^{\mathsf{T}}(\mathbf{X})\mathbf{L} = \overline{\mathsf{\Gamma}}^{\mathsf{T}}(\mathbf{L})\mathbf{X}$
  Because of propositions B.2 and 1, $\mathcal{N}(\boldsymbol{\Sigma}_{\mathrm{AA}}) \subseteq \left(\{\overline{\mathbf{X}}\} \cap \mathcal{I}(\mathsf{\Gamma}^{\mathsf{T}}(\mathbf{L}))\right) \cup \{\mathbf{A}\}$. Note that $\overline{\mathbf{X}}$ is interpreted as a plane with the homogeneous coordinates of the point $\mathbf{X}$. Taking the dual, we obtain $\{\mathbf{X}\} \cap \mathcal{I}(\overline{\mathsf{\Gamma}}^{\mathsf{T}}(\mathbf{L}))$, which is zero if $\mathbf{X} \notin \mathbf{L}$, as $\overline{\mathsf{\Gamma}}^{\mathsf{T}}(\mathbf{L})$ contains the canonical planes of the line $\mathbf{L}$; thus $\boldsymbol{\Sigma}_{\mathrm{AA}} \subseteq \{\mathbf{A}\}$. If $\mathbf{X} \in \mathbf{L}$, then the plane $\mathbf{A}$ is undefined and $\mathrm{rk}(\boldsymbol{\Sigma}_{\mathrm{AA}}) \geq 3$.
  □

To prove the last proposition, we needed the following statement:

Lemma 1 *Assume two uncorrelated uncertain geometric entities* $(\mathbf{x}, \boldsymbol{\Sigma}_{\mathrm{xx}})$ *and* $(\mathbf{y}, \boldsymbol{\Sigma}_{\mathrm{yy}})$. *The homogeneous part of the vectors is larger than the Euclidean part and the nullspaces of the covariance matrices are* $\mathcal{N}(\boldsymbol{\Sigma}_{\mathrm{xx}}) = \mathbf{x}$ *and* $\mathcal{N}(\boldsymbol{\Sigma}_{\mathrm{yy}}) = \mathbf{y}$. *Let* $\mathbf{z} = \mathsf{U}(\mathbf{x})\mathbf{y}$ *be a construction as in table 3.1. Then*

$$\mathcal{N}(\mathsf{U}(\mathbf{x})\boldsymbol{\Sigma}_{\mathrm{yy}}\mathsf{U}^{\mathsf{T}}(\mathbf{x})) \subseteq \mathcal{N}(\mathsf{U}^{\mathsf{T}}(\mathbf{x})) \cup \{\mathbf{z}\} \tag{4.51}$$

Proof: The covariance matrix $\boldsymbol{\Sigma}_{\mathrm{zz}} = \mathsf{U}(\mathbf{x})\boldsymbol{\Sigma}_{\mathrm{yy}}\mathsf{U}^{\mathsf{T}}(\mathbf{x})$ can be decomposed into $\boldsymbol{\Sigma}_{\mathrm{zz}} = \mathsf{U}(\mathbf{x})\mathsf{H}\mathsf{H}^{\mathsf{T}}\mathsf{U}^{\mathsf{T}}(\mathbf{x})$ with H being a $n \times (n-1)$ matrix and nullspace $\mathcal{N}(\mathsf{H}^{\mathsf{T}}) = \mathcal{N}(\boldsymbol{\Sigma}_{\mathrm{yy}}) = \{\mathbf{y}\}$. It is sufficient to consider $\mathsf{H}^{\mathsf{T}}\mathsf{U}^{\mathsf{T}}(\mathbf{x})$ only.
It is obvious that the nullspace $\mathcal{N}(\mathsf{U}^{\mathsf{T}}(\mathbf{x}))$ can be part of $\mathcal{N}(\mathsf{H}^{\mathsf{T}}\mathsf{U}^{\mathsf{T}}(\mathbf{x}))$. On the other hand, if there is a vector $\mathbf{t}$ such that $\mathbf{y} = \mathsf{U}^{\mathsf{T}}(\mathbf{x})\mathbf{t}$, then this vector is also an eigenvector of $\mathsf{H}^{\mathsf{T}}\mathsf{U}^{\mathsf{T}}(\mathbf{x})$. We can show for each operation, that $\mathbf{t} = \mathbf{z}$, for duality reasons we only consider the join operations:

- $\mathbf{l} = \mathbf{x} \wedge \mathbf{y}$
  In this case we consider $\mathbf{m} = \mathsf{S}^{\mathsf{T}}(\mathbf{x})\mathbf{t} = -\mathsf{S}(\mathbf{x})\mathbf{y}$ with $\mathbf{t}$ being interpreted as a 2D point. We need a $\mathbf{t}$, such that $\overline{\mathbf{y}} = \mathbf{m}$, since $\mathbf{y}$ and $\overline{\mathbf{y}}$ are represented by the same vectors. Thus a necessary condition for the existence of $\mathbf{t}$ is that $\mathbf{x} \in \overline{\mathbf{y}} \Leftrightarrow \mathbf{x}^{\mathsf{T}}\mathbf{y} = 0$. Taking the dual of the construction, $(\overline{\mathbf{y}} = -\mathsf{S}(\mathbf{x})\mathbf{t})$ ∘–• $(-\mathsf{S}^{\mathsf{T}}(\mathbf{x})\mathbf{y} = \overline{\mathbf{t}})$, it follows that $\overline{\mathbf{t}} = \mathbf{l}$.
- $\mathbf{L} = \mathbf{X} \wedge \mathbf{Y}$
  Here $\mathsf{U}^{\mathsf{T}}(\mathbf{X}) = \mathsf{\Pi}^{\mathsf{T}}(\mathbf{X})$ and we consider $\mathbf{B} = \mathsf{\Pi}^{\mathsf{T}}(\mathbf{X})\overline{\mathbf{T}}$, where $\mathbf{T}$ is a 3D line. We need a $\mathbf{T}$, such that $\overline{\mathbf{Y}} = \mathbf{B}$ and a necessary condition for the existence of $\mathbf{T}$ is that $\mathbf{X} \in \overline{\mathbf{Y}} \Leftrightarrow \mathbf{X}^{\mathsf{T}}\mathbf{Y} = 0$. Again, $(\overline{\mathbf{Y}} = -\mathsf{\Pi}^{\mathsf{T}}(\mathbf{X})\overline{\mathbf{T}})$ ∘–• $(\mathsf{\Pi}^{\mathsf{T}}(\mathbf{X})\mathbf{Y} = \mathbf{T})$ and thus $\mathbf{T} = \mathbf{L}$.
- $\mathbf{A} = \mathbf{X} \wedge \mathbf{L}$
  (i) For $\mathsf{U}^{\mathsf{T}}(\mathbf{X}) = \overline{\mathsf{\Pi}}(\mathbf{X})$, we consider $\mathbf{M} = \mathsf{\Pi}(\mathbf{X})\mathbf{T}$ or $\overline{\mathbf{M}} = \overline{\mathsf{\Pi}}(\overline{\mathbf{X}})\overline{\mathbf{T}}$, where $\mathbf{T}$ is considered being a 3D point. We need a $\mathbf{T}$ such that $\mathbf{L} = \overline{\mathbf{M}}$ or equivalently $\overline{\mathbf{L}} = \mathbf{M}$. A necessary condition for the existence of $\mathbf{T}$ is

$\mathbf{X} \in \overline{\mathbf{L}}$. From $(\overline{\mathbf{L}} = \overline{\Pi}(\mathbf{X})\overline{\mathbf{T}}) \multimap (\overline{\Pi}^{\mathsf{T}}(\mathbf{X})\mathbf{L} = \mathbf{T})$ it follows that $\overline{\mathbf{T}} = \mathbf{A}$. (ii) For $\mathsf{U}^{\mathsf{T}}(\mathbf{X}) = \overline{\Gamma}(\mathbf{L})$, we consider $\mathbf{B} = \overline{\Gamma}(\mathbf{L})\mathbf{T}$, where $\mathbf{T}$ is interpreted as a 3D point. We need a $\mathbf{T}$ such that $\overline{\mathbf{X}} = \mathbf{B}$, thus a necessary condition for the existence of $\mathbf{T}$ is $\mathbf{L} \in \overline{\mathbf{X}}$. From $(\overline{\mathbf{X}} = \overline{\Gamma}(\mathbf{L})\mathbf{T}) \multimap (\overline{\Gamma}^{\mathsf{T}}(\mathbf{L})\mathbf{X} = \overline{\mathbf{T}})$, it follows that $\overline{\mathbf{T}} = \mathbf{A}$. □

4.4.2 Validation of the Statistical Approach

We validate that the algorithm by Monte Carlo tests in 2D and 3D where all geometric entities are involved. We then compare the sample mean and sample covariance with the computation using first order error propagation as in equation (4.23) on page 111.

In both tests we generate a sufficiently large number of 2D resp. 3D points, where each point is constructed by some join and intersection operations. Liu *et al.* 1996 suggest five hypotheses tests for testing multivariate data: the randomly generated data is assumed to be Gaussian with unknown mean $\boldsymbol{\mu}$ and covariance $\boldsymbol{\Sigma}$ and it may be tested against the approximated mean $\hat{\boldsymbol{\mu}}$ and covariance $\hat{\boldsymbol{\Sigma}}$. For our purpose, we choose the following tests:

$H_1\boldsymbol{\mu} = \hat{\boldsymbol{\mu}}$ with $\boldsymbol{\Sigma}$ being untested.
This test decides whether the generated point data has the computed mean $\hat{\boldsymbol{\mu}}$ using bilinear constructions. The computed covariance is not used.

$H_2\boldsymbol{\Sigma} = \hat{\boldsymbol{\Sigma}}$ with $\boldsymbol{\mu}$ being untested.
This test questions whether the covariance of the generated point cloud has the approximated covariance using first order error propagation. The mean $\boldsymbol{\mu}$ of the point cloud is disregarded.

$H_3\boldsymbol{\mu} = \hat{\boldsymbol{\mu}}$ and $\boldsymbol{\Sigma} = \hat{\boldsymbol{\Sigma}}$.
The last test is a combination of the above, simultaneously testing covariance and mean of the generated point cloud.

Additionally, we define two measures reflecting the relative error of between the sample mean $\bar{\boldsymbol{y}}$ and covariance $\bar{\boldsymbol{\Sigma}}_{yy}$ on one side and the approximated $\hat{\boldsymbol{\mu}}$ and covariance $\hat{\boldsymbol{\Sigma}}$ on the other side.

$$re_E = \sqrt{\frac{||\bar{\boldsymbol{y}} - \hat{\boldsymbol{\mu}}||^2}{\text{trace}(\bar{\boldsymbol{\Sigma}})}} \tag{4.52}$$

$$re_V = \frac{||\bar{\boldsymbol{\Sigma}} - \hat{\boldsymbol{\Sigma}}||}{||\bar{\boldsymbol{\Sigma}}||} \tag{4.53}$$

The number of iterations for the Monte Carlo simulations were fixed to 100'000 for all tests in 2D and 3D.

For 2D, we reconstruct a point by the intersection of two lines, which in turn were each constructed by joining two random points, see table 4.2. Thus both join and intersection operation were used. The three hypothesis tests have

Monte Carlo test for the construction a 2D point
Objective: Test the validity of the algorithm 4.1 by repeatedly constructing a 2D point out of 4 random points.

1. Repeat N times
   a) generate 4 random points $\boldsymbol{x}_i$ according to a Gaussian distribution with σ_x, σ_y and σ_{xy} and $\boldsymbol{\mu}_{x_i}$ and transfer them to homogeneous vectors $\mathbf{x}_i$.
   b) compute 2 lines $\mathbf{l}_1 = \mathbf{x}_1 \wedge \mathbf{x}_2$ and $\mathbf{l}_2 = \mathbf{x}_3 \wedge \mathbf{x}_4$
   c) compute point $\mathbf{y} = \mathbf{l}_1 \cap \mathbf{l}_2$
   d) normalize by $(\boldsymbol{y}, 1)^\mathsf{T} = \mathsf{N}_O(\mathbf{y})$
2. Compute sample mean $\bar{\boldsymbol{y}}$ and sample covariance $\bar{\boldsymbol{\Sigma}}_{yy}$
3. compute point $(\boldsymbol{y}, \boldsymbol{\Sigma}_{yy})$ as in step 1 using the algorithm in table 4.1.
4. compare sample statistics $(\bar{\boldsymbol{y}}, \bar{\boldsymbol{\Sigma}}_{yy})$ with computed statistics $(\boldsymbol{y}, \boldsymbol{\Sigma}_{yy})$

Algorithm 4.2: Monte Carlo test for the construction of a 2D point.

been accepted up to an input noise of 10%, see table 4.2. Only for 25% and 50% input error the approximated covariance matrix was not accepted to be equal to the sample covariance matrix. Similarly, the relative errors of mean and covariance matrix is less than 10% for small noise levels.

Table 4.2. Results of 2D Monte Carlo tests: testing the approximated covariance with the sample covariance matrix according to table 4.2. 100'000 samples have been generated for each error level.

Test-type	1 % error	10% error	25% error	50% error
Test 1 $\mu = \mu_0$ with unknown Σ	0 (+)	0 (+)	0 (+)	0 (+)
Test 2 $\Sigma = \Sigma_0$ with unknown μ	3.25 (+)	4.59 (+)	43.94 (-)	100855.08 (-)
Test 3 $\Sigma = \Sigma_0$ and $\mu = \mu_0$	5.24 (+)	6.52 (+)	45.65 (-)	100903.15 (-)
relative error cov $\|\|\Sigma - \Sigma_0\|\|/\|\|\Sigma\|\|$	0.05 (+)	0.06 (+)	0.21 (-)	0.99 (-)
relative error mean $\|\|\mu - \mu_0\|\|/\sqrt{\mathrm{trace}(\Sigma)}$	0.04 (+)	0.04 (+)	0.04 (+)	0.14 (-)

For 3D, we reconstruct a 3D point given three planes which were constructed by random 3D points. 3D lines were used as intermediate entities and again both join and intersection were used, cf. table 4.3. In this case, the tests have

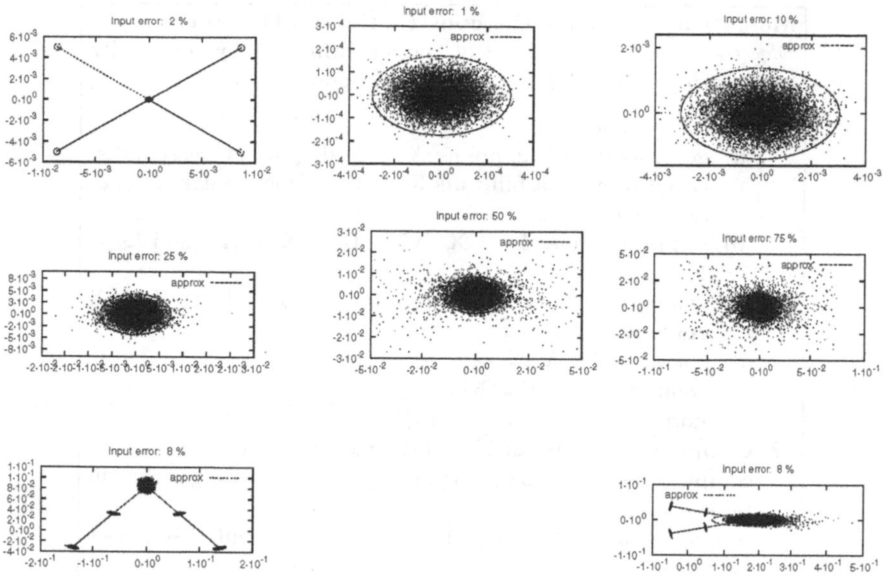

Fig. 4.12. Results of 2D Monte Carlo test using varying σ. Depicted are the points that have been constructed according to the algorithm in table 4.2. The top two rows show one geometric construction with varying input error. In the last row, two other geometric situations are depicted. On the lower right example one sees that if the intersection is not well-defined and the input error becomes larger, the error distribution is not Gaussian anymore.

Table 4.3. Results of 3D Monte Carlo tests: testing the approximated covariance with the sample covariance matrix according to table 4.3. 100'000 samples have been generated for each error level.

Test-type	1 % error	5% error	10% error	25% error
Test 1 $\mu = \mu_0$ with unknown Σ	1.41 (+)	1.33 (+)	2.53 (+)	1261.33 (-)
Test 2 $\Sigma = \Sigma_0$ with unknown μ	4.35 (+)	34.76 (-)	43.94 (-)	978255.37 (-)
Test 3 $\Sigma = \Sigma_0$ and $\mu = \mu_0$	5.78 (+)	36.9 (-)	45.65 (-)	979500.98 (-)
relative error cov $\|\|\Sigma - \Sigma_0\|\|/\|\|S\|\|$	0.06 (+)	0.08 (+)	0.16 (-)	1.00 (-)
relative error mean $\|\|\mu - \mu_0\|\|/\sqrt{\text{trace}(\Sigma)}$	0.04 (+)	0.03 (+)	0.04 (+)	0.98 (-)

Monte Carlo test for the construction a 3D point

Objective: Test the validity of the algorithm 4.1 by repeatedly constructing a 3D point out of 9 random points.

1. Repeat N times
   a) generate 9 random points $\boldsymbol{X}_i$ according to a Gaussian distribution with a full rank 3×3 covariance matrix $\boldsymbol{\Sigma}_{X_i X_i}$ and mean $\boldsymbol{\mu}_{X_i}$.
   b) compute 3 lines $\mathbf{L}_1 = \mathbf{X}_1 \wedge \mathbf{X}_2$, $\mathbf{L}_2 = \mathbf{X}_4 \wedge \mathbf{X}_5$, and $\mathbf{L}_2 = \mathbf{X}_7 \wedge \mathbf{X}_8$.
   c) compute 3 planes $\mathbf{A}_1 = \mathbf{L}_1 \wedge \mathbf{X}_3$, $\mathbf{A}_2 = \mathbf{L}_2 \wedge \mathbf{X}_6$ and $\mathbf{A}_3 = \mathbf{L}_3 \wedge \mathbf{X}_9$
   d) compute line $\mathbf{M} = \mathbf{A}_1 \cap \mathbf{A}_2$
   e) compute point $\mathbf{Y} = \mathbf{M} \cap \mathbf{A}_3$
   f) normalize by $(\boldsymbol{Y}, 1)^\mathsf{T} = \mathsf{N}_O(\mathbf{Y})$
2. Compute sample mean $\bar{\boldsymbol{Y}}$ and sample covariance $\bar{\boldsymbol{\Sigma}}_{YY}$
3. compute point $(\boldsymbol{Y}, \boldsymbol{\Sigma}_{YY})$ as in step 1 using the algorithm in table 4.1.
4. compare sample statistics $(\bar{\boldsymbol{Y}}, \bar{\boldsymbol{\Sigma}}_{YY})$ with computed statistics $(\boldsymbol{Y}, \boldsymbol{\Sigma}_{YY})$

Algorithm 4.3: Monte Carlo test for the construction of a 3D point.

been accepted for 1% and 5% input noise. For 10% input noise, the tests involving the covariance matrix were rejected, though the test-values remain relatively small and the relative error of the covariance matrix was only 8%, see table 4.3.

4.4.3 Construction Using Geometric Transformations

Given a projective transformation H with its covariance matrix $\boldsymbol{\Sigma}_{\mathrm{hh}}$ of $\mathbf{h} = \mathrm{vec}(\mathsf{H}^\mathsf{T})$, the uncertain entity $(\mathbf{x}, \boldsymbol{\Sigma}_{\mathrm{xx}})$ is transformed to a new entity $(\mathbf{x}', \boldsymbol{\Sigma}_{\mathrm{x'x'}})$ with $\mathbf{x}' = \mathsf{H}\mathbf{x}$. The computation of $\boldsymbol{\Sigma}_{\mathrm{x'x'}}$ can be done using error propagation

$$\boldsymbol{\Sigma}_{\mathrm{x'x'}} = \mathsf{H}\boldsymbol{\Sigma}_{\mathrm{xx}}\mathsf{H}^\mathsf{T} + (I_n \otimes \mathbf{x}^\mathsf{T})\boldsymbol{\Sigma}_{\mathrm{hh}}(I_n \otimes \mathbf{x}) \tag{4.54}$$

where I_n is an identity matrix of dimension n, which is equal to the number of rows of H.

It remains to be shown that the resulting entity is a proper uncertain entity as defined in definition 10 on page 104: as a prerequisite we assume that both the transformation and the given entity are proper uncertain entities. In this work we only prove the necessary property of $\boldsymbol{\Sigma}_{\mathrm{x'x'}}$ for homographies in 2D and 3D. For the transformations involving a projective camera, including the fundamental matrix, the proofs are similar but slightly more complicated.

Given a homography H, we consider the first addend in the equation (4.54), $\boldsymbol{\Sigma}^1_{\mathrm{x'x'}} = \mathsf{H}\boldsymbol{\Sigma}_{\mathrm{xx}}\mathsf{H}^\mathsf{T}$. As the homography is assumed to be regular, the possible nullspace $\mathcal{N}(\boldsymbol{\Sigma}_{\mathrm{x'x'}})$ depends only on $\boldsymbol{\Sigma}_{\mathrm{xx}}$. As we assume $\mathcal{N}(\boldsymbol{\Sigma}_{\mathrm{xx}}) = \mathbf{x}$, one obtains $\mathcal{N}(\boldsymbol{\Sigma}^1_{\mathrm{x'x'}}) = \{\mathsf{M}^{-\mathsf{T}}\mathbf{x}\}$. Because adding a second covariance matrix would at most lead to the same nullspace, the condition of in proposition 2 reduces to

$$\mathbf{x}'^{\mathsf{T}}\,\mathsf{M}^{-\mathsf{T}}\mathbf{x} = 0 \;\Leftrightarrow\; \mathbf{x}^\mathsf{T}\mathsf{M}^{-1}\mathbf{x}' = 0 \;\Leftrightarrow\; \mathbf{x}^\mathsf{T}\mathbf{x} = 0$$

The proof for a projective matrix P can be done in a similar way; for P^T, Q and $\overline{\mathsf{Q}}^\mathsf{T}$ one has to take both addends of the error propagation into account. The validity of constructions using the fundamental matrix is then given by the relation $\mathsf{F} = \mathsf{Q}_2\overline{\mathsf{Q}_1}^\mathsf{T}$.

4.5 Testing Geometric Relations

As we have seen in chapter 3 the check of geometric relations between two geometric entities was a simple test whether a distance vector $\mathbf{d}$ or $\mathbf{D}$ resp. distance scalar d was equal to zero. A simple case is the incidence of a 2D line $\mathbf{l}$ and a 2D point $\mathbf{x}$ with $d = \mathbf{l}^\mathsf{T}\mathbf{x}$,, see figure 4.13(a) In practical applications this distance is generally not zero and usually one chooses a small value ϵ and test whether $d < \epsilon$.

The problem is to choose a value ϵ, e.g. for an error-ribbon around the 2D line $\mathbf{l}$, see figure 4.13(b). But even after choosing a fixed ϵ this solution is not optimal as it is not dependent on the position of the points and lines and their confidence regions, see figure 4.13(c). The choice of ϵ becomes even more problematic when testing 3D entities.

This section introduce a method to choose appropriate values for ϵ for the relations introduced in chapter 3 by exploiting simple hypotheses test techniques known from statistics.

4.5.1 Hypothesis Testing

The question "*Is the point* $\mathbf{x}_i$ *incident to the line* $\mathbf{l}$*?*" can be solved by a statistical hypothesis test, cf. Papoulis and Pillai 2002: we want to test the hypothesis H_0: "*The point* $\mathbf{x}_i$ *is incident to* $\mathbf{l}$" against the hypothesis H_1: "*The point* $\mathbf{x}_i$ *is not incident to* $\mathbf{l}$". For general relations, the measure that we want to use is the distance vector $\mathbf{d}$ or $\mathbf{D}$, in the 2D incidence case it collapses to a scalar $d = \mathbf{l}^\mathsf{T}\mathbf{x}$. As we assume that the homogeneous vectors representing the geometric entities are normally distributed, e.g. $\underline{\mathbf{x}} \sim N(\mathbf{x}, \boldsymbol{\Sigma}_{\mathrm{xx}}), \underline{\mathbf{l}} \sim N(\mathbf{l}, \boldsymbol{\Sigma}_{\mathrm{xx}})$, also the distance is normally distributed.

We can now make use the following theorem from statistical testing theory (cf. Koch 1999 sec. 2.7.2 p. 135), which provides a scalar test value with known statistical distribution.

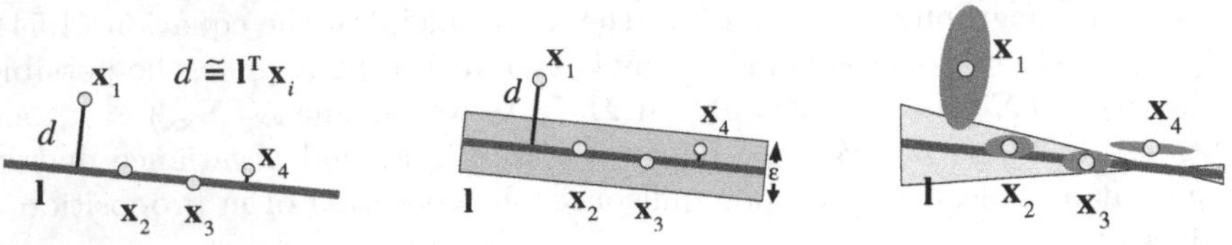

(a) 2D incidence test using dot-product

(b) Test using ϵ-error ribbon

(c) Test using confidence region

Fig. 4.13. How to tell whether a point $\mathbf{x}_i$ is incident to a line $\mathbf{l}$: we use the dot product $d \cong \mathbf{l}^\mathsf{T}\mathbf{x}_i$ to tell (a), but a fixed threshold ϵ as in fig. (b) such that $d < \epsilon$ is not a satisfactory solution: the threshold does not depend on the location and size of the confidence regions of the points $\mathbf{x}_i$ and $\mathbf{l}$.

Proposition 7 (Hypothesis testing) *Given a n-vector* $\mathbf{d}$ *with normal distribution* $\underline{\mathbf{d}} \sim N(\mu_d, \boldsymbol{\Sigma}_{dd})$, *rk*$\boldsymbol{\Sigma}_{dd} = n$. *the optimal test statistic for the hypothesis* $H_0 : \mathbf{d} = \boldsymbol{\mu}$ *is given by*

$$\underline{T} = (\underline{\mathbf{d}} - \boldsymbol{\mu})^\mathsf{T}\boldsymbol{\Sigma}_{dd}^{-1}(\underline{\mathbf{d}} - \boldsymbol{\mu}) \sim \chi^2_n \tag{4.55}$$

where χ^2_r *denotes the* χ^2_n*-distribution with* n *degrees of freedom.*

To perform the test, we need to fix the probability α that we reject H_0 although it is actually true—this situation is called type-I error. The probability α is usually a small number such as 1% or 5% and is called *significance level* of the test.

The critical value ϵ_H such that $P(T > \epsilon_H | H_0) = \alpha$ is given by the (1-α)-quantile of the χ^2 distribution:

$$\epsilon_H := \chi^2_{1-\alpha;n}$$

Now the hypothesis H_0 can be rejected with a significance level of α if

$$T = \underline{\mathbf{d}}^\mathsf{T}\boldsymbol{\Sigma}_{dd}^{-1}\underline{\mathbf{d}} > \epsilon_H = \chi^2_{1-\alpha;n} \tag{4.56}$$

Sometimes it is convenient to use the test-ratio $T_R = \frac{T}{\chi^2_{1-\alpha;n}}$, so that we can compare test-values of different hypothesis tests.

It is crucial to note that a successful hypothesis test $T < \epsilon_H$ does not validate that H_0 is true, it merely states that there is not enough evidence to reject H_0.

4.5.1.1 Testing Geometric Relations We can directly apply the above test method to the 18 relations introduced in table 3.5 on page 78: a relation $R(\mathbf{x}, \mathbf{y})$ can be assumed to hold if the hypothesis

$$H_0: \quad \mathbf{d} = \mathsf{U}(\mathbf{x})\mathbf{y} = \mathsf{V}(\mathbf{y})\mathbf{x} = \mathbf{0} \tag{4.57}$$

can not be rejected, see equation (3.79) on page 79. The covariance matrix $\boldsymbol{\Sigma}_{\mathrm{dd}}$ of $\mathbf{d}$ is given by first order error propagation as

$$\boldsymbol{\Sigma}_{\mathrm{dd}} = \mathsf{U}(\mathbf{x})\boldsymbol{\Sigma}_{\mathrm{yy}}\mathsf{U}^{\mathsf{T}}(\mathbf{x}) + \mathsf{V}(\mathbf{y})\boldsymbol{\Sigma}_{\mathrm{xx}}\mathsf{V}^{\mathsf{T}}(\mathbf{y}) \tag{4.58}$$

In the 2D-incidence case, the last equation reduces to

$$\sigma_d^2 = \mathbf{l}^{\mathsf{T}}\boldsymbol{\Sigma}_{\mathrm{xx}}\mathbf{l} + \mathbf{x}^{\mathsf{T}}\boldsymbol{\Sigma}_{\mathrm{ll}}\mathbf{x}$$

and thus the test-statistic $T = d^2/\sigma_d^2$.

In general $\boldsymbol{\Sigma}_{\mathrm{dd}}$ may be singular, if $\mathbf{d}$ is a $n \times 1$ vector, r is is the degree of freedom of the relation R and $r < n$. The singularity causes a problem, because we have to invert the covariance matrix in equation (4.56). But at least for projective relations we can guarantee that the rank of $\boldsymbol{\Sigma}_{\mathrm{dd}}$ is not less than r, as the following proposition shows:

Proposition 8 (Minimal rank of $\boldsymbol{\Sigma}_{\mathrm{dd}}$) *Let* $(\mathbf{x}, \boldsymbol{\Sigma}_{\mathrm{xx}})$, $(\mathbf{y}, \boldsymbol{\Sigma}_{\mathrm{yy}})$ *be uncertain geometric entities in 2D or 3D, which are conditioned, cf. section 2.4, and the nullspaces for* $\boldsymbol{\Sigma}_{\mathrm{xx}}$ *and* $\boldsymbol{\Sigma}_{\mathrm{yy}}$ *are determined by their homogeneous vectors* $\mathbf{x}, \mathbf{y}$. *The bilinear form* $\mathbf{d} = \mathsf{U}(\mathbf{x})\mathbf{y} = \mathsf{V}(\mathbf{x})\mathbf{y}$ *for a projective relation* $R(\mathbf{x}, \mathbf{y})$ *is defined in table 3.5 on page 78 with* $r = dof(R)$ *being the degree of freedom for* $R(\mathbf{x}, \mathbf{y})$. *Then the covariance matrix* $\boldsymbol{\Sigma}_{\mathrm{dd}}$ *has rank* $rk(\boldsymbol{\Sigma}_{\mathrm{dd}}) \geq r$. *For the reduced distance vector* $\mathbf{d}^{[r]} = \mathsf{U}^{[r]}(\mathbf{x})\mathbf{y} = \mathsf{V}^{[r]}(\mathbf{x})\mathbf{y}$, *the covariance matrix* $\boldsymbol{\Sigma}_{\mathrm{d^{[r]}d^{[r]}}}$ *is regular for all cases mentioned in equations* (3.53), (3.54) *and* (3.55) *on page 69*

Proof: For projective relations, there are three tests with one degree of freedom, which are analogous to the $\mathbb{P}^1$ case in section 4.3.3 on page 117. The other tests mostly stem from bilinear construction forms, which have been analyzed in the proof of proposition 6 on page 122 and the resulting covariance matrices are guaranteed to be at least of rank r. For the special case of the identity of 3D lines we use the bilinear form in equation (3.78), where the matrices $\boldsymbol{\Delta}_i(\mathbf{L})^{\mathsf{T}}$ resp. $\boldsymbol{\Delta}_i(\mathbf{M})^{\mathsf{T}}$ both have the eigenvector $\mathbf{E}_i$. But i is chosen such that $\mathbf{E}_i^{\mathsf{T}}\mathbf{M} > 0$ resp. $\mathbf{E}_i^{\mathsf{T}}\mathbf{L} > 0$, thus with proposition B.1 the rank of $\boldsymbol{\Sigma}_{\mathrm{dd}}$ is at least r.

Given the reduced distance vector of any relation R, the ranks of $\left(\mathsf{U}^{[r]}(\mathbf{x})\right)^{\mathsf{T}}$ and $\left(\mathsf{V}^{[r]}(\mathbf{x})\right)^{\mathsf{T}}$ equals r, if one uses the reduction algorithm 3.1. The nullspaces of $(\mathsf{U}^{[r]}(\mathbf{x}))^{\mathsf{T}}$ and $(\mathsf{V}^{[r]}(\mathbf{x}))^{\mathsf{T}}$ are then zero, thus $\boldsymbol{\Sigma}_{\mathrm{d^{[r]}d^{[r]}}}$ is regular. □

Generally, proposition 8 holds for orthogonality and parallelity relations, too. This time only the sub-covariance matrix of the homogeneous entities, which are basically directional vectors, have to be projected spherically.

But as the conditioning of homogeneous vectors can not be applied to directional vectors, the variance of d can be zero resp. the covariance matrix of $\mathbf{d}$

of can be of rank 1 instead 2, cf. section 4.3.3. These special cases happen only if the relation of the two directional vectors is *exactly* opposite from the tested relations: for example two lines are exactly parallel but are tested for orthogonality. If the vectors are *almost* opposite from the tested relation, the test-value is obviously very large. So in practice the singular cases can be avoided for Euclidean relations by adding a very small amount to the covariance matrix, for example:

$$\boldsymbol{\Sigma}_{\mathsf{d}'\mathsf{d}'} = \boldsymbol{\Sigma}_{\mathsf{d}^{[r]}\mathsf{d}^{[r]}} + 0.0001^2 \,\mathrm{trace}\,(\boldsymbol{\Sigma}_{\mathsf{d}^{[r]}\mathsf{d}^{[r]}})\, I_r \tag{4.59}$$

Thus the new covariance matrix $\boldsymbol{\Sigma}_{\mathsf{d}'\mathsf{d}'}$ is guaranteed to be always regular; this adds only a very small bias to all tests for Euclidean relations, which is negligible as the test-value is very large anyway.

4.5.2 Properties of the Approximated Test-Value T

It is of practical importance to check some properties of the approximated test-value $T(f) = \mathbf{d}^{[r]\mathsf{T}} \boldsymbol{\Sigma}_{\mathsf{d}^{[r]}\mathsf{d}[r]} \mathbf{d}^{[r]}$ from proposition 8 where f is the conditioning factor for $\mathbf{x}$ and $\mathbf{y}$:

- first we discuss the bias between $T(f)$ and the ideal test-value T_E similar to the analysis in section 4.3.3.
- second we check if an important property of the ideal test-value T_E is preserved: is the approximated test-value $T(f)$ monotonously increasing when increasing the distance between the tested entities?

Bias of the Test-value We first discuss the bias of the ideal test value and the computed using first order error propagation. Since the vector $\mathbf{d}$ from equation 4.57 is essentially an entity that was constructed from $\mathbf{x}$ and $\mathbf{y}$, the hypothesis test can be interpreted as a test whether the constructed entity is undefined (null-vector) or not. Thus exactly the same arguments from section 4.3.3 can be applied, where the bias of constructions was discussed and found to be negligible under well-defined conditions.

To demonstrate the existence of the bias, we construct an extreme example, which usually does not occur in practice: assume two uncertain 2D points for $(\mathbf{x}, \boldsymbol{\Sigma}_{\mathsf{xx}})$ and $(\mathbf{l}, \boldsymbol{\Sigma}_{\mathsf{ll}})$ with $\mathcal{N}(\boldsymbol{\Sigma}_{\mathsf{xx}}) = \mathbf{x}$ and $\mathcal{N}(\boldsymbol{\Sigma}_{\mathsf{yy}}) = \mathbf{y}$. The test on identity $\mathbf{x} \equiv \mathbf{y}$ is done using $\mathbf{d} = \mathsf{S}(\mathbf{x})\mathbf{y}$ and

$$\boldsymbol{\Sigma}_{\mathsf{dd}} = \mathsf{S}(\mathbf{y})\boldsymbol{\Sigma}_{\mathsf{xx}}\mathsf{S}(\mathbf{y})^{\mathsf{T}} + \mathsf{S}(\mathbf{x})\boldsymbol{\Sigma}_{\mathsf{yy}}\mathsf{S}(\mathbf{x})^{\mathsf{T}}$$

One may also use the reduced construction matrices $\mathsf{S}^{[2]}(\mathbf{x})$ and $\mathsf{S}^{[2]}(\mathbf{y})$.

An extreme case of the bias is now given when considering $\mathbf{x}^{\mathsf{T}}\mathbf{y} = 0$, see figure 4.14. Note that in practice this case usually does not happen, since $\mathbf{x}^{\mathsf{T}}\mathbf{y} >> 0$ if $\mathbf{x} \approx \mathbf{y}$. Assuming $\mathbf{x}^{\mathsf{T}}\mathbf{y} = 0$, the rank of the covariance matrix $\boldsymbol{\Sigma}_{\mathsf{dd}}$ is less than 2, which can be proven using proposition B.1 on page 183 and the fact that $\mathcal{N}(\boldsymbol{\Sigma}_{\mathrm{xx}}) = \mathcal{N}(\mathsf{S}(\mathbf{y}))^{\perp}$. But $\boldsymbol{\Sigma}_{\mathrm{dd}}$ has to be of at least rank 2, as the test has two degrees of freedom, thus a naive application of the hypothesis testing fails in this example.

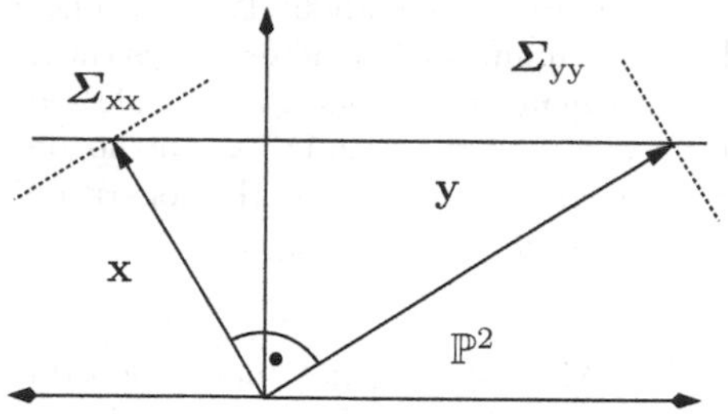

Fig. 4.14. Cross section of the situation of two projective points $\mathbf{x}, \mathbf{y} \in \mathbb{P}^2$, where the homogeneous vectors are orthogonal to each other, $\mathbf{x}^{\mathsf{T}}\mathbf{y} = 0$, the v-coordinate is skipped for visualization purposes. The covariance matrices have nullspaces $\mathcal{N}(\boldsymbol{\Sigma}_{\mathrm{xx}}) = \mathbf{x}$ and $\mathcal{N}(\boldsymbol{\Sigma}_{\mathrm{yy}}) = \mathbf{y}$, thus the respective uncertainty is defined only orthogonal to its homogeneous vector.

The above problem is an extreme example of a bias caused by the approximation of the used error propagation. When we apply a conditioning of the uncertain homogeneous vector with a subsequent orthogonal projection of the covariance matrices with $\mathcal{P}_{\mathbf{x}}$ resp. $\mathcal{P}_{\mathbf{y}}$ (see algorithm in table 4.4), the bias disappears, see also section 4.3.3:

To obtain the true value of the test value, we consider the Euclidean points $\boldsymbol{x} = (x_1, x_2)^{\mathsf{T}}$ and $\boldsymbol{y} = (y_1, y_2)^{\mathsf{T}}$ and their covariances $\boldsymbol{\Sigma}_{xx}$ and $\boldsymbol{\Sigma}_{yy}$. The distance $\boldsymbol{d}$ and the covariance matrix $\boldsymbol{\Sigma}_{dd}$ is given by

$$\boldsymbol{d} = \begin{pmatrix} x_1 - y_1 \\ x_2 - y_2 \end{pmatrix} \tag{4.60}$$

$$\boldsymbol{\Sigma}_{dd} = J \begin{pmatrix} \boldsymbol{\Sigma}_{xx} & 0 \\ 0 & \boldsymbol{\Sigma}_{yy} \end{pmatrix} J^{\mathsf{T}} \qquad \text{with } J = \begin{pmatrix} 1 & 0 & -1 & 0 \\ 0 & 1 & 0 & -1 \end{pmatrix} \tag{4.61}$$

and we obtain the Euclidean test statistic $T_E = \boldsymbol{d}^{\mathsf{T}}\boldsymbol{\Sigma}_{dd}\boldsymbol{d}$. A test proposed with homogeneous vectors as in table 4.4 yields a test-statistic $T(f)$ depending on the conditioning factor f. Figure 4.15 shows the percentage of the bias between T_E and $T(f)$ depending on f: with a factor of $f = 0.1$, the bias is about 1%. If the homogeneous vectors are orthogonal to each other ($f = 1$), the test-statistic is not defined, if they are close to be orthogonal, the test-statistic is very large and is smaller the more collinear the homogeneous vectors are.

As a result of the above example we conclude that the bias can be assumed to be less than 1 % if the conditioning factor $f = 0.1$ is applied. If no conditioning of the geometric entities is used, the bias can be uncontrollably large.

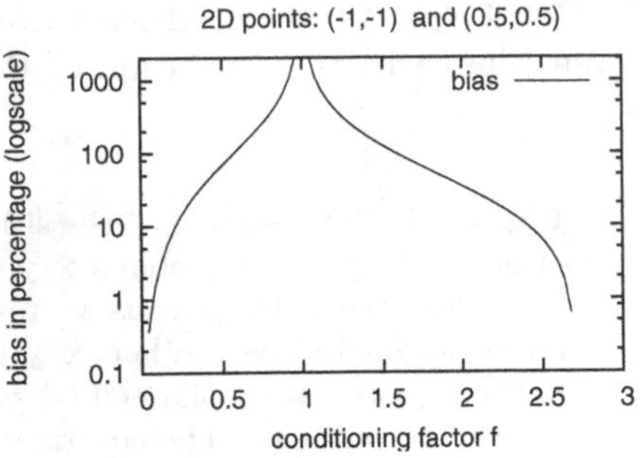

Fig. 4.15. Bias when testing identity of two 2D points $\boldsymbol{x} = (1,1)^\mathsf{T}$ and $\boldsymbol{y} = -\frac{1}{2}(1,1)^\mathsf{T}$ using algorithm in table 4.4. The conditioning factor f varies between 0.1 and 3. With a factor of $f = 0.1$, the bias is about 1%. The bias is maximum with a factor of 1 as the homogeneous vectors $\mathbf{x} = (1,1,1)^\mathsf{T}$ and $\mathbf{y} = (-1/2,-1/2,1)^\mathsf{T}$ are orthogonal to each other. Note the logarithmic scale of the percentage.

Monotony of the Approximated Test-value Although we have identified only a small bias between the Euclidean test-statistic T_E and the approximated test-statistic $T(f)$, it may still be possible that $T(f)$ is not monotonically increasing dependent on the relative positions of the two entities, which would hinder control or guidance of searching algorithms based on $T(f)$. We will discuss the monotony property within $\mathbb{P}^1$, cf. section 4.3.3.

Assume a fixed point $\mathbf{x} = (u,v)^\mathsf{T} = (1,1)^\mathsf{T}$ and a hyperplane $\mathbf{l} = (a,b)^\mathsf{T} = (1,b)^\mathsf{T}$, which solely depends on b. Without loss of generality, we assume $b > 0$. The covariance matrices $\boldsymbol{\Sigma}_{\mathrm{xx}}$ and $\boldsymbol{\Sigma}_{\mathrm{yy}}$ are chosen arbitrarily according to definition 10. We now want to test whether the incidence relation $\mathbf{x} \cap \mathbf{l} \neq \emptyset$ holds, thus $\mathbf{x}^\mathsf{T}\mathbf{l} \stackrel{!}{=} 0$. First, we compute the test statistic T only with error propagation, without prior change of nullspace or conditioning:

$$T = \frac{b^2}{(\sigma_a^2 + 2\,\sigma_{ab} + \sigma_b{}^2) + (\sigma_u^2 + 2\,b\sigma_{uv} + b^2\sigma_v^2)}$$

To check whether T is monotonically increasing dependent on b, we compute the first derivative:

$$T_b = \frac{\partial\,t}{\partial\,b} = 2\,\frac{b\left((\sigma_a^2 + 2\,\sigma_{ab} + \sigma_b^2) + \sigma_u^2 + b\sigma_{uv}\right)}{\left((\sigma_a^2 + 2\,\sigma_{ab} + \sigma_b^2) + (\sigma_u^2 + 2\,b\sigma_{uv} + b^2\sigma_v^2)\right)^2} = 2\,\frac{c_a + \sigma_u^2 + b\,\sigma_{uv}}{(c_a + c_b)^2}$$

The sums c_a and c_b are positive definite forms and thus always positive. To check, whether $T_b > 0$ for all b, it is sufficient to check

$$\sigma_u^2 > -b\,\sigma_{uv} \tag{4.62}$$

If we assume that the nullspace of $\boldsymbol{\Sigma}_{\mathrm{xx}}$ is $\mathbf{x}$ and the nullspace of $\boldsymbol{\Sigma}_{\mathrm{yy}}$ is $\mathbf{y}$, then with equation (4.37) on page 118 the condition (4.62) becomes

$$v^2 > b\,u\,v \quad \text{and thus in our case} \quad 1 > b \tag{4.63}$$

So generally for $b \geq 1$, we can not guarantee the monotony property.

But if we apply the conditioning algorithm 2.1, to $\mathbf{x}$ and $\mathbf{y}$, then $\mathbf{x}^\circ = (f_{min}/b,\, 1)^\mathsf{T}$ and $\mathbf{y}^\circ = (1, f_{min})^\mathsf{T}$ and condition (4.63) becomes $1 > f_{min}$, which is true by definition. Note that condition (4.62) does not change if we allow any v instead of only $v = 1$.

Thus also for ensuring the monotony property of the approximated test-value, the conditioning of the entities is crucial.

Summary As a result of the investigation on the properties of the test-value T we again conclude that it is necessary to condition the uncertain entities prior to applying any algorithm. In section 4.3, the conditioning was shown to be important for construction and normalization. This time, the conditioning keeps bias of the test-value for testing geometric relations sufficiently small; additionally, the test-value is monotonically increasing depending on the relative distance between two entities.

4.5.3 A Statistical Algorithm for Testing Geometric Relations

Summarizing the section, we propose a generic algorithm for statistically testing geometric relations between points, lines and planes in 2D and 3D, cf. table 4.4. The algorithm uses the reduced construction matrices see section 3.2.2, which guarantees the covariance matrix $\boldsymbol{\Sigma}_{\mathrm{d}^{[r]}\mathrm{d}^{[r]}}$ to be regular provided proper uncertain geometric entities, cf. definition 10 on page 104. Thus one can detect non-proper entities within the algorithm if $\boldsymbol{\Sigma}_{\mathrm{d}^{[r]}\mathrm{d}^{[r]}}$ is found to be non-invertible.

An alternative algorithm in table 4.5 does not use the reduced matrices but rather relies on an existing SVD algorithm to compute the pseudo-inverse of $\boldsymbol{\Sigma}_{\mathrm{dd}}$. If an SVD algorithm is available, the latter algorithm may be easier to implement but lacks the feature of detecting non-proper entities.

Opposed to the SVD, an alternative computation of the pseudo-inverse $\boldsymbol{\Sigma}^{+}_{\mathrm{dd}}$ uses the expected nullspace $\mathcal{N}(\boldsymbol{\Sigma}_{\mathrm{dd}}) = \mathsf{H}$ of $\boldsymbol{\Sigma}_{\mathrm{dd}}$. Then the pseudo-inverse is given by

$$\begin{pmatrix} \boldsymbol{\Sigma}^{+}_{dd} & \cdot \\ \cdot & \cdot \end{pmatrix} = \begin{pmatrix} \boldsymbol{\Sigma}_{dd} & H \\ H^{\mathsf{T}} & \mathbf{0} \end{pmatrix}^{-1} \tag{4.64}$$

as it was proposed in FÖRSTNER *et al.* 2000. This is feasible for most cases, but can lead to problems as the nullspace may increase for special cases, such as two exactly identical entities that are tested on identity, see proof for proposition 6 on page 122. Using a SVD is computationally simpler since we know the expected rank of $\boldsymbol{\Sigma}_{dd}$.

4.5.4 Validation of Hypothesis Tests

To validate the algorithm in table 4.4, we generated $n = 200$ random cubes with edge length 1, as in figure 4.16(left) and tested 13 relations between its

Statistically testing uncertain geometric relations (Version 1 with matrix reduction)

Objective: Given two entities $(\mathbf{x}, \boldsymbol{\Sigma}_{\mathrm{xx}})$ and $(\mathbf{y}, \boldsymbol{\Sigma}_{\mathrm{yy}})$ (possibly points, lines or planes) and a relation $R \in \{$*equal, incident, orthogonal, parallel*$\}$. Test the hypothesis that the relation $R(\mathbf{x}, \mathbf{y})$ holds with a given alpha-value α, where r is the degree of freedom for the chosen test, see last column of table 3.5.

1. If necessary, condition $\mathbf{x}$ and $\mathbf{y}$ as in algorithm 2.1 using $f_{min} = 0.1$. Scale the Euclidean parts of the covariance matrices in the same way by first order error-propgation. We obtain $(\mathbf{x}^\circ, \boldsymbol{\Sigma}_{\mathrm{x}^\circ\mathrm{x}^\circ})$ and $(\mathbf{y}^\circ, \boldsymbol{\Sigma}_{\mathrm{y}^\circ\mathrm{y}^\circ})$.
2. Change the nullspaces of $\boldsymbol{\Sigma}_{\mathrm{x}^\circ\mathrm{x}^\circ}$ and $\boldsymbol{\Sigma}_{\mathrm{y}^\circ\mathrm{y}^\circ}$ to $\mathbf{x}^\circ$ resp. $\mathbf{y}^\circ$ using orthogonal projection $\mathcal{P}_{\mathbf{x}^\circ}$ and $\mathcal{P}_{\mathbf{y}^\circ}$, cf. (4.18) on page 111.
3. Obtain Jacobians $\mathsf{U}(\mathbf{x}^\circ)$ and $\mathsf{V}(\mathbf{y}^\circ)$ according to table 3.5, such that $\mathsf{U}(\mathbf{x}^\circ)\mathbf{y}^\circ \stackrel{!}{=} \mathbf{0}$ and $\mathsf{V}(\mathbf{y}^\circ)\mathbf{x}^\circ \stackrel{!}{=} \mathbf{0}$.
4. Obtain the reduced matrices $\mathsf{U}^{[r]}(\mathbf{x}^\circ)$ and $\mathsf{V}^{[r]}(\mathbf{y}^\circ)$, see algorithm 3.1 on page 69. Compute the distance $\mathbf{d}^{[r]} = \mathsf{U}^{[r]}(\mathbf{x}^\circ)\,\mathbf{y}^\circ$ and its covariance matrix
$$\boldsymbol{\Sigma}_{\mathrm{d}^{[r]}\mathrm{d}^{[r]}} = \mathsf{U}^{[r]}(\mathbf{x}^\circ)\boldsymbol{\Sigma}_{\mathrm{y}^\circ\mathrm{y}^\circ}\mathsf{U}^{[r]\mathsf{T}}(\mathbf{x}^\circ) + \mathsf{V}^{[r]}(\mathbf{y}^\circ)\boldsymbol{\Sigma}_{\mathrm{x}^\circ\mathrm{x}^\circ}\mathsf{V}^{[r]\mathsf{T}}(\mathbf{y}^\circ)$$
$\boldsymbol{\Sigma}_{\mathrm{d}^{[r]}\mathrm{d}^{[r]}}$ is guaranteed to be regular. For Euclidean relations (parallelity, orthogonality), one may want to add a infinitesimally small amount to $\boldsymbol{\Sigma}_{\mathrm{d}^{[r]}\mathrm{d}^{[r]}}$ to ensure regularity for some cases, see equation (4.59).
6. Compute chi-square distributed test statistics $T = \mathbf{d}^{[r]\mathsf{T}}\boldsymbol{\Sigma}^{-}_{\mathrm{d}^{[r]}\mathrm{d}^{[r]}}\mathbf{d}^{[r]}$. The hypothesis "$\mathbf{x}$ *and* $\mathbf{y}$ *fulfill the relation* R" can not be rejected if $T < \chi^2_{1-\alpha;n}$. Otherwise the hypothesis is rejected.

Algorithm 4.4: Generic algorithm for testing geometric relations between uncertain geometric entities in 2D or 3D. This version uses the reduction if the construction matrices to obtain an invertible covariance matrix $\boldsymbol{\Sigma}_{\mathrm{dd}}$ of $\mathbf{d}$. Another possibility is to use a SVD, see table 4.5.

entities. Each cube was constructed by first generating 40 random points according to a multi-dimensional Gaussian distribution with a given covariance matrix; for simplicity uncorrelated and identical standard deviations σ were used for the point coordinates.

From these 40 random points, 8 were used for the corners, 24 for the constructions of lines for the 12 edges and 18 for the construction of planes for the 6 surfaces of the cube. For each of the 13 relations in 3D, a set of hypothesis tests was performed: for example for the point-line incidence test, 200 points of a specific corner were tested with 200 lines of an incident edge. Some relations allowed only $200 \cdot 199$ tests as e.g. the test of identity of the same line

Statistically testing uncertain geometric relations (Version 2 with SVD)

Objective: Given two entities $(\mathbf{x}, \boldsymbol{\Sigma}_{xx})$ and $(\mathbf{y}, \boldsymbol{\Sigma}_{yy})$ (possibly points, lines or planes) and a relation $R \in$ {*equal, incident, orthogonal, parallel*}. Test the hypothesis that the relation $R(\mathbf{x}, \mathbf{y})$ holds with a given alpha-value α, where r is the degree of freedom for the chosen test, see last column of table 3.5.

1.-3. see algorithm in table 4.4.
4. Compute the distance $\mathbf{d} = \mathsf{U}(\mathbf{x}^\circ)\, \mathbf{y}^\circ$ and its covariance matrix with the full construction matrices $\mathsf{U}(\mathbf{x}^\circ)$ and $\mathsf{V}(\mathbf{y}^\circ)$

$$\boldsymbol{\Sigma}_{dd} = \mathsf{U}(\mathbf{x}^\circ) \boldsymbol{\Sigma}_{y^\circ y^\circ} \mathsf{U}^\mathsf{T}(\mathbf{x}^\circ) + \mathsf{V}(\mathbf{y}^\circ) \boldsymbol{\Sigma}_{x^\circ x^\circ} \mathsf{V}^\mathsf{T}(\mathbf{y}^\circ)$$

5. Compute the pseudo-inverse $\boldsymbol{\Sigma}^+_{dd}$ for the matrix $\boldsymbol{\Sigma}_{dd}$ from step 4 using singular value decomposition EDG^T: then $\boldsymbol{\Sigma}^+_{dd} = \mathsf{GD}^+\mathsf{E}^\mathsf{T}$, where the first r diagonal elements of matrix D^+ are the inverse of the diagonal elements of D, the other $n - r$ are set to zero.
6. Compute chi-square distributed test statistics $T = \mathbf{d}^\mathsf{T} \boldsymbol{\Sigma}^+_{dd} \mathbf{d}$. The hypothesis "$\mathbf{x}$ *and* $\mathbf{y}$ *fulfill the relation* R" can not be rejected if $T < \chi^2_{1-\alpha;n}$. Otherwise the hypothesis is rejected.

Algorithm 4.5: Generic algorithm for testing geometric relations between uncertain geometric entities in 2D or 3D. This is a variant of the algorithm in table 4.4, the difference is that it does not use reduced construction matrices to ensure regularity of $\boldsymbol{\Sigma}_{dd}$, but rather compute the pseudo-inverse using an SVD.

was forbidden. Altogether 518600 tests were performed with a significance level of $\alpha = 1\%$ and varying σ.

To validate the usefulness of the hypothesis tests, we compare the number of type-I errors with the given $\alpha = 1\%$. The results are summarized in figure 4.16(right): here the test-ratio $T_R = T/\chi^2_{1-\alpha;n}$ was plotted in order to compare test-values between tests of different degrees of freedom. If $T_R < 1$ then the hypothesis H_0 is accepted, for $T_R \geq 1$, H_0 is rejected. For $\sigma = 0.01$, 99 % of the tests have been accepted, thus yielding 1 % percent of type-I errors, as expected for $\alpha = 1$ %. For $\sigma = 0.05$, only a bit less than 99 % of the tests have been accepted. The performance of the test algorithm decreases slowly as σ increases: more type-I errors occur for $\sigma = 0.1$ and $\sigma = 0.5$ which was to be expected, since the estimated covariance matrices are generally too optimistic, cf. discussion on bias on page 4.3.3.

4.5.5 Further Improvements

We shortly discuss further improvements on the testing scheme.

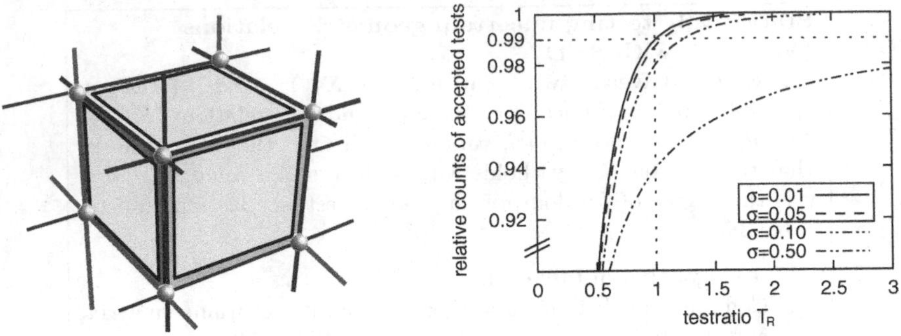

Fig. 4.16. (Left) 3D cube and its associated points, lines and planes. For the Monte Carlo tests 200 random cubes were generated with edge length 1. (Right) Cumulative relative histogram for test-ratios $T_R = T/\chi^2_{1-\alpha;n}$ of 518600 hypothesis tests, each with varying σ and $\alpha = 0.01$; test-ratios of 1 or less indicate that H_0 has been accepted, see text for details.

4.5.5.1 Testability Imagine that we want to test whether one of the tips of the dome in Cologne is incident to the line which is defined by the East Broadway in New York City. We may simplify this problem to a 2D problem and assume that the angle of the street is locally determined with a standard deviation of 1°. Estimating the distance between Cologne and New York at about 6000 km, the displacement error orthogonal to the East Broadway is about 100 km when measured in Cologne, cf. equation (C.3) on page 187. Assuming a confidence region which is 3 times larger than the standard deviation, all points in Germany within a strip of about 600 km width are incident to the East Broadway in New York City according to the hypothesis test. It is easy to see that although the test is accepted for a large number of points in Germany, the result is practically meaningless.

One can argue that the above test failed to reject because of insufficient data. To cope with this problem it is possible to check the *testability* given the uncertain data, as suggested by BAARDA 1968. The idea is to find a lower bound such that testing errors are detectable with a minimum probability, say 80%. For more details, see FÖRSTNER 1987.

4.5.5.2 Assumption of Full-Rank Covariance Matrices For error propagation and normalization we assume that the covariance matrices are of full-rank with respect to the degrees of freedom, in other words we assume that the variance for each Euclidean parameter is non-zero, cf. definition 10 on page 104. Without this assumption, we run into problems when inverting the covariance matrix $\boldsymbol{\Sigma}_{\mathrm{dd}}$ in equation (4.56).

If one wants to weaken this assumption, three approaches may be possible:

- Doing a rank-analysis on covariance matrix $\boldsymbol{\Sigma}_{\mathrm{dd}}$ by computing the eigenvalues and specify a threshold for zero eigenvalues. This has the disadvantage that we have to find a threshold value, which in general is not obvious, especially for arbitrarily scaled homogeneous covariance matrices.

- Specify the additional nullspace in Euclidean space and add it to the representation ($\mathbf{x}$, $\boldsymbol{\Sigma}_{\mathrm{xx}}$) such that we know that an entity is certain in some parameters. The subsequent analysis of the error propagation, a possible bias and the derivation of the new nullspaces may be quite involving.
- Replace the zero variances with very small variances compared to the other non-zero variances, similar to equation (4.59) on page 132. Assuming that the small amount of artificial noise does not influence the final results, we can apply the derived algorithms without any restrictions.

We recommend to implement the third method. All approaches are also valid for the next section, which deals with the optimal geometric estimation of geometric entities.

4.6 Optimal Geometric Estimation

In this last section of the chapter, we will develop a generic algorithm for constructing or estimating a geometric entity or a transformation from a collection of related geometric entities. As for the non-statistical case in chapter 3, the over-constrained construction of uncertain geometric entities benefits from the explorations of geometric relations: the construction is based on the relations between observations and unknowns, which was outlined in the last section. We can use existing techniques such as weighted least-squares estimation to provide a generic algorithm for optimal geometric estimation. The used estimation does not only find an estimate of the unknown entity, but is also capable of correcting the observed entities, cf. also KANATANI 1996.

4.6.1 Statistical Model

We first derive a statistical model for the construction task, where we deduce an estimation $\widehat{\boldsymbol{\beta}}$ of a not observable and hence unknown entity $\tilde{\boldsymbol{\beta}}$ from a set of observations $\mathbf{y}_i$. Here, the tilde denotes that the vector is true but unknown, the hat denotes the vector is an estimated entity. As in section 3.4 on page 80, we assume that the unknown geometric entity $\tilde{\boldsymbol{\beta}}$ is defined by a set of observations $\mathbf{y}_{(R,i)}$, which are related to the unknown by a relation $R(\tilde{\mathbf{y}}_{(R,i)}, \tilde{\boldsymbol{\beta}})$, holding for the true observations $\tilde{\mathbf{y}}_{(R,i)}$.
The statistical model consists of a (a) perturbation model of the observations, (b) an implicit model for the observation process and (c) a model for possible constraints on observations and unknowns. For simplicity in notations, we will only mention 3D entities and their relations explicitly as it is the more general case compared to 2D.

(a) *Perturbation Model of the Observations*
Consider one observed entity $\mathbf{y}_i := \mathbf{y}_{(R,i)}$, the vector $\mathbf{y}$ is obtained by some measurement process and are subject to a perturbation, which we model by a true vector $\tilde{\mathbf{y}}$ with an additional random noise vector $\mathbf{e}$:

$$\underline{\mathbf{y}}_i = \tilde{\mathbf{y}}_i + \underline{\mathbf{e}}_i \quad \text{with } \underline{\mathbf{e}} \sim N(\mathbf{0}, \boldsymbol{\Sigma}_{\mathrm{e_i e_i}}) \tag{4.65}$$

For the estimation of geometric entities, two restrictions have to be made to eq. (4.65):
- though being a homogeneous vector, the observed vector $\mathbf{y}_i$ has a fixed length as it is observed only once. For the sake of simplicity, we may assume that the vectors $\mathbf{y}_i$ and $\tilde{\mathbf{y}}_i$ are of length 1, thus normalized to the unit sphere, $\mathbf{y}_i = \mathsf{N}_s(\mathbf{y}_i)$.
- as we only allow Euclidean perturbations to the observations, we require the pair $(\mathbf{y}_i,\ \boldsymbol{\Sigma}_{\mathrm{y_i y_i}})$ to be a proper uncertain geometric entity, as in definition 10 on page 104. Since in our model $\boldsymbol{\Sigma}_{\mathrm{y_i y_i}} = \boldsymbol{\Sigma}_{\mathrm{e_i e_i}}$, definition 10 applies to the pair $(\mathbf{y},\ \boldsymbol{\Sigma}_{\mathrm{e_i e_i}})$.

The covariance matrix of the noise vector $\mathbf{e}$ can be decomposed in two factors $\boldsymbol{\Sigma}_{\mathrm{e_i e_i}} = \sigma^2 \mathsf{Q}_{e_i e_i}$, cf. KOCH 1999, section 3.2.1: a positive definite symmetric matrix $\mathsf{Q}_{e_i e_i}$ and a variance factor σ^2. $\mathsf{Q}_{e_i e_i}$ can be interpreted as the initial covariance matrix $\boldsymbol{\Sigma}^{(0)}_{\mathrm{y_i y_i}}$, the inverse $\mathsf{Q}^{-1}_{e_i e_i}$ can be interpreted as a weighting matrix, which might be given in advance by the user. The variance factor on the other hand is an unknown variable, which can be estimated. If for example $\mathsf{Q}_{e_i e_i}$ is the covariance matrix which was estimated previously, the result of $\widehat{\sigma}^2$ indicates whether the a-priori uncertainties of the observations were either too pessimistic ($\widehat{\sigma} < 1$) or too optimistic ($\widehat{\sigma} > 1$).

(b) *Implicit Model for the Observation Process*
The geometric relations $R(\tilde{\mathbf{y}}_{(R,i)}, \tilde{\boldsymbol{\beta}})$ express the connection between observations and unknowns, where $R(\cdot,\cdot)$ stands for incidence, identity, parallelity and orthogonality relation (symbols: $\in$, $\equiv$, $\|$ and $\perp$). An algebraic expression for such a set of relations was developed in equation (3.84) on page 83. We now analyse this expression depending on the type of the relations $R(\tilde{\mathbf{y}}_{(R,i)}, \tilde{\boldsymbol{\beta}})$: in general, we may have $n+m+p+q$ relations with

$$\begin{aligned}
\tilde{\mathbf{y}}_{(\in,h)} \cap \tilde{\boldsymbol{\beta}} &\neq \emptyset & h &= 1, \cdots, n \\
\tilde{\mathbf{y}}_{(\equiv,j)} &\equiv \tilde{\boldsymbol{\beta}} & j &= 1, \cdots, m \\
\tilde{\mathbf{y}}_{(\|,k)} &\parallel \tilde{\boldsymbol{\beta}} & k &= 1, \cdots, p \\
\tilde{\mathbf{y}}_{(\perp,l)} &\perp \tilde{\boldsymbol{\beta}} & l &= 1, \cdots, q
\end{aligned}$$

for 3D, the possible relations are again summarized in table 4.4.

Note that $\tilde{\mathbf{y}}_{(R,i)}$ is only a placeholder for a set of different entities: the type of $\tilde{\mathbf{y}}_{(R,i)}$ depend on the unknown entity: e.g. points are not parallel or orthogonal to other entities, but lines an planes may be. Another example

Table 4.4. 3D Relations and algebraic expressions between an unknown $\boldsymbol{\beta}$ and observations $\mathbf{y}_i$. Both $\boldsymbol{\beta}$ and $\mathbf{y}_i$ may be points, lines or planes. See also table 3.5.

3D										
obs.→		$\mathbf{Y}_i$			$\mathbf{M}_i$			$\mathbf{B}_i$		
↓ unk.		R	$\mathsf{U}(\boldsymbol{\beta})$	$\mathsf{V}(\mathbf{Y}_i)$	R	$\mathsf{U}(\boldsymbol{\beta})$	$\mathsf{V}(\mathbf{M}_i)$	R	$\mathsf{U}(\boldsymbol{\beta})$	$\mathsf{V}(\mathbf{B}_i)$
$\boldsymbol{\beta}$	$\mathbf{X}$	$\equiv$	$\mathsf{\Pi}(\mathbf{X})$	$-\mathsf{\Pi}(\mathbf{Y}_i)$	$\in$	$\overline{\overline{\mathsf{\Pi}}}^{\mathsf{T}}(\mathbf{X})$	$\overline{\mathsf{\Gamma}}^{\mathsf{T}}(\mathbf{M}_i)$	$\in$	$\mathbf{X}^{\mathsf{T}}$	$\mathbf{B}_i^{\mathsf{T}}$
	$\mathbf{L}$	$\in$	$\overline{\mathsf{\Gamma}}(\mathbf{L})$	$\overline{\overline{\mathsf{\Pi}}}^{\mathsf{T}}(\mathbf{Y}_i)$	$\in$	$\overline{\mathbf{L}}$	$\overline{\mathbf{M}}_i$	$\in$	$\mathsf{\Gamma}(\mathbf{L})$	$\mathsf{\Pi}(\mathbf{B}_i)$
					$\equiv$	$\Delta_k(\mathbf{L})$	$\Delta_k(\mathbf{M}_i)$			
					$\parallel$	$\mathcal{S}(\boldsymbol{L}_h)$	$-\mathcal{S}((\boldsymbol{M}_i)_h)$	$\parallel$	$\boldsymbol{L}_h^{\mathsf{T}}$	$(\boldsymbol{B}_i)_h^{\mathsf{T}}$
					$\perp$	$\boldsymbol{L}_h^{\mathsf{T}}$	$(\boldsymbol{M}_i)_h^{\mathsf{T}}$	$\perp$	$\mathcal{S}(\boldsymbol{L}_h)$	$-\mathcal{S}((\boldsymbol{B}_i)_h)$
	$\mathbf{A}$	$\in$	$\mathbf{A}^{\mathsf{T}}$	$\mathbf{Y}_i^{\mathsf{T}}$	$\in$	$\mathsf{\Pi}(\mathbf{A})$	$\mathsf{\Gamma}(\mathbf{M}_i)$	$\equiv$	$\overline{\overline{\mathsf{\Pi}}}(\mathbf{A})$	$\overline{\overline{\mathsf{\Pi}}}(\mathbf{B}_i)$
					$\parallel$	$\boldsymbol{A}_h^{\mathsf{T}}$	$(\boldsymbol{M}_i)_h^{\mathsf{T}}$	$\parallel$	$\mathcal{S}(\boldsymbol{A}_h)$	$-\mathcal{S}((\boldsymbol{B}_i)_h)$
					$\perp$	$\mathcal{S}(\boldsymbol{L}_h)$	$-\mathcal{S}((\boldsymbol{B}_i)_h)$	$\perp$	$\boldsymbol{L}_h^{\mathsf{T}}$	$(\boldsymbol{B}_i)_h^{\mathsf{T}}$

is the incidence relation, where an unknown point $\mathbf{X}$ can be related to observed lines $\mathbf{M}_{\in,h}$ and planes $\mathbf{B}_{\in,h}$; but an unknown line $\mathbf{L}$ may have incident points $\mathbf{Y}_{\in,h}$, lines $\mathbf{M}_{\in,h}$ and planes $\mathbf{B}_{\in,h}$. In general, $n = n_Y + n_M + n_B$ where n_Y denotes the number of incident points, n_M the number of incident lines and n_B the number of incident planes. Likewise, $m = m_Y + m_M + m_B$, $p = p_M + p_B$ and $q = q_M + q_B$.

Practically, for unknown points $\mathbf{X}$ we can have up to three different blocks of observations: $\mathbf{X} \equiv \mathbf{Y}_j$, $\mathbf{X} \in \mathbf{M}_i$ and $\mathbf{X} \in \mathbf{B}_i$. Likewise for unknown lines $\mathbf{L}$ there exists up to 8 blocks and for unknown planes $\mathbf{A}$ up to 7 blocks of observations, see also table 4.4.

We can reformulate the relationships to bilinear expressions, cf. section 3.3 on page 70 and obtain and implicit algebraic model for the observation process

$$\mathbf{w}_{R,i}(\tilde{\mathbf{y}}_{(R,i)}, \tilde{\boldsymbol{\beta}}) = \mathbf{0} \quad \Leftrightarrow \quad \mathsf{U}(\tilde{\mathbf{y}}_{(R,i)})\tilde{\boldsymbol{\beta}} = \mathsf{V}(\tilde{\boldsymbol{\beta}})\tilde{\mathbf{y}}_{(R,i)} = \mathbf{0} \tag{4.66}$$

where the matrices U and V are determined by the relation R, the type of the observations $\tilde{\mathbf{y}}_{(R,i)}$ and of the unknowns β, as defined in table 3.5.

We can simplify the observation model equation in (4.66) by replacing $\mathsf{U}(\tilde{\mathbf{y}}_{(R,i)})$ and $\mathsf{V}(\tilde{\boldsymbol{\beta}})$ with the reduced construction matrices $\mathsf{U}^{[r_i]}(\tilde{\mathbf{y}}_{(R,i)})$ and $\mathsf{V}^{[r_i]}(\tilde{\boldsymbol{\beta}})$, where r_i is the degree of freedom of relation $R(\mathbf{y}_i, \boldsymbol{\beta})$. The reduction can be done by the algorithm 3.1 on page 69, where the first step of coordinate transformation can safely be skipped as we assume that all relations are approximately true. The resulting reduced model

$$\mathbf{w}'_{R,i}(\tilde{\mathbf{y}}_{(R,i)}, \tilde{\boldsymbol{\beta}}) = \mathbf{0} \quad \Leftrightarrow \quad \mathsf{U}^{[r_i]}(\tilde{\mathbf{y}}_{(R,i)})\tilde{\boldsymbol{\beta}} = \mathsf{V}^{[r_i]}(\tilde{\boldsymbol{\beta}})\tilde{\mathbf{y}}_{(R,i)} = \mathbf{0} \tag{4.67}$$

is equivalent to (4.66) , cf. section 3.3 The redundancy of the model can be calculated using table 3.7, as previously for non-statistical case on page 83.

(c) *Model of Constraints on Unknowns and Observations*
The observations and the unknown have to fulfill the homogeneous constraints and possibly the Plücker condition, We introduce two functions $\mathbf{h}(\tilde{\boldsymbol{\beta}})$ and $\mathbf{g}(\tilde{\mathbf{y}})$ with their Jacobians H and $\mathsf{G}_{R,i}$ such that

$$\mathbf{h}(\tilde{\boldsymbol{\beta}}) = \mathbf{0} \Leftrightarrow \begin{cases} |\tilde{\boldsymbol{\beta}}|^2 = \tilde{\boldsymbol{\beta}}^\mathsf{T}\tilde{\boldsymbol{\beta}} = \mathsf{H}\,\tilde{\boldsymbol{\beta}} = 1 & \text{for non-3D lines,} \\ \begin{pmatrix} \tilde{\boldsymbol{\beta}}^\mathsf{T} \\ \frac{1}{2}(\mathsf{C}\tilde{\boldsymbol{\beta}})^\mathsf{T} \end{pmatrix} \tilde{\boldsymbol{\beta}} = \mathsf{H}\,\tilde{\boldsymbol{\beta}} = \begin{pmatrix} 1 \\ 0 \end{pmatrix} & \text{for 3D lines} \end{cases} \tag{4.68}$$

$$\mathbf{g}(\tilde{\mathbf{y}}_{(R,i)}) = \mathbf{0} \Leftrightarrow \begin{cases} |\tilde{\mathbf{y}}_{(R,i)}|^2 = \tilde{\mathbf{y}}^\mathsf{T}_{(R,i)}\tilde{\mathbf{y}}_{(R,i)} = \mathsf{G}_{(R,i)}\,\tilde{\mathbf{y}}_{(R,i)} = 1 \\ \qquad \text{for non-3D lines,} \\ \begin{pmatrix} \tilde{\mathbf{y}}^\mathsf{T}_{(R,i)} \\ \frac{1}{2}(\mathsf{C}\tilde{\mathbf{y}}_{(R,i)})^\mathsf{T} \end{pmatrix} \tilde{\mathbf{y}}_{(R,i)} = \mathsf{G}_{(R,i)}\,\tilde{\mathbf{y}}_{(R,i)} = \begin{pmatrix} 1 \\ 0 \end{pmatrix} \\ \qquad \text{for 3D lines} \end{cases} \tag{4.69}$$

The matrix C denotes the dual operator for 3D lines, see equation (2.31) on page 44. The second condition for 3D lines ensures the Plücker condition $\boldsymbol{L}_h^\mathsf{T}\boldsymbol{L}_O = 0$, cf. equation (2.20) on page 33.

Equations (4.65), (4.66) and (4.68) together form a statistical model for the observation process using the observed entities $\mathbf{y}_{(R,i)}$ and the unknown entity $\tilde{\boldsymbol{\beta}}$. In general such a model is called Gauß-Helmert model, cf. HELMERT 1872, FÖRSTNER 2000. A special case is the Gauß-Markoff-model, where equation (4.66) is replaced by an explicit relation between observations and unknowns: $\tilde{\mathbf{y}} = \mathbf{w}^*(\tilde{\boldsymbol{\beta}})$.

Block Structure of the Model Note that the observation model equation (4.66) is actually a set of equations, defined for each block of observations. We could stack up all observations in one big vector $\mathbf{y}$,

$$\mathbf{y} = (\mathbf{y}_\in,\ \mathbf{y}_\equiv,\ \mathbf{y}_\parallel,\ \mathbf{y}_\perp)^\mathsf{T} \quad \text{with}$$
$$\mathbf{y}_R = \left(\mathbf{Y}_{(R,1)}, \cdots, \mathbf{Y}_{(R,s_Y)}, \mathbf{M}_{(R,1)}, \cdots, \mathbf{M}_{(R,s_M)}, \mathbf{B}_{(R,1)}, \cdots, \mathbf{B}_{(R,s_B)}\right)^\mathsf{T} \tag{4.70}$$

As all observations are assumed to be independent from each other, the corresponding covariance matrix $\boldsymbol{\Sigma}_{\mathrm{yy}}$ would be a block-diagonal matrix. Similarly, the Jacobians for $\mathbf{w}(\mathbf{y}, \boldsymbol{\beta})$ would be of diagonal structure, consisting of the matrices $\mathsf{U}(\tilde{\mathbf{y}}_{(R,i)})$ and $\mathsf{V}(\tilde{\boldsymbol{\beta}})$.

4.6.2 Iterative Estimation

Using the previously derived statistical model, we now discuss the problem to estimate the best fit of the vectors $\widehat{\boldsymbol{\beta}}$, $\widehat{\mathbf{y}}$ and $\hat{\sigma}$ given an observation vector $\mathbf{y}$, a covariance matrix $\boldsymbol{\Sigma}_{\mathrm{yy}}$. We use a linearized iterative method as described in Mikhail and Ackermann 1976, which is covered in more detail in appendix C.2.

We assume that there exists an initial guess for the unknown $\widehat{\boldsymbol{\beta}}^{(0)}$, the observations are given as $\widehat{\mathbf{y}}^{(0)}_{(R,i)} = \widehat{\mathbf{y}}_{(R,i)}$. The initial guess for the unknown can be obtained by using the non-statistical algorithm 3.3 on page 95. The proposed solution is an iterative two-step procedure.

1. we first estimate the vectors using $\mathbf{w}(\tilde{\mathbf{y}}, \tilde{\boldsymbol{\beta}}) = \mathbf{0}$ and $\mathbf{h}(\tilde{\boldsymbol{\beta}}) = \mathbf{0}$ by finding the minimum of

$$\Phi(\tilde{\mathbf{y}}, \boldsymbol{\beta}, \boldsymbol{\lambda}, \boldsymbol{\mu}) = \frac{1}{2}(\mathbf{y} - \tilde{\mathbf{y}})^{\mathsf{T}} \mathsf{Q}_{yy}^{-1} (\mathbf{y} - \tilde{\mathbf{y}}) + \boldsymbol{\lambda}^{\mathsf{T}} \mathbf{w}(\tilde{\mathbf{y}}, \tilde{\boldsymbol{\beta}}) + \boldsymbol{\mu}^{\mathsf{T}} \mathbf{h}(\tilde{\boldsymbol{\beta}}) \rightarrow \min \tag{4.71}$$

   where $\boldsymbol{\lambda}$ and $\boldsymbol{\mu}$ are Lagrangian multipliers. One can solve this minimization problem by a linearization at $\tilde{\mathbf{y}}^{(0)}$ and $\tilde{\boldsymbol{\beta}}^{(0)}$, obtaining the Jacobians $\mathsf{U}(\tilde{\mathbf{y}}^{(0)})$ and $\mathsf{V}(\tilde{\boldsymbol{\beta}}^{(0)})$. A solution of the optimization problem with a complete algorithm can be found in the appendix C.2. If the observations $\mathbf{y}$ are normally distributed, the solution represents a maximum-likelihood estimate. As the problem has a block-structure, the normal equation matrix of the solution is given by

$$\mathsf{N} = \begin{pmatrix} \bar{\mathsf{N}} & \mathsf{H}^{\mathsf{T}} \\ \mathsf{H} & 0 \end{pmatrix}$$
$$\text{with } \bar{\mathsf{N}} = \sum_{R,i} \mathsf{U}^{\mathsf{T}}(\widehat{\mathbf{y}}^{(0)}_{(R,i)}) (\mathsf{V}(\widehat{\boldsymbol{\beta}}) \boldsymbol{\Sigma}_{\mathrm{y}_{(R,i)}\mathrm{y}_{(R,i)}} \mathsf{V}^{\mathsf{T}}(\widehat{\boldsymbol{\beta}}))^{-1} \mathsf{U}(\widehat{\mathbf{y}}^{(0)}_{(R,i)}) \tag{4.72}$$

   After the first step, we obtain an estimate $\widehat{\mathbf{y}}^{(1)}_{(R,i)}$, $\widehat{\boldsymbol{\beta}}^{(1)}$ and $\hat{\sigma}^{(1)}$.

2. in a second step, we make sure, that for each $\widehat{\mathbf{y}}^{(1)} = \widehat{\mathbf{y}}^{(1)}_{(R,i)}$ fulfills the constraint (4.69). This can be done again using an iteration process, where the closest vector $\widehat{\widehat{\mathbf{y}}}^{(1)}$ to the given $\widehat{\mathbf{y}}^{(1)}$ is computed, that fulfills $\mathbf{g}(\widehat{\widehat{\mathbf{y}}}^{(1)}) = 0$. As described in the non-statistical case in section 3.4.3 on page 84, we can approximate the constraint $\mathbf{g}(\widehat{\widehat{\mathbf{y}}}^{(1)}) = 0$ locally and therefore iteratively apply for

$$\widehat{\mathbf{y}}^{(1)} = \mathbf{y} - \boldsymbol{\Sigma}_{\mathrm{yy}}\mathsf{G}^{\mathsf{T}}(\mathsf{G}\boldsymbol{\Sigma}_{\mathrm{yy}}\mathsf{G}^{\mathsf{T}})^{-1}\mathbf{g}(\widehat{\mathbf{y}}^{(0)}) \quad \text{with } \widehat{\mathbf{y}}^{(0)} := \widehat{\mathbf{y}}^{(1)}$$

cf. KANATANI 1996 or MATEI AND MEER 2000. Here, the matrix G is the Jacobian $\frac{\partial \mathbf{g}}{\partial \mathbf{y}}$ computed at the position $\widehat{\mathbf{y}}^{(1)}$, cf. equation (4.69). It is sufficient to use only a couple of iterations, as we expect that $\widehat{\mathbf{y}}^{(1)}$ *almost* fulfills the constraint.

Note that for the implementation, we recommend to use the actual Jacobians derived in the last chapter instead of a numerical differentiation: not only is this solutions probably more aesthetical, we also expect a better convergence and stability compared to obtaining the Jacobians numerically. A complete algorithm can be found in table 4.6.

4.6.2.1 Estimating Projective Transformations So far we have focused on unknown geometric entities. Now the task is to estimate an unknown projective transformation denoted by $\tilde{\mathsf{M}}$ given a set of observed corresponding entity pairs $(\mathbf{x}_i, \mathbf{x}'_i)$ with given covariance matrices. Fortunately, the above estimation method needs only small adjustments to solve this problem: as described in section 3.5 on page 85, the corresponding entities and the unknown transformation form a triple relationship $R(\mathbf{x}'_i; \mathsf{M}, \mathbf{x}_i)$, for which we know trilinear algebraic expressions, cf. tables 3.8 and 3.9, thus we have explicit forms for the Jacobians of the relation.

To use the iterative estimation scheme, we only have to turn the observed pairs of entities $(\mathbf{x}_i, \mathbf{x}'_i)$ into one observation vector $\mathbf{y}_i^{\mathsf{T}} = (\mathbf{x}_i^{\mathsf{T}}, \mathbf{x}_i'^{\mathsf{T}})$. The covariance matrix and Jacobians for $\mathbf{w}(\tilde{\mathbf{y}}_i, \tilde{\boldsymbol{\beta}})$ and $\mathbf{g}(\tilde{\mathbf{y}}_i)$ have to be combined accordingly, e.g. for the observation equations we get with $\mathbf{m} = \mathrm{vec}(\mathsf{M})$

$$\mathsf{V}_i(\mathbf{m}) = \frac{\partial \mathbf{w}(\mathbf{y}_i, \mathbf{m})}{\partial \mathbf{y}} = \frac{1}{2}\begin{pmatrix} \frac{\partial \mathbf{w}(\mathbf{y}_i, \mathbf{m})}{\partial \mathbf{x}_i} & 0 \\ 0 & \frac{\partial \mathbf{w}(\mathbf{y}_i, \mathbf{m})}{\partial \mathbf{x}'_i} \end{pmatrix}$$

$$= \frac{1}{2}\begin{pmatrix} \mathsf{V}_i(\mathbf{x}'_i, \mathbf{m}) & 0 \\ 0 & \mathsf{V}_i(\mathbf{x}_i, \mathbf{m}) \end{pmatrix}$$

With these preparations, the algorithm in table 4.6 can be applied for geometric transformations, too. One only has to take into account that the reduction $\mathsf{V}_i^{[r]}(\mathbf{m})$ of the matrices $\mathsf{V}_i(\mathbf{m})$ have to be done separately for the underlying sub matrices.

4.6.3 Generic Algorithm for Optimal Geometric Estimation

The complete algorithm see table 4.6. It can handle both geometric entities such as points, lines and planes in 3D and projective transformations, such as a projective camera or homographies in 2D and 3D. Applications of this algorithm are manifold, see section 3.4 for some examples, that can be optimally solved using the algorithm in table 4.6.

Statistical optimization of geometric entities
Objective: find an optimal estimation for an unknown geometric entity $\left(\tilde{\beta},\ \boldsymbol{\Sigma}_{\tilde{\beta}\tilde{\beta}}\right)$ using a set of observed entities $\left(\mathbf{y}_{(R,i)},\ \boldsymbol{\Sigma}_{\mathbf{y}(\mathrm{R,i})\mathbf{y}(\mathrm{R,i})}\right)$, that are supposed to be related to the unknown entity by the geometric relations $R(\mathbf{y}_{(R,i)}, \tilde{\beta})$. Additionally, compute the fitted observations $\widehat{\mathbf{y}}_{(R,i)}$ and the estimated variance factor $\hat{\sigma}$.

If the unknown $\tilde{\beta}$ is a transformation, then the observed entities are corresponding pairs $(\mathbf{y}_{(R,i)}, \mathbf{y}'_{(R,i)})$.

1. If necessary, condition the observations $\mathbf{y}_{(R,i)}$ with a conditioning factor f as determined in algorithm 2.1 using $f_{min} = 0.1$. For transformations, obtain two different factors f, f'. Scale the Euclidean parts of the covariance matrices in the same way.
2. Change the nullspaces of $\boldsymbol{\Sigma}_{\mathbf{y}(\mathrm{R,i})\mathbf{y}(\mathrm{R,i})}$ to $\mathbf{y}_{(R,i)}$ using the orthogonal projection $\mathcal{P}_{\mathbf{y}_{(R,i)}}$, cf. (4.18) on page 111.
3. Group the observations in b blocks depending on the type of relation R and type of the observation vector $\mathbf{y}_{(R,i)}$.

   For each block $i = 1, \cdots, b$, define a function $\mathsf{A}_i(\mathbf{y}^*, \beta^*)$, $\mathsf{B}_i(\mathbf{y}^*, \beta^*)$ that return the Jacobians $\partial\mathbf{w}(\mathbf{y}, \beta)/\partial\beta$ at a given position $(\mathbf{y}^*, \beta^*)$:
   a) Obtain Jacobians $\mathsf{U}(\mathbf{y}^*)$ and $\mathsf{V}(\beta^*)$ according to table 3.5
   b) Select those r rows from $\mathsf{U}(\mathbf{y}^*)$ and $\mathsf{V}(\beta^*)$ which have the largest $L1$-norm with respect to one of the matrices, see the second step in algorithm 3.1 on page 69. One obtains the reduced matrices $\mathsf{U}^{[r]}(\mathbf{x}')$ and $\mathsf{V}^{[r]}(\mathbf{y}')$.

   Similarly, for each block i, define a function $\mathsf{G}(\mathbf{y}^*)$, that returns the Jacobian of $\mathbf{g}(\mathbf{y}^*)$ as defined in equation (4.69).

   If the unknown is a transformation, the observation vector consists of stacked pairs of homologous vectors, cf. section 4.6.2.1.

   Lastly, define a function for $\mathsf{H}(\beta^*)$, cf. (4.68).
4. Compute an approximate solution for $\widehat{\beta}^{(0)}$ using the direct algorithm 3.3.
5. Compute the iterative solution of Gauß-Helmert model for block structures as outlined in the algorithm in table C.3 on page 195.
6. If necessary, recondition the estimated $\left(\widehat{\beta},\ \widehat{\boldsymbol{\Sigma}}_{\widehat{\beta},\widehat{\beta}}\right)$ and the fitted observations by $1/f$; for a transformation $\widehat{\beta}$ use equation (2.25) with f, f'.

Algorithm 4.6: Algorithm for optimally estimating geometric entities using weighted least squares.

Alternative Solutions for Iterative Minimization As described in FÖRSTNER 2001A it is possible to extend the eigenvector solution described in section 3.4 to take the uncertainty of the observations in to account. The solution is similar to LEEDAN AND MEER 2000 and MATEI AND MEER 2000 and exploits the fact, that the problem is bilinear in each pair of unknowns and observations. One can minimize

$$\Omega = \mathbf{w}^\mathsf{T} \boldsymbol{\Sigma}_{\mathbf{ww}} \mathbf{w}$$
$$= \boldsymbol{\beta}^\mathsf{T} \left(\sum_{(R,i)} \mathsf{U}^\mathsf{T}(\widehat{\mathbf{y}}^{(0)}_{(R,i)})(\mathsf{V}(\widehat{\boldsymbol{\beta}})\boldsymbol{\Sigma}_{\mathbf{y}(\mathrm{R},\mathrm{i})\mathbf{y}(\mathrm{R},\mathrm{i})}\mathsf{V}^\mathsf{T}(\widehat{\boldsymbol{\beta}}))^{-1}\mathsf{U}(\widehat{\mathbf{y}}^{(0)}_{(R,i)}) \right) \boldsymbol{\beta} \rightarrow \min$$

under the constraint $\boldsymbol{\beta}^\mathsf{T}\boldsymbol{\beta} = 1$. An additional constraint such as the Plücker condition has to be imposed to the unknown in a second step. Likewise, any constraints on the observations have to be imposed manually. The advantage of this estimation is that it is supposed to converge faster than other iterative methods, Matei and Meer 2000.

Another method for iterative estimation is the commonly used Levenberg-Marquardt iteration, for example described in Press *et al.* 1992. Compared to the previous methods it adds a regularization to avoid local minima. The reason for using an iterative Gauß-Helmert model is not only the fact, that it is a classical geodesic estimation method. It also fits nicely into the geometric framework, since relations between entities are algebraically expressed by implicit equations.

4.6.4 Example: Estimating Entities of a Cube

We now test the estimation algorithm using an estimation of cube entities by a large set of different observations. For example, assume that a 3D line $\mathbf{L}$ representing one edge of the cube is unknown; the observed entities are: (1) incident points $\mathbf{X}_i^{\in}$, (2) incident lines $\mathbf{L}_i^{\cap}$, (3) collinear (i.e. equal) lines $\mathbf{L}_i^{\equiv}$, (4) parallel lines $\mathbf{L}_i^{\parallel}$, (5) orthogonal lines $\mathbf{L}_i^{\perp}$, (6) incident planes $\mathbf{A}_i^{\ni}$, (7) parallel planes $\mathbf{A}_i^{\parallel}$ and (8) orthogonal planes $\mathbf{A}_i^{\perp}$. All observed entities are derived by the other 9 edges, 8 corners and 6 surfaces of the cube. We obtain the implicit observation model $\mathbf{w}(\mathbf{L}; \mathbf{X}_i^{\in}, \mathbf{L}_i^{\cap}, \mathbf{L}_i^{\equiv}, \mathbf{L}_i^{\parallel}, \mathbf{L}_i^{\perp} \mathbf{A}_i^{\ni}, \mathbf{A}_i^{\parallel}, \mathbf{A}_i^{\perp}) = \mathbf{0}$ and the Jacobians listed in table 3.5. In the same manner we can also construct a point representing a cube corner or a plane representing a cube surface.

For testing purposes we used artificial data to validate the estimation method with the ground-truth. We estimated a 3D point, line and plane being part of translated and rotated cubes of size 1. For each of the 8 corners, 12 edges and 6 surfaces of the cube, 5 observations have been generated randomly with a Gaussian error of 0.1% and 1%. When taking into account all possible constraints, the redundancy of the estimations are 57 for the unknown point, 176 for the unknown line and 162 for the unknown plane. We drew 1000 sample cubes and then estimated 10000 points, lines and planes. In fig. 4.17, the first row shows the cumulative histogram for the estimated $\widehat{\sigma}^2$ for the points $\widehat{\mathbf{X}}$, $\widehat{\mathbf{Y}}$ and planes $\widehat{\mathbf{A}}$, over all estimates we get an average of $\overline{\widehat{\sigma}}^2 = 0.989$. The second row depicts the scatter diagram for the estimated point $\widehat{\mathbf{Y}}$, projected onto the three principal planes. The 10000 covariance matrices obtained from the estimation have been averaged and its confidence region has been plotted against the confidence region of the sample covariance matrix of the results. The estimated covariances are too optimistic compared to the sample covariances, which is conform with the theoretical explorations of the bias in section 4.3.

For the given datasets, all estimations converged within 3 to 4 iterations. The stopping criteria for the iteration was defined as follows: the corrections to the values of $\boldsymbol{\beta}$ should be less than 1% with respect to its standard deviation. The algorithm was implemented in the scripting language Perl and takes about 1 [sec] for 50 point resp. plane observations with a factor 1.5 slower for lines.

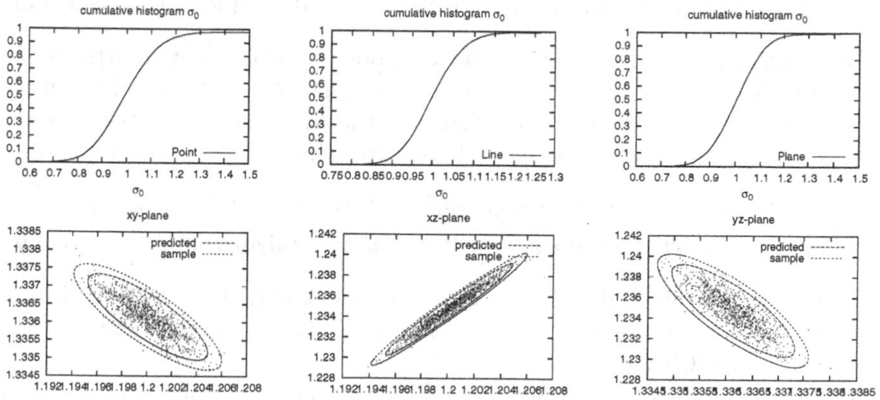

Fig. 4.17. Results of Monte Carlo test for point estimations. First row: cumulative histograms of the estimated variance factor $\widehat{\sigma}^2$ for the estimated point, line and plane. Second row: scatter diagram of the estimated point $\widehat{\mathbf{Y}}$ projected on the $xy-, xz-, yz-$planes. Only 1000 out of the 10000 estimated points are shown. The inner confidence ellipse is averaged over all 10000 estimated ellipses, the outer one is the sample confidence ellipse based on all estimated points. The empirical confidence ellipse is larger by a factor of about 1.2; this validates the theoretical result that the estimated covariance matrix is too optimistic compared with the true one, cf. section 4.3.

4.7 SUGR: a Library for Statistical Uncertain Geometric Reasoning

As a final result of this chapter, we compile the introduced algorithms in a library called called SUGR, standing for *Statistically Uncertain Geometric Reasoning*. The library is capable of

- Representing, combining and estimating uncertain points, lines and planes in 2D and 3D
- Establishing geometric relationships between the entities
- Representing and estimating projective transformations including projective camera transformations.

For the representation of the entities, we choose the approximate representation, defined in definition 10 on page 104.

Only the following algorithms are necessary to accomplish the mentioned tasks:

- *Join and intersection with uncertainty* (algorithm 4.1 on page 123)
- *Statistically testing uncertain geometric relations (Version 1)* (algorithm 4.4 on page 136)
- *Statistical optimization of geometric entities* (algorithm 4.6 on page 145)

For testing uncertain geometric relations, both versions using matrix reduction or SVD can be used (algorithms 4.4, 4.4). We recommend to implement version 1 with matrix reduction, see the discussion in section 4.5.3 on page 135. To implement the algorithms, two helper routines are required:

- *Conditioning a set of geometric entities* (algorithm 2.1 on page 40)
- *Consistent Reduction of construction matrices* (algorithm 3.1 on page 69)

At the time of this writing, two implementations exist for `SUGR`, one in Perl and one in Java, both can be found at `http://www.ipb.uni-bonn.de/ipb/projects/SUGR/index.html`.

In this chapter, we have shown only some artificial examples using `SUGR`. A practical application is demonstrated in the next chapter in the context of Polyhedral Object Reconstruction.

5 Polyhedral Object Reconstruction

As mentioned in the first chapter, the automated reconstruction of polyhedral objects from multiple images is an on-going and challenging task within the field of Computer Vision: from a given set of different images of a polyhedral objects and some orientation information about the camera positions, the aim is to optimally reconstruct the shape of the polyhedral object. Different approaches exist, see section 1.5.3, combining various cues from geometrical, radiometrical and topological observations and possibly exploiting pre-defined knowledge about the object.

The goal of this chapter is not to propose another automatic approach to the general reconstruction problem, we rather demonstrate the capabilities of the SUGR procedures for statistical geometric reasoning proposed in the last chapter, see 4.7. To accomplish this, we almost exclusively exploit geometric cues and don't use any other knowledge about the object other than its polyhedral shape. It is not intended to promote the idea that solely relying on geometric cues will solve the reconstruction problem. But we want to demonstrate the power of the given geometric constraints of the scene combined with a statistical framework. For a successful reconstruction system, it is recommended to integrate other cues e.g. coming from radiometry or topology. An appropriate approach for the integration of cues could be the application of Bayesian networks, cf. Pearl 1988, Sarkar and Boyer 1994.

To show the capabilities of the SUGR procedures we apply them to reconstruction systems with varying degrees of automation, cf. section 1.3.3 on page 7. In particular, the two main goals of using statistical geometric reasoning for object reconstructions in this work are:

- pushing the automation limits of user-assisted systems,
- enhancing the performance characterization of automated methods

In the beginning of this chapter, we describe the principle workflow of polyhedral object extraction and summarize preliminary steps such as obtaining camera orientation parameters and automatic feature extraction from images. As an example for user-assisted systems we describe two commercial modeling systems and demonstrate how to include the proposed techniques of statistical geometric reasoning. To completely eliminate the amount of user interaction for object reconstruction, a method for automatically establish-

S. Heuel: Uncertain Projective Geometry, LNCS 3008, pp. 149-172, 2004.

ing corresponding 2D line segments is described, which is completely based on geometric cues. As in all automated approaches, there are some deficiencies in this method; the final part of this chapter shows ways to overcome these deficiencies by including a minimal amount of user interaction. Combining techniques of automated systems with minimal and simplified user interaction are steps on the way from exhausting manual user input toward semi-automatic or even automatic systems as outlined in the first chapter, see table 1.1 on page 8.

5.1 Principle Workflow

We assume the following principle workflow for a polyhedral object extraction system, cf. figure 5.1: given a set of multiple images, one first has to determine the camera orientation and possibly automatically extract geometric features such as points or line segments from the images. These preliminary steps will be covered within the remainder of this section. After these preliminary steps, one could invoke a user-assisted system to manually extract the geometric information that is necessary to reconstruct the shape of the pictured polyhedra, see section 5.2. Alternatively, an automatic system may replace the manual work of the user, see section 5.3. One may also use automated techniques in user-assisted systems, cf. section 5.4. In all approaches one can make use of the SUGR procedures.

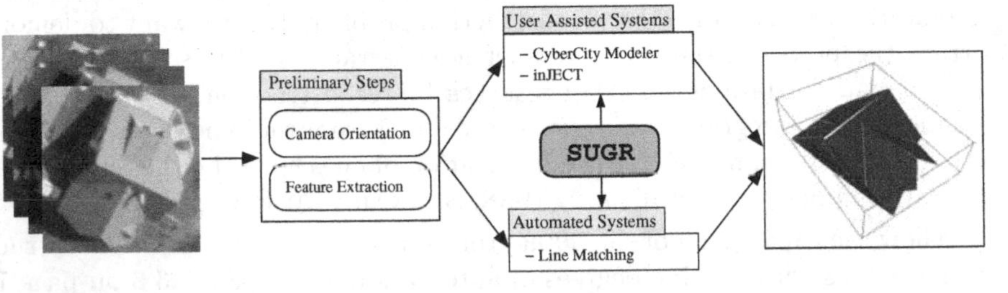

Fig. 5.1. Principle workflow of polyhedral object extraction: from a set of images, one first has to determine the camera orientation parameters of each image (sec. 5.1.2). Additionally one extracts geometric features from the images (sec. 5.1.1), such as points and line segments. Then either a user-assisted system (sec. 5.2) or an automatic system (sec. 5.3) is used to compute a 3D description of the scene. One may also combine interactive and automatic methods (sec. 5.4). Both automatic and interactive approaches can make use of the SUGR procedures.

5.1.1 Feature Extraction

The first step from digital images to a 3D object is the segmentation of the images, i.e. the extraction of image features that are relevant to the subsequent tasks. For the extraction of polyhedral objects, we are especially interested in points, straight line segments and eventually image regions that correspond to polyhedral surfaces.

The extraction of geometric features in the images can be done manually by simply clicking on a point for point measurements. In case of line measurements on may click on two points on the image, which are supposed to be on the image line. Both points and line measurements require the user to be quite precise in her clicks with respect to the orthogonal distance of the line.

Instead of manual extraction one can also apply image segmentation algorithms to automatically extract 2D points and line segments. The automatically extracted features are not necessarily corresponding to 3D object features, but are solely based on the observed image intensities.

Note that intensities are measurements of the physical real world and consequently are only precise up to some degree. Obviously the automatically extracted geometric features are also uncertain and their precision depend on the measurement of the intensities. Therefore the segmentation algorithm should not only deliver the geometric entities but additionally some information on how precise these entities are to be expected.

In this work we used the feature extraction system `FEX` for image segmentation (FÖRSTNER 1994, FUCHS 1998). `FEX` simultaneously extracts a set of points, line segments and homogeneous regions from one or multi-channel digital images. The extraction process of this system can be divided in three steps, cf. figure 5.2: first, the image is completely subdivided into non-intersection regions, in which either points, line segments or regions are supposed to be extracted (*ternary image*). To do this, each pixel is labelled as one of the three feature types according to the gradient of the image intensities in its neighborhood. To decide on the feature type a hypotheses testing is invoked which is based on a homogeneity or a shape criterion of the image intensities. For reliable decisions one needs to estimate the noise of the image intensities (BRÜGELMANN AND FÖRSTNER 1992).

In a second step, the geometric features are precisely localized within each of the regions — again using the intensity gradients within a specified area. Finally, in a third step, the neighborhood (or topological) relationships between the extracted features is determined using a Voronoi-Diagram. An overview of the segmentation process is found in figure 5.2.

We assume that the feature extraction yields realistic estimates for the uncertainty of the position of the points and line segments. Otherwise one has to calibrate the variance factors, for example based on the residuals of a bundle adjustment.

5.1.2 Acquiring the Camera Parameters

In the last chapter we have outlined an algorithm for the computation of the camera parameters of one image. The determination of camera parameters

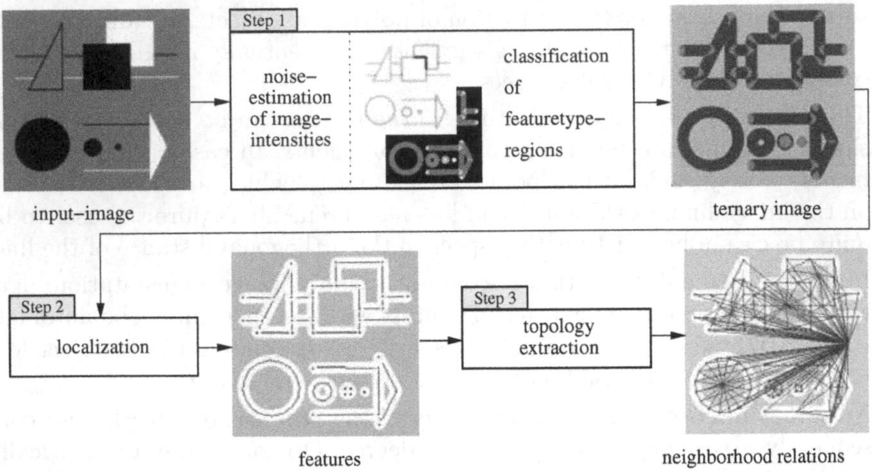

Fig. 5.2. Overview of the segmentation process of FEX, divided in three steps: 1. segmentation of the image in three feature types: points, line-segments and blob regions; 2. localization of the features within their region; 3. extraction of the neighborhood relations between the localized features.

for a set of images is a central and well-researched topic in Photogrammetry (KRAUS 1997) as well as in Computer Vision (HARTLEY AND ZISSERMAN 2000). A large number of different approaches for the computation of camera parameters exist for varying situations, but the classical approach of bundle adjustment – the joint optimization of camera parameters and observed 3D entities – is the most general and commonly recommended solution, for a review see TRIGGS *et al.* 2000. In classical photogrammetric tasks, bundle adjustment is formulated as nonlinear least-squares problem, which is especially suitable for problems with no gross outliers. In case outliers exists one may pre-label them prior to the optimization or use different cost functions, see HUBER 1981, KLEIN AND FÖRSTNER 1984, TRIGGS *et al.* 2000.

One does not necessarily have to compute the full projective camera matrix in order to compute the inherent geometric structure of a scene: FAUGERAS 1992 and HARTLEY *et al.* 1992 showed that one can use the fundamental matrix only to obtain a scene reconstruction up to an unknown homography H. Simply stated, one chooses an arbitrary projective camera matrix for the first camera. Then the camera parameters for the other images are completely determined by the given fundamental matrices. One can now do any geometric reasoning with the assumed projection matrices – but without using affine or Euclidean relations. At the very end one can find the homography H to obtain the real Euclidean projective camera matrices ("stratifying the reconstruction"); for example, this can be done using control points.

In all cases, we assume a camera, which preserves straight lines. If this assumption is violated, one can apply a rectification prior to the image analysis, for example as described in ABRAHAM AND FÖRSTNER 1997.

5.2 Enhancing User Assisted Reconstruction Systems

As a first application, we demonstrate how to use statistical geometric reasoning for interactive reconstruction of generic polyhedra. We start with two existing systems for user-assisted reconstruction of man-made objects and describe how they may benefit from the techniques for uncertain geometric reasoning. We will then explicitly show the usefulness of SUGR for two specific tasks: the reconstruction of building edges and the automatic grouping of these edges to surface patches.

5.2.1 Existing User Assisted Systems

There exists a number of different user assisted reconstruction systems for man-made structures, see BALTSAVIAS *et al.* 2001 for a recent overview. We choose two existing systems, especially suited for building reconstruction: CYBERCITY MODELER and INJECT, which both started as research projects and are now commercialized, see CYBERCITY 2002 and INPHO 2002. The system CYBERCITY MODELER, introduced by GRÜN AND WANG 1999, is embedded in a classical photogrammetric environment and contains a method for fitting planar faces given a set of well-defined point-clouds. The acquisition of the 3D data only involves the measurement of points: a single object, which is identified by the user, is measured by first clicking the boundary points of the object in each image. then the interior points are specified in arbitrary sequence. After the user completed the acquisition, the software starts a relaxation process to identify planar surfaces from the given set of labelled points. In a final step a global least-squares adjustment is performed in order to obtain an optimal estimate of the reconstructed object-

The INJECT system GÜLCH *et al.* 2000, uses a different measurement technique: first the user chooses a building primitive from an extensible set of pre-generated primitives, such as saddle-roof buildings or simple boxes. Then, the position and shape of the chosen primitive is not measured by point clicks, but rather by adjusting a projected version of the 3D-primitive in the image with a mouse. A set of building primitives are then combined to a complex building by following a set of CSG (Constructive Solid Geometry) operations (ENCARNACAO *et al.* 1997).

Both systems could benefit from the developed procedures for uncertain geometric reasoning:

- *Integration of points, lines and planes.* One may enhance the computation of the optimal geometric shape of a building by simultaneously using points and line measurements in the images, contrary to only use point measurements: as explicated in section 3.1.2, the transfer between image and object space can be done simultaneously for points and lines by using the backward and forward projection with the matrices P and Q, see table 3.2 on page 61. Additionally in 3D, all possible constructions involving any number of related geometric entity can be implemented in a statistically optimal manner, cf. section 3.4 and 4.6.

- *Uncertainty analysis.* Throughout the system process it is important to check whether the accuracy specifications are fulfilled for both intermediate and final results. Using the SUGR procedures, it is easy to obtain information about the uncertainty of reconstructed objects: assume the measurement errors in the images and the precision of the camera parameters are known, then one can propagate this information directly to the reconstructed entities, such as corner points or building edges. The obtained covariance matrices can be analyzed and compared with the expected error; one may show the user the quality of the obtained reconstructions by traffic-light indicators (PREGIBON 1986, FÖRSTNER 1995).
- *Ease of extensibility*: for the development of object extraction system it is important that it is easy to continuously create extensions to the existing system. Due to the consistency of the representation of points, lines, planes and their transformation from chapter 3 it is easy to derive new methods and still use the same framework. For example, a geometric regularization of a geometric shape may be done under coplanarity, parallelity or orthogonality constraints, cf. GRÜN AND WANG 2001. These constraints can be selectively imposed on points, lines and planes using simple bilinear expressions for the corresponding relations, cf. table 3.5.

In the following we will demonstrate the above mentioned benefits of SUGR with two examples: the manual reconstruction of building edges and the automatic grouping of these edges to polyhedral surface patches. Later on in section 5.4, an enhancement using automation techniques is proposed.

It is possible to add the enhancements to existing systems in order to acquire additional polyhedral structure to the already generated reconstructions. This is especially interesting for the INJECT system: its simple user interaction of measuring primitives instead of only points is gained by restrictions of the existing primitive models. But a very complicated polyhedral structure may be hard to model by using a sequence of CSG operations on primitives. In this case it may be simpler to specify the polyhedral surfaces directly and glue these parts to a polyhedral shape, which in turn can be combined with existing primitives. This type of reconstruction is an example of combining specific, semantic models (set of building primitives) with generic models (polyhedron) – a discussion of such a combination can be found in HEUEL AND KOLBE 2001 where two approaches of automatic building reconstruction are compared.

5.2.2 User Assisted Constructions of Building Edges

The first example of enhancing user-assisted systems is the task of reconstruction of 3D building edges and corners. The geometric situation in 3D is depicted in figure 5.3. We will see that both point and line segments in the images can be simultaneously used to obtain optimal reconstruction of geometric primitives. In the following we will concentrate on the reconstructions of edges, as it is the more interesting case.

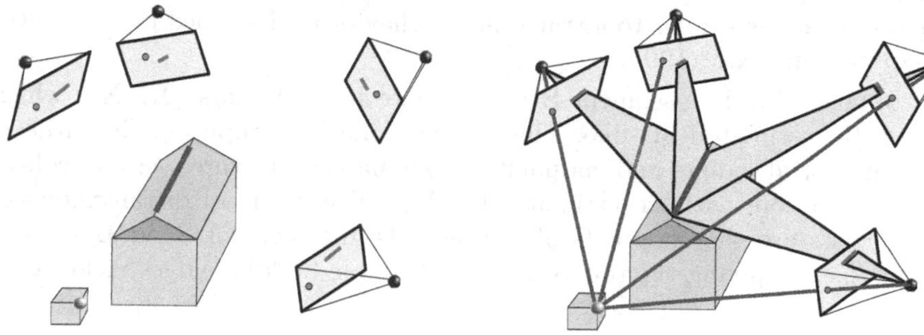

Fig. 5.3. Geometric situation for reconstruction of building corners and edges including observed image points and line segments (*left*) and the same scene with the corresponding viewing rays and planes (*right*).

Assume that the image has been preprocessed by an image segmentation algorithm providing points and line segments, cf. section 5.1.1. The user now specifies a corresponding points and line segments by clicking on the overlaid geometric features, see figure 5.5 One obtains a set $\mathcal{M}$ of corresponding features for at least two different images. Note that using automatically extracted images features has the advantage that (i) only one click is needed for specifying line segments and (ii) that the precision in the location of the extracted image features is expected to be higher compared to non-guided measurements.

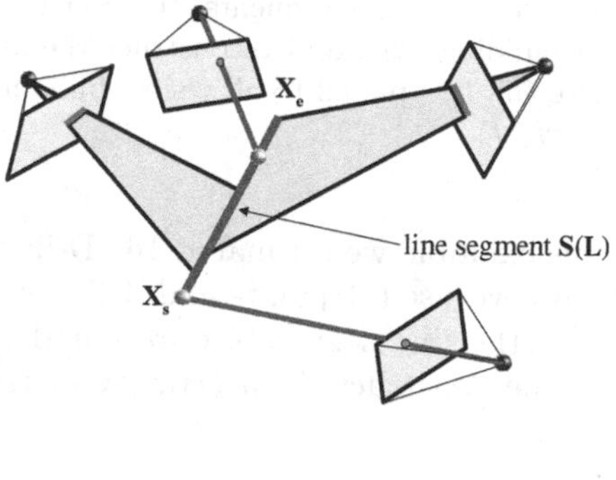

Fig. 5.4. Optimal estimation of a 3D line segment using image points and image line segments: first, estimate the infinite line **L**, then compute the endpoints of the 3D line segment by the most distant image points resp. endpoints, see text for details.

Now a 3D line $\hat{\mathbf{L}}_{\mathcal{M}}$ is estimated from the corresponding points and line segments using the algorithm 4.6. It is advisable to check the validity of the estimation by analyzing the estimated variance factor $\hat{\sigma}^2$. If the a-priori uncertainties from the feature extraction `FEX` are correct, the expected value is equal to 1. Assuming $\sigma_0 = 1$, the variance factor is Fisher distributed; for the result $\hat{\sigma}^2$ of our estimation, we allow a factor of up to 2, which basically is equivalent to an hypotheses test with a sufficiently large significance level. Thus we can alert the user for possible problems such as mismatched features, if the estimated variance factor exceeds a value of 2. It is even an option to refuse storing the construction if the value has a very high value. This feedback is similar to a traffic-light behavior as described in PREGIBON 1986 or FÖRSTNER 1995.

To obtain a 3D line segment $\hat{\mathbf{S}}$ we compute its endpoints $(\mathbf{X}_s, \mathbf{X}_e)$ which lie on the estimated 3D line. This can be done by computing the projecting lines for all points and endpoints of the matched features $\mathcal{M}$ and collect the intersections with the estimated line $\hat{\mathbf{L}}_{\mathcal{M}}$. The two most distant intersection points are the endpoints of the new 3D line segment $\hat{\mathbf{S}}(\mathcal{M}, \hat{\mathbf{L}}_{\mathcal{M}})$, also cf. fig. 5.4. Applying the procedure for all visible building edges yields a set

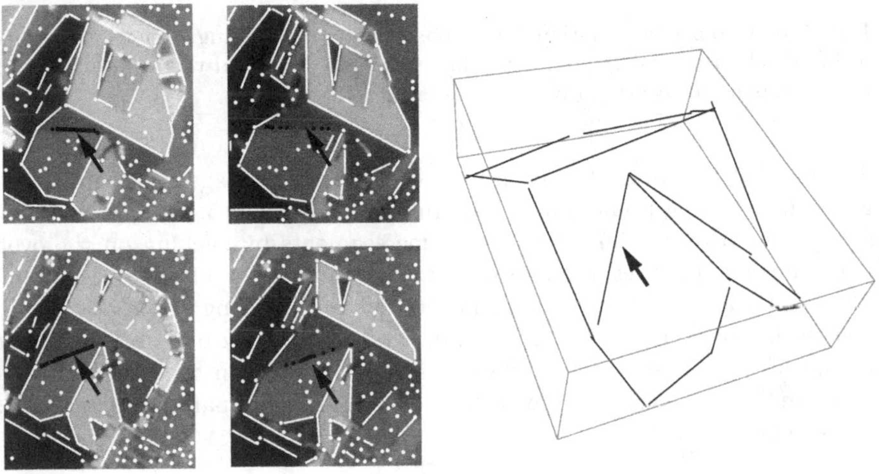

Fig. 5.5. Interactive reconstruction using points and line segments: the user may click on image features corresponding to a building edge and reconstruct the 3D edge using the algorithm 4.6. This procedure can be applied to all visible building features obtaining 16 building edges in 3D (*right*).

of 3D line segments, cf. figure 5.5: for this example we estimated 16 3D-lines corresponding to object edges, on average we used 4 point- and 4 line observation for each 3D line. The length of the line segments were found to be between 2 [m] and 12 [m]. To asses the constructed uncertainty of the line-segments, we checked the error orthogonal to the line directions at the endpoints of each line segment, see fig. 5.6. The average standard deviation at the line endpoints orthogonal to the 3D line segments were between 0.02 [m] and 0.16 [m].

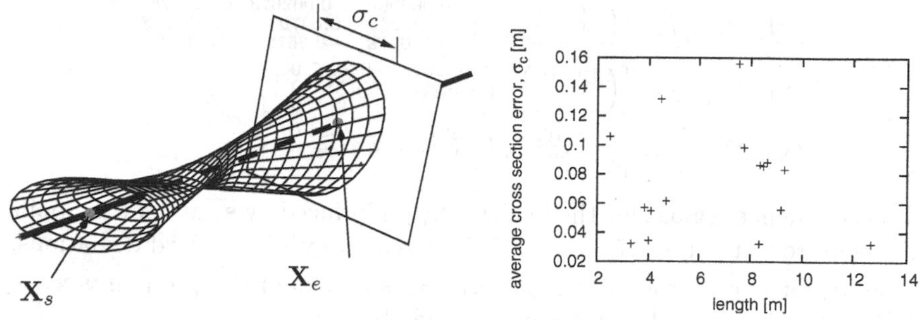

Fig. 5.6. *Left:* The uncertainty of a line segment can be expressed by using its infinite line. Given the start and endpoints $\mathbf{X}_s, \mathbf{X}_e$ of the line segments, the uncertainty of the position of the line is hence only considered in between these points. One can use the error σ_c on the plane orthogonal to the line direction to get interpretable measures for the uncertainty of the line. *Right*: using the 16 reconstructed line segments from figure 5.5, this diagram shows a plot of the errors σ_c orthogonal to the measured lines at the endpoints.

Note that the 3D line can not be constructed if all 2D line segments $\mathbf{l}'_i$ lie on the same epipolar plane. Then all viewing planes $\mathbf{A}'(\mathbf{l}'_i)$ are identical and the intersection is not defined. This may happen the camera was moved along only one axis, and the corresponding object edge is parallel to the moving direction.

Numerical Example We now want to show some numerical examples of construction and error propagation on real data, easing the check of a re-implementation. Consider one of the images in figure 5.5: we choose one line segment $\mathbf{s}_1$ and two points $\mathbf{x}_1, \mathbf{x}_2$, that are supposed to lie on the same line, see figure 5.7(a). The parameters of the features are extracted as follows

$\mathbf{s}_1$		$\mathbf{x}_1$		$\mathbf{x}_2$	
r_m	83.09 [pel]	r	104.79 [pel]	r	99.162 [pel]
c_m	112.95 [pel]	c	110.38 [pel]	c	130.526 [pel]
φ	18.62°	σ_r^2	0.297 [pel^2]	σ_r^2	0.72 [pel^2]
σ_d	0.5 [pel]	σ_c^2	0.9792 [pel^2]	σ_c^2	0.4149 [pel^2]
σ_φ	5.38°	σ_{rc}	-0.2367 [pel^2]	σ_{rc}	-0.1224 [pel^2]
l	13.11 [pel]				

where the coordinates are in the row-column coordinate system with the origin on the top-left corner. The line segment $\mathbf{s}_1$ is given using the 5-tuple from equation 4.11 on page 107 and its length l. From these parameters one obtains the following uncertain geometric entities (see section 4.2.2 on page 105):

$$(\mathbf{l}_1, \boldsymbol{\Sigma}_{\mathbf{l}_1\mathbf{l}_1}) = \left[\begin{pmatrix} 0.947 \\ 0.319 \\ -133.56 \end{pmatrix}, \begin{pmatrix} 0.000900 & -0.002673 & 0.120378 \\ -0.002673 & 0.007933 & -0.357248 \\ 0.120378 & -0.35724 & 16.3369 \end{pmatrix}\right]$$

$$(\mathbf{x}_1, \boldsymbol{\Sigma}_{\mathbf{x}_1\mathbf{x}_1}) = \left[\begin{pmatrix} 104.79 \\ 110.38 \\ 1 \end{pmatrix}, \begin{pmatrix} 0.297 & -0.2367 & 0 \\ -0.2367 & 0.9792 & 0 \\ 0 & 0 & 0 \end{pmatrix}\right]$$

$$(\mathbf{x}_2, \boldsymbol{\Sigma}_{\mathbf{x}_2\mathbf{x}_2}) = \left[\begin{pmatrix} 99.16 \\ 130.52 \\ 1 \end{pmatrix}, \begin{pmatrix} 0.72 & -0.1224 & 0 \\ -0.1224 & 0.4149 & 0 \\ 0 & 0 & 0 \end{pmatrix}\right]$$

The confidence regions for the line $(\mathbf{l}_1, \boldsymbol{\Sigma}_{\mathbf{l}_1\mathbf{l}_1})$ induced by $\mathbf{s}_1$ and for the points are depicted in figure 5.7(b) resp. 5.7(c), scaled by 10 standard deviations. One may now construct a line $\mathbf{l}_2 = \mathbf{x}_1 \wedge \mathbf{x}_2$ as a join of the two points $\mathbf{x}_1, \mathbf{x}_2$. Using the algorithm 4.1 on page 123, one obtains

$$\begin{aligned}(\mathbf{l}_2, \boldsymbol{\Sigma}_{\mathbf{l}_2\mathbf{l}_2}) &= (\mathbf{x}_1, \boldsymbol{\Sigma}_{\mathbf{x}_1\mathbf{x}_1}) \wedge (\mathbf{x}_2, \boldsymbol{\Sigma}_{\mathbf{x}_2\mathbf{x}_2}) \\ &= \left[\begin{pmatrix} -0.963 \\ -0.269 \\ 130.65 \end{pmatrix}, \begin{pmatrix} 0.000900 & -0.002673 & 0.120378 \\ -0.002673 & 0.007933 & -0.357248 \\ 0.120378 & -0.35724 & 16.3369 \end{pmatrix}\right]\end{aligned}$$

Transforming back to a Euclidean interpretation yields

$$(x_m, y_m) = (103.32\,\texttt{[pel]}, 115.63\,\texttt{[pel]}) \;,\; \varphi = 15.5^\circ \;,$$
$$\sigma_d = 0.044287\,\texttt{[pel]} \;,\; \sigma_\varphi = 2.53^\circ)$$

To check whether the line $\mathbf{l}_2$ is identical to the line $\mathbf{l}_1$, we first condition both uncertain entities with $f = 7.6 \cdot 10^{-4} = 7.6e - 4$ and obtain

$$(\mathbf{l}_1^\circ, \boldsymbol{\Sigma}_{\mathbf{l}_1^\circ\mathbf{l}_1^\circ}) = \left[\begin{pmatrix} 0.947 \\ 0.319 \\ -0.10222 \end{pmatrix}, \begin{pmatrix} 0.000900 & -0.002673 & 9.21341e{-}05 \\ -0.002673 & 0.007933 & -0.000273427 \\ 9.21341e{-}05 & -0.000273427 & 9.57006e{-}06 \end{pmatrix}\right]$$

$$(\mathbf{l}_2^\circ, \boldsymbol{\Sigma}_{\mathbf{l}_2^\circ\mathbf{l}_2^\circ}) = \left[\begin{pmatrix} -0.963 \\ -0.269 \\ 0.1 \end{pmatrix}, \begin{pmatrix} 0.000900 & -0.002673 & 3.37749e{-}05 \\ -0.002673 & 0.007933 & -0.000120754 \\ 3.37749e{-}05 & -0.000120754 & 8.11294e{-}06 \end{pmatrix}\right]$$

We now compute the distance vector $\mathbf{d}$ similar to to (3.63) on page 72 with the conditioned vectors:

$$\mathbf{d} = \mathsf{S}(\mathbf{l}_1^\circ)\,\mathbf{l}_2^\circ = \begin{pmatrix} 0 & 0.10222 & 0.319 \\ -0.10222 & 0 & -0.947 \\ -0.319 & 0.947 & 0 \end{pmatrix} \begin{pmatrix} -0.963 \\ -0.269 \\ 0.1 \end{pmatrix} = \begin{pmatrix} 0.0043955058 \\ 0.0036850488 \\ 0.052256559 \end{pmatrix}$$

As the test of identity of 2D lines has two degrees of freedom, we only need to check 2 of the three entries in $\mathbf{d}$ The reduction algorithm 3.1 on page 69 deletes the first row of $\mathsf{S}(\mathbf{l}_1^\circ)$, thus we obtain

$$\mathbf{d}^{[2]} = \mathsf{S}^{[2]}(\mathbf{l}_1^\circ)\,\mathbf{l}_2^\circ = \begin{pmatrix} -0.10222 & 0 & -0.947 \\ -0.319 & 0.947 & 0 \end{pmatrix} \begin{pmatrix} -0.963 \\ -0.269 \\ 0.1 \end{pmatrix} = \begin{pmatrix} 0.0036850488 \\ 0.052256559 \end{pmatrix}$$

For a reliable decision whether the two lines are perform an hypothesis test with $\alpha = 0.95$ according to algorithm 4.4. The test is not rejected and the test-value resp test-ratio are

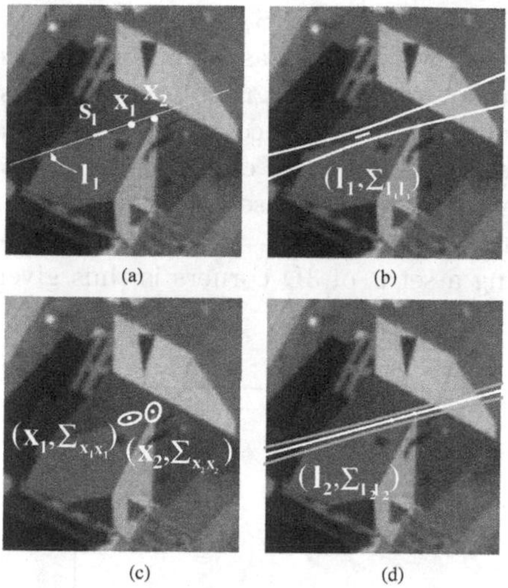

Fig. 5.7. Picture (a) shows an aerial image with selected geometric features extracted by FEX: a line segment $\mathbf{s}_1$ and two points $\mathbf{x}_1, \mathbf{x}_2$. The confidence regions of these features are shown in (b) and (c), scaled by 10 standard deviations. The uncertain line $(\mathbf{l}_2,\ \boldsymbol{\Sigma}_{\mathbf{l}_2\mathbf{l}_2})$ is constructed by a join of $(\mathbf{x}_1,\ \boldsymbol{\Sigma}_{\mathbf{x}_1\mathbf{x}_1})$ and $(\mathbf{x}_2,\ \boldsymbol{\Sigma}_{\mathbf{x}_2\mathbf{x}_2})$, see (d).

$$T = 0.268 \quad ; \quad T_R = 0.0450$$

as $T < T_{1-\alpha,2} \approx 5.9$.

Using the identity test of 2D lines and incidence test of a 2D line and a 2D point, we can test collinearity of all 2D features that are supposed to belong to the same building edge. Doing this for all images, we can compute 3D line segments by using a joint estimation as described above. For the indicated image features in figure 5.5 one obtains the following 3D line:

$$\mathbf{L} = \begin{pmatrix} -0.449 \\ -0.751 \\ 0.483 \\ -535.76 \\ 460.05 \\ 216.83 \end{pmatrix}$$

$$\boldsymbol{\Sigma}_{\mathrm{LL}} = \begin{pmatrix} 1.83118e{-}05 & 3.14034e{-}06 & 2.18747e{-}05 & 0.00223648 & -0.00674255 & -0.00090315 \\ 3.14034e{-}06 & 5.91207e{-}06 & 1.20933e{-}05 & 0.00421620 & 0.00125156 & -0.00170701 \\ 2.18747e{-}05 & 1.20933e{-}05 & 3.90813e{-}05 & 0.00862158 & -0.00431638 & -0.00348846 \\ 0.00223648 & 0.00421620 & 0.00862158 & 3.0073 & 0.8935 & -1.2174 \\ -0.00674255 & 0.00125156 & -0.00431638 & 0.8935 & 3.5619 & -0.3628 \\ -0.00090315 & -0.00170701 & -0.00348846 & -1.2174 & -0.3628 & 0.4929 \end{pmatrix}$$

The length of the underlying 3D line segment is 8.34 [m]. The maximal cross section error is $\max \sigma_c = 5$ [cm], the minimal error $\min \sigma_c = 1.5$ [cm].

5.2.3 Grouping 3D Line Segments to Surface Patches

In the previous step 3D line segments were obtained by manually establishing the correspondences between 2D line segments. Using the testing algorithm 4.4 it is now possible to group the computed 3D line segments to more complex aggregates.

First, the line segments are aggregated to 3D corners, i.e. corner-points associated with two 3D line segments. This means we have to find line segment pairs

$\mathbf{S}_1, \mathbf{S}_2$ in object space $\mathcal{O}$ such that the lines $\mathbf{L}(\mathbf{S}_1), \mathbf{L}(\mathbf{S}_2)$ are incident and non-parallel. These tests can be performed with some reasonable significance level such as $\alpha = 1\%$. Furthermore we do not want to group line segments, that are too far away. If we know the units in object space, we may specify a threshold T_d for the maximal distance between the two closest endpoints of the line segments $\mathbf{S}_1$ and $\mathbf{S}_2$. After identifying two such segments, the corner point $\mathbf{X}_c$ is estimated and the corner is defined by $\mathbf{C} = (\mathbf{X}_c, \mathbf{S}_1, \mathbf{S}_2)$. The resulting grouping algorithm for finding a set $\mathcal{C}$ of 3D corners is thus given as:

Grouping 3D line segments

Objective: Grouping 3D line segments $\mathbf{S}_i \in \mathcal{O}$ based on incidence and distance criteria.

for $\mathbf{S}_1, \mathbf{S}_2 \in \mathcal{O}$ do
  $\mathbf{L}_1 := \mathbf{L}(\mathbf{S}_1), \mathbf{L}_2 := \mathbf{L}(\mathbf{S}_2)$
  if $\mathbf{L}_1 \cap \mathbf{L}_2 \neq \emptyset$ and not $(\mathbf{L}_1 \parallel \mathbf{L}_2)$
  then $\mathbf{X}_c :=$ Estimate-Incident$(\mathbf{L}_1, \mathbf{L}_2)$
    skip if distance$(\mathbf{X}_c, \mathbf{S}_1, \mathbf{S}_2) > T_d$
    $\mathcal{C} := \mathcal{C} \cup \{(\mathbf{X}_c, \mathbf{S}_1, \mathbf{S}_2)\}$
done

Algorithm 5.1: Simple pseudo-code algorithm for grouping 3D line segments based on incidence and distance criteria.

This is a first and simple example of how to apply the SUGR-library, cf. section 4.7 on page 147 to automatically obtain new hypotheses about the object. The new corner hypotheses are regarded as the best explanation of the observations, thus the underlying inference is essentially an abduction.

Finally the corners can be grouped to planar surface patches: one can find all those corners $\mathbf{C}_i, \mathbf{C}_j$, that define identical planes $\mathbf{A}(\mathbf{C}_i) \equiv \mathbf{A}(\mathbf{C}_j)$. The planar patches are then defined by the convex hull of all identical corners, ignoring a possible concave shape of the surface at this stage of reasoning.

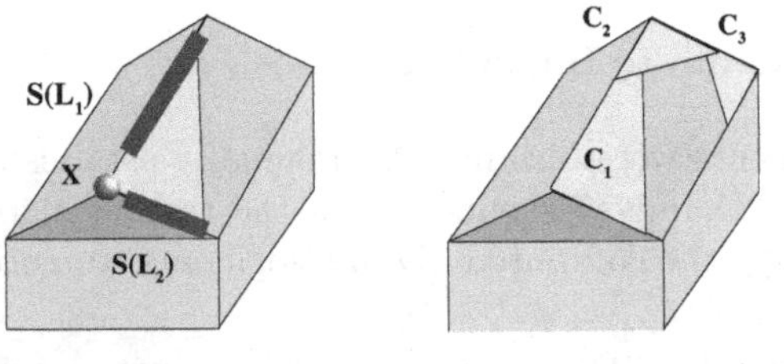

Fig. 5.8. Grouping of 3D line segments to planar patches: first, pairs of line segments are grouped when being coplanar and non-parallel, together with their intersection $\mathbf{X}_c$, they form a corner. Several corners $\mathbf{C}_i$ may be grouped to one planar surface patch if the underlying planes are identical.

5.3 Automated Reconstruction

The enhancements proposed in the previous section are an improvement of the quality of user interactions compared to pure point measurements, but it still contained a considerable amount of user interaction. We now describe an approach to replace the manual establishment of image line correspondences to an automatic matching system solely based on geometric constraints (HEUEL AND FÖRSTNER 2001). Combining the matching result with the reconstruction and grouping, we have automated the following tasks:

- matching 2D line segments from multiple oriented images
- optimally reconstructing 3D line segments
- grouping 3D line segments to corners

5.3.1 Matching of Corresponding Line Segments

The matching of image line segments is considerably more complex than the matching of image points: first of all, line segments observed by a segmentation algorithm are imperfect: the location of the end-points are not reliable; additionally, an ideal line segment might be broken into two or more small line segments, that are not connected to each other. A second reason for the complexity of line matching is due to the epipolar geometry: the search of a corresponding point in a second image to the search on the epipolar line, thus restricting the search-space from 2D to 1D. For lines, there is no strict geometric constraint of the location of the line in the other images.

The now following matching algorithm 5.2, is essentially a heuristic search for finding hypotheses of corresponding line segments; these hypotheses are constructed and tested using the SUGR procedures.

5.3.1.1 Selection of Segment Pairs In the first step of the algorithm 5.2, a pair of image line segments is selected, that are supposed to be caused by a building edge. Although the epipolar constraint does not hold for infinite lines, we can make use of the epipolar lines of the endpoints of a line segment: given a line segment $\mathbf{s}$ with start- end endpoint $\mathbf{x}_s, \mathbf{x}_e$, the epipolar lines of $\mathbf{x}_s$ and $\mathbf{x}_e$ define a region (called "epipolar beam") in the other images, cf. ZHANG 1994 and figure 5.9(1.). Therefore we define a set $\mathcal{E}(\mathbf{s})$ for each line segment $\mathbf{s}$, which contains all line segments *and* points being in the epipolar beam. For the epipolar beam, the uncertainties of $\mathbf{x}_s, \mathbf{x}_e$ have a small, negligible influence on the result and safely can be ignored.

5.3.1.2 Construction and Verification of 3D Lines (Steps 1 and 2). For a pair of 2D lines $\mathbf{l}_1, \mathbf{l}_2$ we compute the 3D line $\mathbf{L}_{12}$ by intersecting the projecting planes $\mathbf{A}'(\mathbf{l}_1), \mathbf{A}'(\mathbf{l}_2)$. This apparently causes problems for specific configurations: the uncertainty of the constructed 3D line will be very large

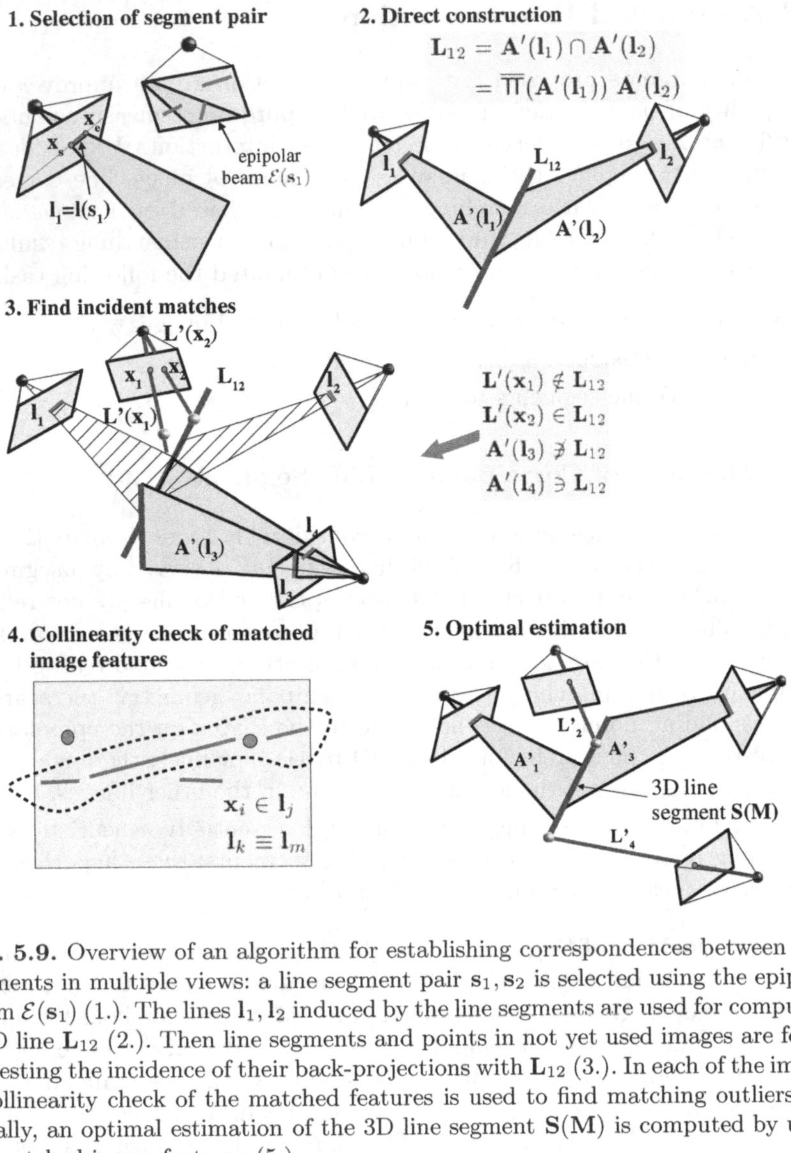

Fig. 5.9. Overview of an algorithm for establishing correspondences between line-segments in multiple views: a line segment pair $\mathbf{s}_1, \mathbf{s}_2$ is selected using the epipolar beam $\mathcal{E}(\mathbf{s}_1)$ (1.). The lines $\mathbf{l}_1, \mathbf{l}_2$ induced by the line segments are used for computing a 3D line $\mathbf{L}_{12}$ (2.). Then line segments and points in not yet used images are found by testing the incidence of their back-projections with $\mathbf{L}_{12}$ (3.). In each of the images a collinearity check of the matched features is used to find matching outliers (4). Finally, an optimal estimation of the 3D line segment $\mathbf{S}(\mathbf{M})$ is computed by using all matched image features (5.)

when the line segments are close to the epipolar line, see figure 5.10 for an example. In these cases the chosen pair has to be rejected: a possible criteria for large uncertainty is the error $sigma_c$ orthogonal to the line direction, cf. figure 5.6(left). The error σ_c should not exceed a maximal error σ_c^{max} which has the advantage that the maximal error σ_c^{max} can be expressed in object dimensions. Therefore it is not possible that a 3D edge is considered, where *all* image lines are on one epipolar line.

Fig. 5.10. Assume a 3D line is constructed by two image line segments, that are almost on the same epipolar line, see images on the right. Then the uncertainty of the 3D line is quite large, which can be checked by projecting the 3D line segment back to the image, the confidence hyperbola (scaled by 3 standard deviations) is significantly larger than expected for the images on the left.

Find Incident Matches (Steps 3 and 4) After constructing a 3D line $\mathbf{L}_{12}$, the next step involves the search of matching features in the other $N-2$ images. Again we can use the epipolar beam to reduce the number of candidates for the match. To test the candidates we can test the incidences of the back-projected image features $\mathbf{L}'(\mathbf{x}'_i)$ and $\mathbf{A}'(\mathbf{l}'_j)$ with the 3D line $\mathbf{L}_{12}$, see figure 5.9(3.), again using the SUGR algorithm for testing relations, cf. algorithm 4.4. In case of a successful test we may measure the closeness of each image feature feature to the 3D line $\mathbf{L}_{12}$ by using the test-values T of the incidence tests. All matched features are collected in a set $\mathcal{M}$. Note that we formulated the tests in 3D, but it is equivalent to project $\mathbf{L}_{12}$ into the images and test on collinearity in 2D with the feature candidates.

The set $\mathcal{M}$ of matched image features is now checked for consistency within each image, figure 5.9(4.): all matched 2D points $\mathbf{x}_i$ and 2D line segments $\mathbf{l}_k$ in one image should be collinear to each other. We pairwise test collinearity between the image features $\mathbf{x}_i \in \mathbf{l}_j$ and $\mathbf{l}_k \equiv \mathbf{l}_m$. A feature is then removed from the set $\mathcal{M}$ of matches, if

- the collinearity test was not successful, i.e. $\mathbf{x}_i \notin \mathbf{l}_j$ and $\mathbf{l}_k \not\equiv \mathbf{l}_m$, and
- the feature has a larger test-value T for the incidence with $\mathbf{L}_{12}$ than its counterpart.

Note that with the second criterion we take advantage of the monotony property of the approximated test values (section 4.5.2): we assume a feature to be further away from the line $\mathbf{L}_{12}$ if the test-value is larger than the one for its counterpart.

Optimal Estimation (Step 5) With the feature-set $\mathcal{M}$ it is now possible to estimate a 3D line $\hat{\mathbf{L}}_{\mathcal{M}}$ with all back-projected point and line segment matches using the algorithm 4.6. This is done accordingly to the estimation of the user-assisted case and we may reject the feature-sets that give a variance factor $\hat{\sigma}$ higher than a suitable threshold, see discussion above. As a final test of the estimated line $\hat{\mathbf{L}}_{\mathcal{M}}$, we check again if all matches in $\mathcal{M}$ are still incident to $\hat{\mathbf{L}}_{\mathcal{M}}$.

The last part of the algorithm is the computation of the endpoints $(\mathbf{X}_s, \mathbf{X}_e)$ of the 3D line segment $\hat{\mathbf{S}}$ which lie on the estimated 3D line, again done as in the user-assisted case, see figure 5.9(5.).

It is possible that multiple 3D line segments $\hat{\mathbf{S}}_j$ have been extracted for the same true line segment $\tilde{\mathbf{S}}$. Then all $\hat{\mathbf{S}}_j$ should be equivalent to each other, which again is tested statistically. We merge all equivalent 3D line segments by taking all 2D feature-sets and recompute the 3D line $\hat{\mathbf{L}}_{\mathcal{M}'}$.

The approach of computing a 3D line segment solely based on geometric constraints is summarized in the algorithm 5.2.

5.3.2 Examples

The matching algorithm has been tested on an artificial scene and aerial imagery.

Artificial Example The artificial scene (figure 5.11) consisted of cubes with edge-length 1 that were stacked upon each other. Eight views of the 3D scenes were produced and a bundle adjustment was computed to obtain optimal projective camera matrices. Then the images were corrupted by a Gaussian noise of 1 [pel] and segmented using the FEX-system. From these features, the

---

Matching and Reconstruction of Line Segments
Objective: Compute a 3D line segment $\hat{\mathbf{S}}$ from image feature sets $\mathcal{F}^i, i = 1, \ldots, N$ for N images.

```
 1  for s₁ ∈ 𝓕¹ ∪ ⋯ ∪ 𝓕ᴺ do
 2     for s₂ ∈ ℰ(s₁) do l₁ := l(s₁), l₂ := l(s₂)
 3        // unique construction using first two lines
 4        (L₁₂, Σ_L₁₂L₁₂) := A′(l₁) ∩ A′(l₂)
 5        𝓜 := ∅ // initial match-set 𝓜
 6        // search projections of L₁₂ in other image
 7        for x, m ∈ ℰ(l₁) \ (𝓕(l₁) ∪ 𝓕(l₂)) do
 8           if L′(x) ∩ L₁₂ ≠ ∅ then x ∈ 𝓜
 9           if A′(m) ∋ L₁₂ then m ∈ 𝓜
10        done
11        // pairwise collinearity check of matched features
12        for i := 1, …, N; x, m₁/₂ ∈ 𝓜 ∩ 𝓕ⁱ do
13           if (x ∉ m₁/₂ or not(m₁ ≡ m₂))
14              then 𝓜 := 𝓜 \ {x or m₁ or m₂}
15        done
16        skip 𝓜 if ∃i : 𝓕ⁱ ∩ 𝓜 = ∅
17        // estimate 3D line by incident matches 𝓜
18        (L̂_𝓜, Σ̂_LL, σ̂) := Estimate-Incident(𝓜)
19        skip 𝓜 if σ̂ > T_σ // check estimation
20        if (∀x, m ∈ 𝓜 : A′(m) ∈ L̂_𝓜 and L′(x) ∩ L̂_𝓜 ≠ ∅)
21           then compute 3D line segment Ŝ(𝓜, L̂_𝓜)
22     done
23  done
```

Algorithm 5.2: Pseudo-code algorithm to compute a 3D line segment $\hat{\mathbf{S}}$ from feature sets $\mathcal{F}^i, i = 1, \ldots, N$ for N images. A feature set consists of image line segments $\mathbf{s}_{1,2}, \mathbf{m}_{1,2}$ and points $\mathbf{x}$. The set $\mathcal{E}(\mathbf{s})$ is the set for a line segment $\mathbf{s}$, which contains all line segments and points being in the epipolar beam of $\mathbf{s}$. The threshold T_σ is chosen according to an hypothesis test, see text for details. All tests and constructions are done using the SUGR-algorithms, cf. section 4.7 on page 147.

matching algorithm 5.2 was applied. It was required that matches should exist in at least 6 images. Only true line matches are found, but the endpoints of the line segments are sometimes extended due to accidental matches in special viewpoints. Additionally some line segments have not been found because either they were not visible in 6 or more images or the FEX has not found an edge due to low contrast.

Aerial Imagery The algorithm has been applied to aerial images, where the projection matrices had been computed by bundle adjustment. The datasets we used were made available within the context of a workshop on automatic extraction of man-made objects, cf. GRÜN *et al.* 1995and consisted of four overlapping aerial images. By selecting image patches with one small-sized

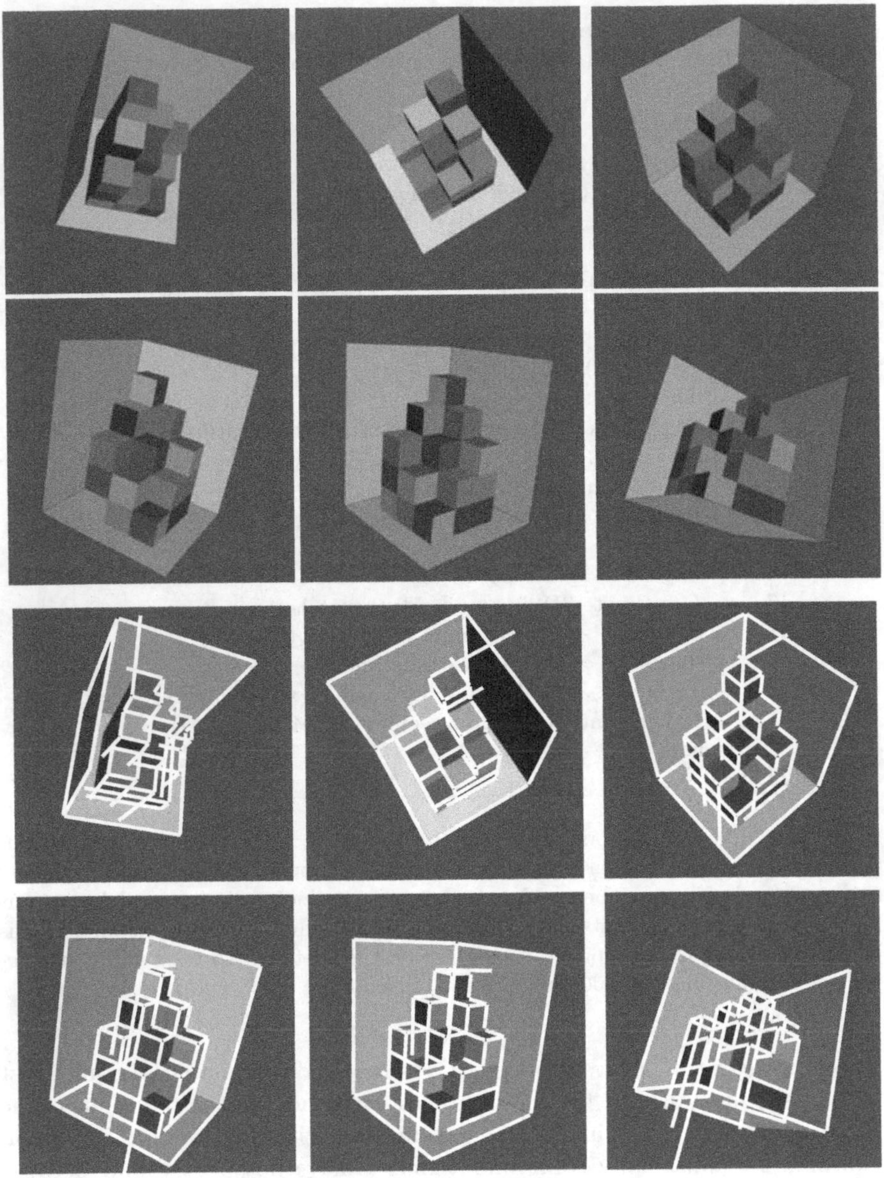

Fig. 5.11. Six out of eight images of an artificial seen (*top two rows*) and the reconstructed 3D line segments projected in the images (*bottom rows*). Only true line matches are found, but the endpoints of the line segments are sometimes extended due to accidental matches in special viewpoints. Additionally some line segments have not been found because either they were not visible in 6 or more images or the feature extraction program FEX has not found an edge due to low contrast.

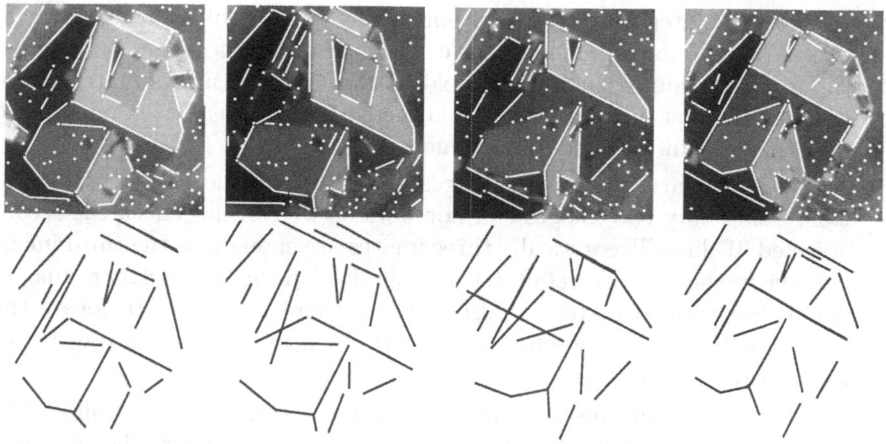

Fig. 5.12. first row: extracted line segments and points superimposed on three out of four images; second row: projected 3D line segments computed with the proposed algorithm using line segments and points.

building, the algorithm was able to reconstruct most of the 3D lines, see figure 5.12, while some false matches were detected and some edges were missing. Since the purpose here is to demonstrate the feasibility of SUGR, we don't provide an extensive examination of the building reconstruction results. The next section contains a discussion of the properties of the results.

5.3.3 Discussion

The sketched algorithm is a first proposal for the extensive use of SUGR for matching algorithm. The overall strategy was to build up a set of hypotheses and filter out those hypotheses, which do not fit to the geometric constraints. To reliably use these geometric constraints, the algorithms from the last chapter have been extensively used. There are some interesting properties of the approach:

- There are no parameters to be set other than significance levels for the hypotheses tests, which was fixed to 95% and a distance threshold T_d for the grouping of 3D line segments to corners.
- The step of optimal reconstruction of a 3D line segments with respect to the observations is an essential part of the matching algorithm, as it can reject bad hypotheses based on the result of the joint estimation.
- Image points can be integrated into the matching and may contribute to the final 3D line segment. After a pair of 2D line segments have been chosen, back-projected points in the other images can be tested for incidence to the match hypothesis. One may even allow the observation of two points in an image, which can replace a line observation.

However the integration of image points may have problems if too many of the extracted image points do not correspond to polyhedral corners. Then the consideration of points can yield to more false-positives matches: the accidental situation that a false image edge is consistent with one or two false image points becomes much more likely.

- A sufficiently large number of images is crucial for a successful reconstruction. Using only two images can not work as we can not check the reconstructed 3D line. Theoretically three images are enough, as the third image can verify the reconstructed 3D line obtained from the first two images. But if the verified match is false-positive, there is no way to locate the erroneous feature. Only when using four or more images it is possible to locate errors in the match.

  Practically one can observe that the geometric constraints are only sufficient when having four or more images. Using only 3 images, the chance of accidental matches is quite high, depending on the scene. But this chance reduces significantly by an increasing number of images.

  More than three images are also required when two image lines in different images lie on the same epipolar plane as in figure 5.10. Then all lines in a third image could be a potential match and one needs a fourth image to check the validity of the matching hypotheses.
- The success of the algorithm depends on the precision of the involved entities: if the uncertainty of the image features or the uncertainty of the projective parameters become large, more hypotheses will be accepted and the search space becomes larger. This is a natural relationship between the observations and the unknowns: the less we know about the observations, the less we can infer about the unknown.
- The algorithm was applied using known camera matrices; but since all relations that are used are invariant under projective transformation. Only the grouping of corners required a test on parallelity. Note that it is possible to do the same reasoning using uncalibrated cameras: then we only use the relative orientation of the images and from there artificial cameras, as outlined in section 5.1.2.
- The transfer from matches of the first two views to the third one is done via 3D by computing the associated 3D line. One may also perform the transfer of the matches only in 2D using the trifocal tensor, cf. SHASHUA AND WERMAN 1995, HARTLEY 1995A.

As mentioned above, the algorithm 5.2 solely depends on geometric constraints and first of all serves as a test-case for the validity of the SUGR algorithms. It cannot resolve all issues of the matching problem since other sources of information such as radiometry or topology are not used. In detail, the inherent problems of the matching algorithm are as follows:

- *Lack of notion for proximity*: the search for matching features given a hypothesized 3D line segment is done by projecting the hypothesis into the images and then find incident features for this projected line. This criteria does not take into account any kind of proximity criteria. Additionally the probability that a feature is incident to the hypothesis becomes larger the further the feature is away from the point of gravity of the line. Thus the matching algorithm can only be applied to small image patches so that the proximity condition can be neglected. A possible remedy is to use neighborhood relationships between image features, see below.
- *Lack of notion for radiometry*: another important problem is that the matching algorithm ignores any image intensities that are on the left or on the right side of an image edge. This constraint may also filter some of the false matches. Note however that even radiometric constraints cannot solve all problems: consider repetitive structures such as a crosswalk with parallel white stripes (figure 5.13): even a matching algorithm taking a local neighborhood image intensities into account could not distinguish between the second and the fourth stripe.
- *Computational Complexity*: Not taking the epipolar beam constraint into account, the above algorithm has a complexity of $O(m \cdot n^2)$, where n is the number of image line segments and m is the average number of image features in $N - 2$ images. In worst case the epipolar beam does not have an effect on the complexity.

  On our prototype system, implemented with language Perl, a pass of one pair of line segment (lines 3-21 in algorithm 5.2) takes on average 5 seconds, using the epipolar beam. This number may vary depending on the geometry of the scene. While the speed of the system can be greatly reduced with an optimized, the complexity issue still has to be resolved by introducing more constraints to the solution.

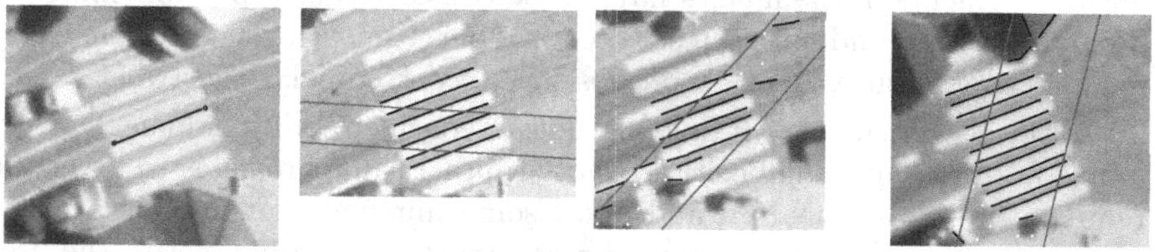

Fig. 5.13. In some situation, there might arise problems when matching line segments because of parallel lines: the pedestrian crossing pictured in the image allows for multiple matches, even when considering left and right image intensities.

5.3.4 Effect of Topological Selection to Matching

As pointed out in the previous discussion, using only geometric cues such as incidence and identity does not incorporate any proximity criterion. To overcome this problem, one may use the neighborhood relationship between image features to infer the neighborhood relationship in 3D. An example for this approach is described by HEUEL *et al.* 2000 who proposed a system for the reconstruction of polyhedra, which uses a strategy for matching 2D corners instead of 2D lines, see LANG 1999. The aggregation of basic features to corners does already start in the image domain rather than in the 3D domain. The early aggregation has the advantage that the subsequent matching process becomes simpler. On the other hand, the identification of corners in the image is more involving. The result of the corner matching reconstruction is then used within a 3D grouping stage, that is similar to the one proposed in this work, the result being again polyhedral 3D patches. The difference is here that topological constraints are used in additional to the geometric constraints: the system infers the neighborhood relationships between 3D features from the underlying 2D neighborhoods, which have been generated by FEX. The use of topological constraints has the advantage that it is computationally much simpler and consists only of a sequence of lookups in the 2D neighborhood. HEUEL *et al.* 2000 demonstrated that the application of topological criteria prior to geometric tests greatly reduces the number of hypotheses for the more involving geometric tests. Furthermore the topology selection prevents problems of proximity outlined in the last section.

5.4 Combining Interaction and Automation

We now describe two simple methods to combine techniques of automated systems with minimal and simplified user interaction in order to overcome problems in the automated system. The previous section has demonstrated an automated approach for matching 2D line-segments across multiple images. One of the disadvantages of this approach is the size of the search-space: first, all line segments in the images have to be considered as a possible image of a polyhedron; second, for each selected 2D line segments, the only constraint for matching features in other images was a weak epipolar constraint.

To solve the search-space problem, one can consider additional cues, for example from the image intensities. Another practical approach is to lower the requirements of automation and allow some minimal user interaction that may steer the automated process of reconstruction. In other words, the user is supposed to interactively constrain the search space by some simple clicks, see also table 1.1 on page 8.

Choosing a Line Segment One first may enhance the approach described in section 5.2.2: there matching image line segments have to be specified in all existing images in order to obtain an optimal result. An apparent way of using the automated approach is to let the user specify only *one* line segment in *one* image that refers to a polyhedral edge. Then the outer loop of the algorithm 5.2 can be skipped and likely results in a 3D line segment, provided there are enough matching features extracted in the other images. If additionally one corresponding second line segment is clicked in a second image, the inner loop is skipped too and it becomes almost certain that an optimal match is chosen automatically. In cases such as in figure 5.10 this reasoning might fail, but the system is aware of this situation and can warn the user to choose a different line segment.

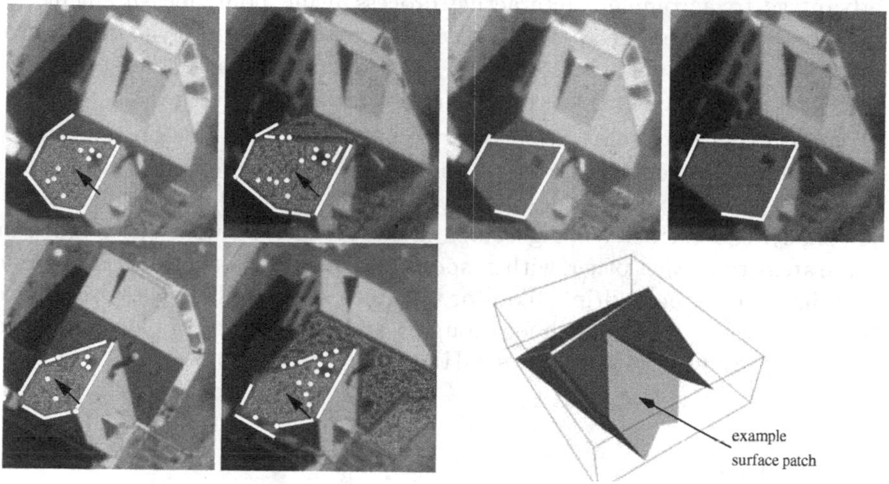

Fig. 5.14. Example of interactively reducing the search-space of the matching algorithm. The left four images show the four blobs that have been selected by the user by four clicks. The top right images show the projections of the 3D line segments found by matching neighboring features of the blob, the bottom right shows the set of reconstructed surfaces by clicking on matching surface areas. Note that no single corners are shown and the different surfaces are not glued to one polyhedral part.

Choosing Matching Blobs A more effective example of constraining the search-space exploits the neighborhood-relationships between the image features that were extracted by FEX. The idea is that the user specifies matching areas in each image, which belong to the same polyhedral surface: only line segments within this areas should be considered for matching. One can outline the acquisition process as follows, see also figure 5.14:

- In each image, the user clicks on an area that belong to the same polyhedral surface.
- For each click in each image, the system finds the closest blob, that was extracted by FEX. The blob is supposed to correspond to the area that the user clicked on.

- All line segments and points that are neighbored to a selected blob are collected in a set of image features $\mathcal{F}'$.
- Start the matching algorithm 5.2 with the reduced set of features $\mathcal{F}'$.
- Compute a surface patch from all found 3D line segments, that are coplanar to each other.

The process above only needs user interaction at the first step, namely clicking somewhere in an area that refers to a polyhedral surface. The output is at best the complete polyhedral surface. This acquisition process is quite an advantage to a complete interactive process using point measurement: (i) only four clicks were necessary to reconstruct a complete polyhedral surface and (ii) no precision is required when clicking on the images, one just has to hit the area belonging to the polyhedral surface.

Of course there are some drawbacks: the approach depends on the quality of the image segmentation result - if no image edges were found, the system can not reconstruct any edges, which is true for all applications above. Additionally, the extracted image blobs might be merged with other blobs or separated from each other with respect to the ideal image blob. Thus one may has to click on multiple blobs or take into account that the search space becomes larger. An approach for grouping and partitioning of blobs might be helpful to resolve this problem, see Heuel and Förstner 1998.

5.5 Summary

In this chapter we have demonstrated the feasibility of the SUGR library by applying it to the task of polyhedral object reconstruction. It was shown that for both user-assisted systems and automated systems, the proposed approach of including uncertainty in the reasoning process can be used. The statistical approach enabled the reconstruction to take over the precision measures from the image segmentation and made it possible to characterize the reconstruction results in a consistent manner.

6 Conclusions

This chapter summarizes the main aspects of this work, mentions its original contributions and finally gives an outlook on possible future work based on our approach.

6.1 Summary

We have introduced a unified approach for *representing*, *combining* and *estimating* points, lines and planes in 2D and 3D taking uncertainty information into account. Additionally we have shown how to *establish geometric relations* between entities using a simple hypothesis test. These tools enable us to perform geometric reasoning with these entities in a statistical manner.

The motivation of this work was driven by the task of polyhedral object reconstruction. Polyhedral objects can be represented by linear subspaces – namely points, lines and planes – which in turn can be treated in a nice mathematical way using the Grassmann-Cayley algebra. We have chosen a homogeneous vector representation for these entities, thus representing the subspaces by projective elements. By subdividing each homogeneous vectors into two parts we have established a consistent connection to the underlying Euclidean entities.

Using homogeneous vectors, we have derived simple bilinear expressions for the join and intersection operators by using three basic types of matrices: the skew-symmetric matrices $\mathsf{S}(\cdot)$, $\Gamma(\cdot)$ and $\Pi(\cdot)$. These matrices enable us to perform a manipulation of an entity by a simple matrix-vector product (table 3.2 on page 61).

A close look on the construction matrices shows that the rows and columns contain special joins and intersections with canonical elements, as suggested by the Grassmann-Cayley algebra. This observation can be used to easily express incidence and identity relations. Together with parallelity and orthogonality, we obtain a set of simple tests to establish relationships between entities, cf. table 3.5 on page 78.

Using the established relationships between entities, it is possible to formulate a generic yet simple estimation scheme for fitting an unknown entity to a set of observations, see table 3.3 on page 95. This scheme can also be used

S. Heuel: Uncertain Projective Geometry, LNCS 3008, pp. 173-177, 2004.

to estimate an unknown transformation from a set of observations, which is known as general direct linear transformation (DLT). With the proposed framework, the idea of a DLT can be reformulated as a problem of establishing relationships. This interpretation allows a quick way to establish the algebraic expressions for a DLT, which is demonstrated for general homographies and the projective cameras.

For real applications, the proposed algebraic framework has limitations, as it takes no uncertainties into account. For the usual case of noisy observations we proposed a representation based on homogeneous covariance matrices. The choice of covariance matrices constitutes an approximation of the real underlying probability density function. It was shown that it is easily possible to limit the amount of error that is induced by this choice. The only requirement on the observed data is that the amount of noise should be small, for example less than 10%. Based on this approximation, statistical versions of the algebraic methods were described, which are combined into a library for geometric tasks called SUGR (Statistically Uncertain Geometric Reasoning).

To demonstrate the feasibility of the SUGR procedures, the tools were applied to the task of polyhedral object reconstruction. It was shown that it is easily possible to propagate uncertainty information in the images directly to 3D reconstructions, which in turn can be used for establishing hypotheses about relations in 3D. The tools can be used for user-assisted systems as well as automated systems. From the analysis of the automated approach it was shown that it is possible to include partial solutions of the automation to user-assisted systems.

6.2 Contribution of Thesis

The main contribution of the thesis is the development of a methodology for doing statistical geometric reasoning on points, lines and planes in 2D and 3D. We have provided detailed procedures for the core methods, which are generic enough to solve a wide range of geometric problems in Computer Vision and Photogrammetry yet always taking the uncertainty into account. In detail, the important achievements are:

- We provided a *consistent description* of points, lines, planes and their transformations in the context of projective geometry. All entities can represented as homogeneous vectors including transformations. The homogeneous vectors were subdivided into a Euclidean and a homogeneous part, such that the distance of the entity to the origin is equal to the ratio of the norm of the parts.

  We *translated the join and intersection* operators from Grassmann-Cayley algebra into a matrix-vector notation. For this we used the already known 3×3 skew-symmetric matrix $\mathsf{S}(\cdot)$ for 2D lines and points and the 4×4

Plücker matrix $\Gamma(\cdot)$ for lines; additionally a new matrix called $\Pi(\cdot)$ was introduced, which plays the same role for 3D points and planes as the Plücker matrices for lines and dual lines. The three matrices may also serve as a representation for geometric entities or they can be used for construction or testing of entities. The interpretation of the matrices in table 3.3 on page 66 allow a clarified geometric viewpoint of the structure of the algebraic expressions in full analogy to the Grassmann-Cayley algebra.

With the interpretation of the matrices, it is easy to formulate expressions for projectively invariant relations such as *incidence* and *identity*. Additionally, with our choice of a Euclidean and homogeneous part of a vector, it is even possible to express projectively non-invariant relations such as *parallelity* or *orthogonality*. Also distances between entities can be expressed in simple formulas, see table 3.5 on page 78.

Using the projective relations, it is possible to *re-interpret the well-known DLT method* in terms of relations between the corresponding entities and the transformation consistently for points, lines and planes. With these interpretations, the derivation of algebraic expressions of the DLT is very simple, see tables 3.8, 3.9.

Finally, a generic estimation algorithm of geometric entities and transformation by minimizing the algebraic error, similar to the traditional DLT. This is an acceptable method for fitting unknown geometric parameters to observations as outlined in HARTLEY 1998. One only has to *condition* (or normalize) the entities to minimize a bias in the solution. As an approximation of the conditioning operation we proposed to ensure a minimum scale difference between the homogeneous and the Euclidean part of the vectors.

- We described how to *integrate statistics* into the above algebraic framework: homogeneous covariance matrices have been shown to be sufficient to represent and propagate the uncertainty of entities, assuming a reasonably realistic amount of error. Although the representation of the uncertainty using homogeneous covariance matrices is not unique, one can define simple properties to check the validity of an arbitrary covariance matrix with respect to its homogeneous vector, cf. section 4.1.2 on page 102. The property merely states that a homogeneous vector is unconstrained except of homogeneity and a possible Plücker constraint. For example we do not allow points with one or two degrees of freedom, in other words points that are only determined on a specific line or plane. Using this property a construction algorithm can be formulated for which the first-order error propagation is always possible and yields a valid covariance matrix, see table 4.1.

  Using the proposed uncertainty representation, it was possible to formulate *hypotheses tests for geometrical relations* in a consistent manner, cf table 4.4. Finally, the generic estimation scheme was extended to exploit the uncertainty information using a Gauß-Helmert model, where the initial values are computed with the non-statistical version. This provides a simpler solution to a wide range of geometrical tasks than the one described by KANATANI 1996, see table 3.3

The proposed statistical algorithms have been checked using Monte Carlo simulations and enable an efficient way to perform geometric reasoning including uncertainty information.

- As a test-case for the feasibility of the approach, we applied the SUGR-algorithms to the task of *polyhedral object reconstruction.* It was possible to integrate the algorithms to systems with varying scales of automation, demonstrating that the methods are capable of pushing the automation limits of user-assisted systems or enhancing the performance characteristics of automated methods.

  Within the automated approach, it was also demonstrated how to build up hypotheses of 3D features and aggregates using relation tests and how to compute the best estimate given a set of hypotheses. These were the main operations for geometric reasoning mentioned in the introduction.

6.3 Outlook

The SUGR procedures provide a good basis for further research in the domain of geometric reasoning in Computer Vision and Photogrammetry. There are a number of paths one may follow to enhance the existing methods or apply them to other applications.

- *Constrained geometric entities:* as pointed out above, we assume unconstrained geometric entities, thus a 3D point has to be uncertain in all 3 degrees of freedom. Though a work-around was suggested by adding relatively small noise to constrained parameters, a theoretically sound solution is desirable. Especially because the constraints have to be known to the testing and estimation schemes, one could think of adding a rank number r to the representation of an uncertain geometric object $(\mathbf{x}, \boldsymbol{\Sigma}_{\mathrm{xx}})$. This would be sufficient when using a singular value decomposition. In general though, the theoretical and practical impact of this addition has not been covered yet.
- *Correlation between entities:* we assumed that a correlation exists only between the parameters of one entity, but not between two or more entities. It remains to be investigated in which cases the assumption of no correlation seriously affects the result. Essentially, the error propagation in equation (4.23) has to be replaced by (4.22). It seems to be reasonable to investigate this issue in the context of constrained vectors.
- *Estimation of aggregates of entities:* related to the last point, it would be extremely useful to extend the estimation to entity-aggregates, instead of only one entity at a time. One has to develop a unified representation of

aggregate models such as corners, rectangles, cuboids, etc. using geometric entities and a minimal set constraints between them. For example, such a method would allow to optimally estimate a 3D cuboid from a set of image points and edges.

- *Include conics and quadrics:* it would be interesting to see how other entities such as conics and quadrics could fit into the proposed framework. Conics and quadrics can be represented by matrices C, for which a quadratic constraint for all points on the conic resp. quadric holds: $\mathbf{x}^\mathsf{T}\mathsf{C}\mathbf{x} = 0 \Leftrightarrow (\mathbf{x}^\mathsf{T} \otimes \mathbf{x}^\mathsf{T})\text{vec}(\mathsf{C}) = 0$.
- *Use other cues for matching:* with respect to polyhedral object reconstruction, immediate extensions are possible: for a successful automated reconstruction, especially the matching problem, it is inevitable to use more than only geometric knowledge to obtain predictable, robust results. Such knowledge could be obtained by assuming only subclasses of a polyhedron, or by combining other cues such as image intensities close to the entities, which has been successfully demonstrated to be useful by researchers. One promising approach would be to extend the example from section 5.4, where just the identification of corresponding regions yield a considerable reduction of the search-space for finding matching line-segments. The "plane–sweep" approach of BAILLARD AND ZISSERMAN 1999 is a good example for including image intensities in a similar framework
- *Uncalibrated cameras:* in our work we assume the projective camera matrix to be completely known, thus knowing internal and external camera parameters. The vision community has been extensively working on cases with uncalibrated cameras and shown that a reconstruction is possible up to a projective transformation. If one uses only projective relations, all tools for statistical geometric reasoning are applicable for the uncalibrated case. In practice, it remains to be investigated what impact the assumption of uncalibrated cameras has for the reasoning algorithms. We expect though that the reasoning is not affected if the projective reconstruction is close to the Euclidean reconstruction (i.e. the Euclidean infinite plane is far away from the origin of the projective reconstruction with respect to the scene).

A Notation

As stated in section 2.1, we do not distinguish between projective points and and its coordinate vectors compared to FAUGERAS AND LUONG 2001, p. 78. HARTLEY AND ZISSERMAN 2000 take a similar approach, see p.3f.

$x, y, z, a, b, c,$ X, Y, Z, L_i, U, V, W, T	coordinates from $\mathbb{R}$
$\boldsymbol{x}, \boldsymbol{y}$	vectors $\mathbb{R}^2$ points from $\mathbb{R}^2$
$\boldsymbol{x}', \boldsymbol{y}'; \boldsymbol{x}'', \boldsymbol{y}''; \boldsymbol{x}^{(i)}, \boldsymbol{y}^{(i)}$	vectors $\mathbb{R}^2$ points from the first, second or i-th image
$\boldsymbol{X}, \boldsymbol{Y}, \boldsymbol{Z}$	vectors $\mathbb{R}^3$ points from $\mathbb{R}^3$
$\boldsymbol{e}_i, \boldsymbol{E}_i,$	canonical basis $\mathbb{R}^3$ resp. $\mathbb{R}^2$ points, lines or planes from $\mathbb{R}^3$ resp. $\mathbb{R}^2$
$\wedge$	intersection of two projective subspaces, yielding a new projective subspace constructor yielding new points, lines and planes, cf. 3.1
$\cap$	join of two projective subspaces, yielding a new projective subspace constructor yielding new points, lines and planes, cf. 3.1
$\otimes$	Kroneckerproduct
$\mathbf{x}, \mathbf{y}$	homogeneous vectors from $\mathbb{R}^3$ points from $\mathbb{P}^2$
$\mathbf{x}', \mathbf{y}'; \mathbf{x}'', \mathbf{y}''; \mathbf{x}^{(i)}, \mathbf{y}^{(i)}$	homogeneous vectors from $\mathbb{R}^3$ points from the first, second or i-th image

S. Heuel: Uncertain Projective Geometry, LNCS 3008, pp. 179-181, 2004.

continued from previous page

$\mathbf{l}, \mathbf{m}$	homogeneous vectors from $\mathbb{R}^3$ lines from $\mathbb{R}^2$
$\mathbf{l}', \mathbf{m}'; \mathbf{l}'', \mathbf{m}''; \mathbf{l}^{(i)}, \mathbf{m}^{(i)}$	homogeneous vectors from $\mathbb{R}^3$ points from the first, second or i-th image
$\mathsf{N}_s(\mathbf{x})$	spherical normalization of a homogeneous vector
$\mathsf{N}_O(\cdot)$	Euclidean normalization of a homogeneous vector, cf. eq. (2.14) on page 26
$\mathsf{S}(\cdot)$	construction matrices for points and lines from $\mathbb{P}^2$
$\mathbf{X}, \mathbf{Y}, \mathbf{Z}$	homogeneous vectors $\mathbb{R}^4$ points from $\mathbb{P}^3$
$\mathbf{A}, \mathbf{B}, \mathbf{C}, \mathbf{D}$	homogeneous vectors $\mathbb{R}^4$ planes from $\mathbb{P}^3$
$\mathbf{L}, \mathbf{M}$	homogeneous vectors $\mathbb{R}^{5+1}$ lines from $\mathbb{P}^3$
$\overline{\mathbf{L}}, \overline{\mathbf{M}}$	homogeneous vectors $\mathbb{P}^{5+1}$ dual lines from $\mathbb{P}^3$ with respect to $\mathbf{L}, \mathbf{M}$
$\mathbf{e}_i, \mathbf{E}_i,$	canonical basis of $\mathbb{R}^n$ points, lines or planes from $\mathbb{P}^3$ resp. $\mathbb{P}^2$
$\mathsf{\Pi}(\cdot)$	construction matrix for points and planes
$\overline{\mathsf{\Pi}}(\cdot)$	dual construction matrix: $\overline{\mathsf{\Pi}}(\mathbf{A}) = C\mathsf{\Pi}(\cdot)$
$\mathsf{\Gamma}(\mathbf{L})$	construction matrix of a line (Plückermatrix)
$\overline{\mathsf{\Gamma}}(\mathbf{L})$	dual construction matrix $\overline{\mathsf{\Gamma}}(\mathbf{L}) = \mathsf{\Gamma}(\overline{\mathbf{L}})$
Δ_i	6×6 matrix for the identity of lines, see eq. (3.77) on page 76
$\mathsf{M}^{[r]}$	reduction of a matrix M, cf. section 3.2.2 on page 67.
$\boldsymbol{X}_0$, A_0, $\boldsymbol{L}_0$	euclidean parts of homogeneous point, line and plane vectors
X_h, $\mathbf{A}_h$, $\boldsymbol{L}_h$	homogeneous parts of homogeneous point, line and plane vectors
H	3×3 or 4×4 homogeneous matrix homography $\mathbb{P}^2 \to \mathbb{P}^2$ or $\mathbb{P}^3 \to \mathbb{P}^3$
$\mathsf{P} = (\mathbf{A}, \mathbf{B}, \mathbf{C})^\mathsf{T}$	3×4 homogeneous matrix projective camera for points $\mathbb{P}^3 \to \mathbb{P}^2$
$\mathsf{Q} = (\mathbf{L}, \mathbf{M}, \mathbf{N})^\mathsf{T}$ $\mathbf{L} = \mathbf{B} \cap \mathbf{C}$, $\mathbf{M} = \mathbf{C} \cap \mathbf{A}$, $\mathbf{N} = \mathbf{A} \cap \mathbf{B}$	3×6 homogeneous matrix projective camera for mapping 3D lines to 2D lines

continued from previous page	
$\mathbf{p} = \text{vec}(\mathsf{P}), \mathbf{q} = \text{vec}(\mathsf{Q})$, $\mathsf{H}=\text{vec}(\mathsf{H})$	vector representation of transformations $\mathsf{P}, \mathsf{Q}, \mathsf{H}$.
J_{qp}	Jacobian $\frac{\partial \mathbf{q}(\mathbf{p})}{\partial \mathbf{p}}$, cf. eq. (3.51) on page 59
C	dual or hodge operator for 3D lines, see eq. (2.31) on page 44
F	fundamental matrix, see eq. (2.23)
I_n	$n \times n$ identity matrix
$\mathsf{W}(f)$	conditioning matrix for the conditioning factor f, eq. (2.25) on page 39
$\mathcal{I}(\mathsf{M})$	image or column space of a matrix M
$\mathcal{N}(\mathsf{M})$	nullspace or kernel of a matrix M
$\Psi_{A,W}$	orthogonal projection onto the space $\{A\}^{\perp}$ with respect to a scalar product $< \cdot, \cdot >_W$.

$\underline{x}$	random variable
$\underline{\mathbf{x}}$	random homogeneous vector
p_{x}	probability density function of sqx
$\boldsymbol{\mu}_{\mathrm{x}}$	mean value of a random homogeneous vector $\underline{\mathbf{x}}$
$\boldsymbol{\Sigma}_{\mathrm{xx}}$	covariance matrix of a random homogeneous vector $\underline{\mathbf{x}}$
$\tilde{\beta}$	true unknown entity
$\tilde{\beta}$	estimation for an unknown entity
$\mathbf{y}$	observations

B Linear Algebra

B.1 Ranks and Nullspaces

For the analysis of homogeneous covariance matrices in chapter 4 we need some statements on the rank of matrix products and the nullspace of matrix sums.

Proposition B.1 (Rank of matrix products) *Let A be a $n \times m$ matrix and B a $m \times r$ matrix. Then*

$$\begin{aligned} &rk(AB) < min(rk(A), rk(B)) \\ &\qquad \Leftrightarrow \quad dim(\mathcal{N}(A) \cap Image(B))) > max(rk(B) - rk(A), 0) \end{aligned} \tag{B.1}$$

Especially, if $rk(B) \leq rk(A)$ then

$$rk(AB) < min(rk(A), rk(B)) \quad \Leftrightarrow \quad dim(\mathcal{N}(A) \cap \mathcal{N}((B^{\mathsf{T}})^{\perp})) > 0$$

Proof: The rank $rk(AB)$ of the matrix multiplication is at most equal to $\min(rk(A), rk(B))$. If $rk(AB) < \min(rk(A), rk(B))$ or $rk(AB) = \min(rk(A), rk(B))$ depends on the relative position of the nullspace $\mathcal{N}(A)$ to the image or column space $\mathcal{I}(B)$. Equality holds if and only if $\mathcal{N}(A)$ is *transversal* to the image $\mathcal{I}(B)$, which is defined as

$$dim(\mathcal{N}(A) \cap \mathcal{I}(B)) = \max(rk(B) - rk(A), 0)$$

One obtains

$$rk(AB) < \min(rk(A), rk(B)) \Leftrightarrow dim(\mathcal{N}(A) \cap \mathcal{I}(B)) > \max((rk(B) - rk(A), 0)$$

The second equation follows from the fact that $\mathcal{I}(B) = (\mathcal{N}((B^T)^{\perp}))$ $\square$

Proposition B.2 (Nullspace of sum of matrices) *Let A and B be two symmetric, positive semidefinite $n \times n$ matrices. Then*

$$\mathcal{N}(A + B) = \mathcal{N}(A) \cap \mathcal{N}(B)$$

S. Heuel: Uncertain Projective Geometry, LNCS 3008, pp. 183-185, 2004.

Proof: If $\mathbf{x} \in \mathcal{N}(A) \cap \mathcal{N}(B)$, then $A\mathbf{x} + B\mathbf{x} = \mathbf{0}$ or $(A + B)\mathbf{x} = \mathbf{0}$, thus $\mathbf{x} \in \mathcal{N}(A + B)$. To show the other direction of the equivalence, assume $\mathbf{x} \in \mathcal{N}(A + B)$, then $\mathbf{x}^\mathsf{T}(A\mathbf{x} + B\mathbf{x}) = 0$ which is $\mathbf{x}^\mathsf{T}A\mathbf{x} + \mathbf{x}^\mathsf{T}B\mathbf{x} = 0$. Since A and B are positive semi-definite, the last equation holds only if $\mathbf{x} \in \mathcal{N}(A)$ and $\mathbf{x} \in \mathcal{N}(B)$. □

B.2 Orthogonal Projections

Orthogonal projections map the vector space $\mathbb{R}^n$ onto a subspace U or its orthogonal complement $U^\perp$.

Proposition B.3 (Orthogonal projection) *Let A be a $n \times m$ matrix. The orthogonal projection Ψ of $\mathbb{R}^n$ onto the space $\{A\}^\perp$ orthogonal to the column space A is given by*

$$\Psi_A = I - A(A^\mathsf{T}A)^{-1}A^\mathsf{T} \tag{B.2}$$

The introduced orthogonal projection was introduced according to the scalar product $\boldsymbol{x}^\mathsf{T}\boldsymbol{y}$, but may also be defined using a weighted scalar product $\boldsymbol{x}^\mathsf{T}W\boldsymbol{y}$ with a positive semidefinite $n \times n$ matrix W.

Proposition B.4 (General orthogonal projection) *Let A be a $n \times m$ matrix and. The orthogonal projection Ψ of $\mathbb{R}^n$ which maps onto the space $\{A\}^\perp$ orthogonal to the column space A according to the scalar product $\boldsymbol{x}^\mathsf{T}W\boldsymbol{y}$ is given by*

$$\Psi_{A,W} = I - A(A^\mathsf{T}WA)^{-1}A^\mathsf{T}W \tag{B.3}$$

B.3 Kronecker Product and vec(·) Operator

The Kronecker product and the vec(·)-operator are important to derive trilinear relations between geometric entities and their transformations.

Definition B.1 (Kronecker product) *Let $A = (a_{ij})$ be a $m \times n$ matrix and $B = (b_{ij}$ be a $p \times q$ matrix. Then the* Kronecker product *$A \otimes B$ of A and B defines a $mp \times nq$ matrix*

$$A \otimes B := \begin{pmatrix} a_{11}B & \cdots & a_{1n}B \\ \cdots & \ddots & \cdots \\ a_{m1}B & \cdots & a_{mn}B \end{pmatrix}$$

Definition B.2 (vec() operator) *Let $A = (a_{ij})$ be a $m \times n$ matrix, then* vec(A) *is a $mn \times 1$ vector, that is formed by stacking the columns of the matrix A:*

$$\text{vec}(A) := (a_{11}, \cdots, a_{m1}, a_{12}, \cdots, a_{mn})^\mathsf{T}$$

The following important proposition establishes a relationship between the vec(·)-operator and the Kronecker product, see e.g. KOCH 1999.

Proposition B.5 *Let A be a $n \times m$ matrix, B a $n \times p$ matrix, C a $p \times s$ matrix and $\boldsymbol{x}$ an n-vector. Then*

$$\operatorname{vec}(ABC) = (C^\top \otimes A)\operatorname{vec}(B) = (A \otimes C^\top)\operatorname{vec}(B^\top) \tag{B.4}$$

$$A\boldsymbol{x} = \operatorname{vec}(A\boldsymbol{x}) = (\boldsymbol{x}^\top \otimes I_m)\operatorname{vec}(A) \tag{B.5}$$

$$= \operatorname{vec}(\boldsymbol{x}^\top A^\top) = (I_m \otimes \boldsymbol{x}^\top)\operatorname{vec}(A^\top) \tag{B.6}$$

C Statistics

C.1 Covariance Matrices for 2D Lines

C.1.1 Uncertainty of a 2D Line

We adopt the representation of FÖRSTNER 1992 for the uncertainty of a straight line and start with the angle-distance form of a line

$$x\, cos(\underline{\varphi}) + y\, sin(\underline{\varphi}) = \underline{d} \tag{C.1}$$

The angle φ and the distance d are uncertain and with the covariance $\sigma_{\varphi d}$ we obtain the covariance matrix of the vector $(\varphi, d)^{\mathsf{T}}$

$$\Sigma_{\varphi d} = \begin{pmatrix} \sigma_\varphi^2 & \sigma_{\varphi d} \\ \sigma_{\varphi d} & \sigma_d^2 \end{pmatrix} \tag{C.2}$$

We first assume the simplest case of a line $\mathbf{l}_x(\underline{\varphi}, \underline{d})$ being the x-axis in the Euclidean plane with $\mu(\underline{\varphi}) = 90°$ and $\mu(\underline{d}) = 0$. We further assume both parameters to be uncorrelated, thus $\sigma_{\varphi d} = 0$. We can reformulate the line $\mathbf{l}_x$ in the slope-intercept form $y = \underline{m}\,x + \underline{b}$, where $\mu(\underline{m}) = \mu(\underline{b}) = 0$. In this case $\underline{b} = \underline{d}$ and $\underline{m} = cos(\underline{\varphi})$ with σ_b and σ_m being uncorrelated. Then the displacement error orthogonal to the line is by simple and rigorous error propagation

$$\sigma_y^2 = \sigma_m^2\, x^2 + \sigma_b^2 \tag{C.3}$$

This clearly gives us an hyperbola as a confidence region, cf. figure C.1: the line is most precise at the origin with an error of σ_d^2 orthogonal to the line. The further we move along the line by x, the larger is the displacement error. Using first order error propagation, $\sigma_m^2 \approx sin(\mu(\varphi)) \cdot \sigma_\varphi^2 = \sigma_\varphi^2$, thus for this special case we can identify $\sigma_b = \sigma_d$ and $\sigma_\varphi = \sigma_m$. The last approximation is valid if the $\sigma_\varphi < 15°$, as $\mathrm{d}sin(x) \approx \mathrm{d}x$ for small x.
Expressing $\mathbf{l}_x(\underline{\varphi}, \underline{d})$ in homogeneous coordinates $\underline{\mathbf{l}}_x = (\underline{u}, \underline{v}, \underline{w})^{\mathsf{T}}$ can be easily done some by the following mapping: $\underline{u} = \underline{m} = \cos(\underline{\varphi})$, $\underline{v} = \sin(\underline{\varphi})$ and $\underline{w} = \underline{d}$, yielding:

S. Heuel: Uncertain Projective Geometry, LNCS 3008, pp. 187-195, 2004.

$$(\mathbf{l}_x, \boldsymbol{\Sigma}_{\mathrm{l_x l_x}}) = \left(\begin{pmatrix} 0 \\ 1 \\ 0 \end{pmatrix}, \begin{pmatrix} \sigma_\varphi^2 & 0 & 0 \\ 0 & 0 & 0 \\ 0 & 0 & \sigma_d^2 \end{pmatrix} \right) \tag{C.4}$$

We may now represent any 2D line $\mathbf{l}$ based on the above representation $\mathbf{l}_x$ of the $x - axis$ by rotating the line by some angle α and and translating by a vector $\mathbf{x}_t = (t_x, t_y)$.

$$\begin{aligned} \mathbf{l} &= \mathsf{T}(\mathbf{x}_t)\mathsf{R}(\alpha)\mathbf{l}_x \\ &= \begin{pmatrix} 1 & 0 & 0 \\ 0 & 1 & 0 \\ -t_x & -t_y & 1 \end{pmatrix} \begin{pmatrix} \sin(\alpha) & \cos(\alpha) & 0 \\ -\cos(\alpha) & \sin(\alpha) & 0 \\ 0 & 0 & 1 \end{pmatrix} \begin{pmatrix} 0 \\ 1 \\ 0 \end{pmatrix} \end{aligned}$$

The line $\mathbf{l}$ has now a normal with mean angle $\mu(\underline{\varphi}_l) = \alpha$ and a distance $\mu(\underline{d}_l) = t_x \cos(\alpha) + t_y \sin(\alpha)$. The covariance matrix $\boldsymbol{\Sigma}_{\mathrm{ll}}$ of the line $\mathbf{l}$ is given by

$$\boldsymbol{\Sigma}_{\mathrm{ll}} = \mathsf{T}(\mathbf{x}_t)\mathsf{R}(\alpha)\boldsymbol{\Sigma}_{\mathrm{l_x l_x}}\mathsf{R}^{\mathsf{T}}(\alpha)\mathsf{T}^{\mathsf{T}}(\mathbf{x}_t) \tag{C.5}$$

Equation (C.5) yields a rank-2 matrix, where in general all entries are non-zero. In particular, this means that we have a correlation between the angle and the distance, introduced by the translation. This is depicted in figure C.1. Thus another valid representation of an uncertain 2D line is the 5-tuple

$$\boldsymbol{l}_e = (x_m, y_m, \varphi, \sigma_d, \sigma_\varphi) \tag{C.6}$$

The point $(x_m, y_m)^{\mathsf{T}}$ is the center point or point of gravity of the line, where the displacement error orthogonal to the line is minimal.

C.1.2 Euclidean Interpretation of Homogeneous Covariances

Assume a line $\mathbf{l} = (a, b, c)^{\mathsf{T}}$ is given with its covariance matrix

$$\boldsymbol{\Sigma}_{\mathrm{ll}} = \begin{pmatrix} \sigma_a^2 & \sigma_{ab} & \sigma_{ac} \\ \sigma_{ab} & \sigma_b^2 & \sigma_{bc} \\ \sigma_{ac} & \sigma_{bc} & \sigma_c^2 \end{pmatrix}$$

either by construction or by error propagation from some other representation, cf. C.1.1. The covariance matrix $\boldsymbol{\Sigma}_{\mathrm{ll}}$ may have rank 2 or 3. We are interested in the Euclidean interpretation of the homogeneous covariance matrix is sought as in (C.6). Among others, CLARKE 1998 or FAUGERAS AND

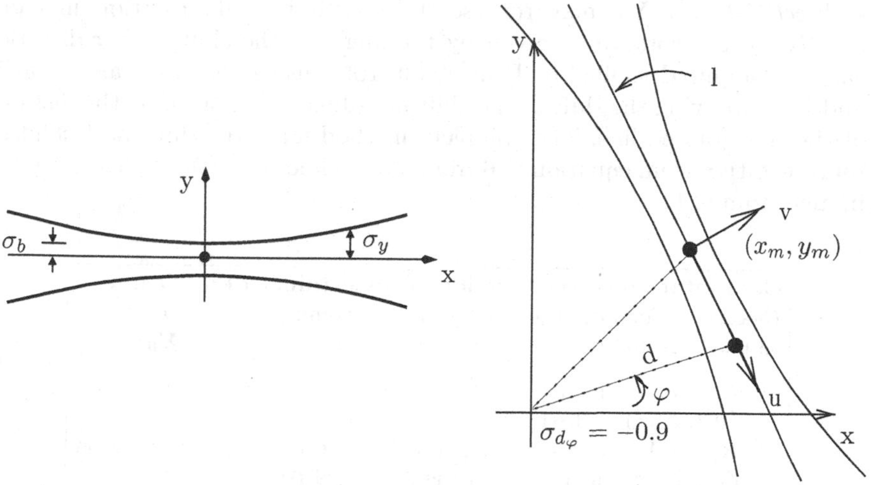

Fig. C.1. The uncertainty of a 2D line can be represented by a hyperbola (*left*). The point $(x_m, y_m)^\mathsf{T}$ with smallest uncertainty is called the center point or point of gravity of the line. Its position depends on the covariance $\sigma_{\varphi d}$ (*right*).

LUONG 2001 suggest to use the duality between the point space $\mathbb{P}^2$ and the line space $(\mathbb{P}^2)^*$, where an ellipse in $(\mathbb{P}^2)^*$ yields an hyperbola in $\mathbb{P}^2$.
An alternative method is to use the Euclidean interpretation of $\mathbf{l} = (a, b, c)^\mathsf{T}$ and $\boldsymbol{\Sigma}_{\mathrm{ll}}$ as outlined in chapter 2 and 4. As proposed in section 2.2.1, a normalization $\mathsf{N}_O(\mathbf{l})$ yields the angle-distance form (C.1), but we still do not know the center point $\boldsymbol{x}_m = (x_m, y_m)$ of the line. The center point is determined by the covariance $\sigma_{\varphi d}$ or alternatively σ_{ac}, σ_{bc} which is supposed to be zero for the representation (C.6).

Ad hoc Solution To determine the center point, one has to find a line translation $\mathsf{T}(\boldsymbol{x}_m)$, such that $\sigma_{ac} = \sigma_{bc} = 0$, thus we have to find a unique translation such that for the translated line $\mathbf{l}_t$ the covariance matrix

$$\boldsymbol{\Sigma}_{\mathrm{l_t l_t}} = \mathsf{T}(\boldsymbol{x}_m)\boldsymbol{\Sigma}_{\mathrm{ll}}\mathsf{T}^\mathsf{T}(\boldsymbol{x}_m)$$

has zero entries for the elements $(1, 3)$ and $(2, 3)$. It turns out that the solution is

$$\boldsymbol{x}_m = \begin{pmatrix} \frac{\sigma_{ab}\cdot\sigma_{bc} - \sigma_{ac}\cdot\sigma_b^2}{\sigma_a^2\cdot\sigma_b^2 - \sigma_{ab}^2} \\ \frac{\sigma_a^2\cdot\sigma_{bc} - \sigma_{ab}\cdot\sigma_{ac}}{\sigma_a^2\cdot\sigma_b^2 - \sigma_{ab}^2} \end{pmatrix}$$

and therefore is undefined if $\sigma_a^2\sigma_b^2 = \sigma_{ab}^2$, which means that $\underline{a}$ and $\underline{b}$ are correlated by factor 1. But after a Euclidean normalization, the first two coordinates of a line vector refer to the normal of the line and thus $\underline{a}$ and $\underline{b}$ are indeed completely correlated. Therefore a direct determination of the center point fails.

Indirect Solution We may reverse the translation and rotation in equation (C.5): the rotation is given by the angle φ, the shift u in x-direction can be computed similar to FÖRSTNER 1992 using the covariance matrix and after applying the shift u, the shift in y-direction is given by the distance of the transformed line. The complete method for extracting the Euclidean representation from equation C.6 from a 3×3 line covariance matrix is given in algorithm C.1.

Computation of the Euclidean uncertainty of a 2D line
Objective: Compute the Euclidean uncertainty, cf. equation (C.6) on page 188 from an uncertain homogeneous 2D line $(\mathbf{l}, \boldsymbol{\Sigma}_{\mathrm{ll}})$.

1. Normalize the line
   $(\mathbf{l}^{(1)}, \boldsymbol{\Sigma}_{\mathrm{ll}}^{(1)}) = \mathsf{N}_O(\mathbf{l}, \boldsymbol{\Sigma}_{\mathrm{ll}})$
2. Rotate $\mathbf{l}_1$ such that it is parallel to the x-axis, R is a rotation defined by the angle φ of the normal of $\mathbf{l}_1$:
   $\left(\mathbf{l}^{(2)}, \boldsymbol{\Sigma}_{\mathrm{ll}}^{(2)}\right) = \left(\mathsf{R}\mathbf{l}^{(1)},\ \mathsf{R}\boldsymbol{\Sigma}_{\mathrm{ll}}^{(1)}\mathsf{R}^{\mathsf{T}}\right)$
3. Determine the shift u of the center of the rotated line $\mathbf{l}^{(2)}$ to the y-axis parallel to the x -axis. As in FÖRSTNER 1992, $u = -\sigma_{(1,2)}^{(2)}/\sigma_{(1,1)}^{(2)}$, where $\sigma_{(i,j)}^{(2)}$ is the (i,j)-th element of $\boldsymbol{\Sigma}_{\mathrm{ll}}^{(2)}$.
   $\left(\mathbf{l}^{(3)}, \boldsymbol{\Sigma}_{\mathrm{ll}}^{(3)}\right) = \left(\mathsf{T}(u)\mathbf{l}^{(2)},\ \mathsf{T}(u)\boldsymbol{\Sigma}_{\mathrm{ll}}^{(2)}\mathsf{T}(u)^{\mathsf{T}}\right)$
4. Now determine the shift v along the y-axis, such that the line $\mathbf{l}^{(3)}$ is equal to the x-axis. The shift v is simply the distance of $\mathbf{l}^{(2)}$ to the origin, thus the third coordinate of $\mathbf{l}^{(2)}$.
   $\left(\mathbf{l}^{(4)}, \boldsymbol{\Sigma}_{\mathrm{ll}}^{(4)}\right) = \left(\mathsf{T}(v)\mathbf{l}^{(3)},\ \mathsf{T}(v)\boldsymbol{\Sigma}_{\mathrm{ll}}^{(3)}\mathsf{T}(v)^{\mathsf{T}}\right)$
5. The covariance matrix $\boldsymbol{\Sigma}_{\mathrm{ll}}^{(4)}$ has two non-zero entries being σ_φ^2 and σ_d^2 as in equation (C.4). The point of gravity is contained in the motion $\mathsf{H} = \mathsf{T}(v)\mathsf{T}(u)\mathsf{R}$, the angle φ was determined in step 2.

Algorithm C.1: Algorithm for the computation of the Euclidean uncertainty of a 2D line given a homogeneous covariance matrix

C.2 Gauss Helmert Estimation

We will shortly describe the Gauss-Helmert model and its iterative solution as it is used in section 4.6. A reference for the Gauß-Helmert model is MIKHAIL AND ACKERMANN 1976, the description here follows the one in FÖRSTNER 2000. We also give an efficient variation of the Gauss-Helmert model, which can be applied to the estimation problems discussed in chapter 4.

C.2.1 General Gauss Helmert Model

The Gauss-Helmert-model is assumed to be given by the following entities (cf. MIKHAIL AND ACKERMANN 1976): The W constraints $\boldsymbol{w}(\tilde{\boldsymbol{\beta}}, \tilde{\boldsymbol{y}}) = \mathbf{0}$ between the unknown parameters $\boldsymbol{\beta}$ and the true observations $\tilde{\boldsymbol{y}}$ are assumed to hold. Also the H constraints $\boldsymbol{h}(\boldsymbol{\beta}) = \mathbf{0}$ between the unknown parameters alone are given.

Additionally, we assume G constraints $\mathbf{g}(\tilde{\boldsymbol{y}}) = \mathbf{0}$ only for the observations $\boldsymbol{y}$. First, we will neglect this constraint and impose it later on onto the fitted observation $\widehat{\boldsymbol{y}}$, see also section 4.6.2.

The true observation vector is related to the observed N-vector by $\tilde{\boldsymbol{y}} = \boldsymbol{y} - \boldsymbol{e}$, where the errors $\boldsymbol{e}$ are unknown. The observations $\underline{\mathbf{y}}$ are assumed to be distributed normally with $\underline{\boldsymbol{y}} \sim N(\tilde{\boldsymbol{y}}, \sigma^2 Q_{yy})$. Finding optimal estimates for $\boldsymbol{\beta}$ and $\tilde{\boldsymbol{y}}$ can be done by minimizing

$$\Phi(\tilde{\boldsymbol{y}}, \tilde{\boldsymbol{\beta}}, \boldsymbol{\lambda}, \boldsymbol{\mu}) = \frac{1}{2}(\boldsymbol{y} - \tilde{\boldsymbol{y}})^T Q_{yy}^{-1}(\boldsymbol{y} - \tilde{\boldsymbol{y}}) + \boldsymbol{\lambda}^T \boldsymbol{g}(\tilde{\boldsymbol{\beta}}, \tilde{\boldsymbol{y}}) + \boldsymbol{\mu}^T \boldsymbol{h}(\tilde{\boldsymbol{\beta}}) \qquad \text{(C.7)}$$

with the W-vector $\boldsymbol{\lambda}$ and the H-vector $\boldsymbol{\mu}$ are Lagrangian multipliers.

For solving this nonlinear problem in an iterative manner we need approximate values $\widehat{\boldsymbol{\beta}}^{(0)}$ and $\widehat{\mathbf{y}}^{(0)}$ for the unknowns $\widehat{\boldsymbol{\beta}} = \widehat{\boldsymbol{\beta}}^{(0)} + \widehat{\boldsymbol{\Delta\beta}}$ and $\widehat{\boldsymbol{y}} = \widehat{\boldsymbol{y}}^{(0)} + \widehat{\boldsymbol{\Delta}\mathbf{y}}$. The corrections for the unknowns and the observation are obtained iteratively. With the Jacobians

$$A = \left(\frac{\partial \boldsymbol{w}(\boldsymbol{\beta}, \boldsymbol{y})}{\partial \boldsymbol{\beta}}\right)\bigg|_{\widehat{\boldsymbol{\beta}}^{(0)}, \boldsymbol{y}=\widehat{\boldsymbol{y}}^{(0)}} \quad B = \left(\frac{\partial \boldsymbol{w}(\boldsymbol{\beta}, \boldsymbol{y})}{\partial \boldsymbol{y}}\right)\bigg|_{\widehat{\boldsymbol{\beta}}^{(0)}, \boldsymbol{y}=\widehat{\boldsymbol{y}}^{(0)}} \quad H = \left(\frac{\partial \boldsymbol{h}(\boldsymbol{\beta})}{\partial \boldsymbol{\beta}}\right)\bigg|_{\widehat{\boldsymbol{\beta}}^{(0)}} \qquad \text{(C.8)}$$

and the relation $\widehat{\boldsymbol{\Delta}\mathbf{y}} = (\boldsymbol{y} - \widehat{\boldsymbol{y}}^{(0)}) - \widehat{e}$ we obtain the linear constraints $\boldsymbol{w}(\widehat{\boldsymbol{\beta}}, \widehat{\boldsymbol{y}}) = \boldsymbol{w}(\widehat{\boldsymbol{\beta}}^{(0)}, \widehat{\boldsymbol{y}}^{(0)}) + A\widehat{\boldsymbol{\Delta\beta}} + B\widehat{\boldsymbol{\Delta y}}$ or

$$\boldsymbol{w}(\widehat{\boldsymbol{\beta}}, \widehat{\boldsymbol{y}}) = \boldsymbol{c}_w + A\widehat{\boldsymbol{\Delta\beta}} - B\widehat{e} \qquad \text{(C.9)}$$

$$\boldsymbol{h}(\widehat{\boldsymbol{\beta}}) = \boldsymbol{c}_h + H\widehat{\boldsymbol{\Delta\beta}} \qquad \text{(C.10)}$$

with

$$\boldsymbol{c}_w = \boldsymbol{w}(\widehat{\boldsymbol{\beta}}^{(0)}, \widehat{\boldsymbol{y}}^{(0)}) + B(\boldsymbol{y} - \widehat{\boldsymbol{y}}^{(0)}) \quad \text{and} \quad \boldsymbol{c}_h = \boldsymbol{h}(\widehat{\boldsymbol{\beta}}^{(0)}) \qquad \text{(C.11)}$$

are the contradictions between the approximate values for the unknown parameters and the given observations and among the approximate values for the unknowns.

Setting the partials of Φ from (C.7) zero yields

$$\frac{\partial \Phi}{\partial \widehat{\boldsymbol{y}}^\mathsf{T}} = -Q_{yy}^{-1}\widehat{e} + B^\mathsf{T}\boldsymbol{\lambda} = \mathbf{0} \qquad \frac{\partial \Phi}{\partial \widehat{\boldsymbol{\beta}}^\mathsf{T}} = A^\mathsf{T}\boldsymbol{\lambda} + H^\mathsf{T}\boldsymbol{\mu} = \mathbf{0} \qquad \text{(C.12)}$$

$$\frac{\partial \Phi}{\partial \boldsymbol{\lambda}^\mathsf{T}} = \boldsymbol{c}_w + A\widehat{\boldsymbol{\Delta\beta}} - B\widehat{e} = \mathbf{0} \qquad \frac{\partial \Phi}{\partial \boldsymbol{\mu}^\mathsf{T}} = \boldsymbol{c}_h + H\widehat{\boldsymbol{\Delta\beta}} = \mathbf{0} \qquad \text{(C.13)}$$

From (C.12a) follows the relation

$$\widehat{e} = Q_{yy} B^{\mathsf{T}} \boldsymbol{\lambda} \tag{C.14}$$

When substituting (C.14) into (C.13a), solving for $\boldsymbol{\lambda}$ yields

$$\boldsymbol{\lambda} = (B Q_{yy} B^{\mathsf{T}})^{-1} (c_w + A \widehat{\Delta\beta}) \tag{C.15}$$

Substitution in (C.12b) yields the symmetric normal equation system

$$\begin{pmatrix} A^{\mathsf{T}} (B Q_{yy} B^{\mathsf{T}})^{-1} A & \boldsymbol{H}^{\mathsf{T}} \\ H & \mathbf{0} \end{pmatrix} \begin{pmatrix} \widehat{\Delta\beta} \\ \mu \end{pmatrix} = \begin{pmatrix} -A^{\mathsf{T}} (B Q_{yy} B^{\mathsf{T}})^{-1} c_w \\ -c_h \end{pmatrix} \tag{C.16}$$

The Lagrangian multipliers can be obtained from (C.15) which then yields the estimated residuals $\widehat{e}$ in (C.14). Using the residuals, the fitted observations are given by $\widehat{\boldsymbol{y}} = \boldsymbol{y} - \widehat{\mathbf{e}}$. To impose the constraints $\mathbf{g}(\widehat{\boldsymbol{y}}) = \mathbf{0}$, we solve a reduced Gauß-Helmert model with no unknowns. then $\mathbf{w}'(\tilde{\mathbf{y}}, \tilde{\boldsymbol{\beta}}) = \mathbf{g}(\tilde{\boldsymbol{\beta}})$ and of course $\mathbf{h}'(\tilde{\boldsymbol{\beta}})$ undefined. Then the new estimation $\widehat{\widehat{\boldsymbol{y}}}$ is

$$\widehat{\widehat{\boldsymbol{y}}} = \boldsymbol{y} - \boldsymbol{\Sigma}_{\mathrm{yy}} \mathrm{G}^{\mathsf{T}} (\mathrm{G} \boldsymbol{\Sigma}_{\mathrm{yy}} \mathrm{G})^{-1} \mathbf{g}(\widehat{\boldsymbol{y}})$$

The estimated variance factor $\widehat{\sigma}^2$ is given by

$$\widehat{\sigma}^2 = \frac{\widehat{e}^{\mathsf{T}} Q_{yy}^{-1} \widehat{e}}{G + H - U} \tag{C.17}$$

The number R of constraints above the number $U - H$, which is necessary for determining the unknown parameters, the *redundancy* is the denominator $R = G - (U - H)$. We finally obtain the *estimated* covariance matrix

$$\widehat{\boldsymbol{\Sigma}}_{\widehat{\beta}\widehat{\beta}} = \widehat{\sigma}^2 Q_{\widehat{\beta}\widehat{\beta}} \tag{C.18}$$

of the estimated parameters, where $Q_{\widehat{\beta}\widehat{\beta}}$ results from the inverted reduced normal equation matrix using $\overline{N} = A^{\mathsf{T}} (\boldsymbol{B} Q_{yy} B^{\mathsf{T}})^{-1} A$

$$\begin{pmatrix} Q_{\widehat{\beta}\widehat{\beta}} & \boldsymbol{S}^{\mathsf{T}} \\ \boldsymbol{S} & \boldsymbol{T} \end{pmatrix} = \begin{pmatrix} \overline{N} & H^{\mathsf{T}} \\ H & \mathbf{0} \end{pmatrix}^{-1} \tag{C.19}$$

We have summarized the approach in algorithm in table C.2. For a criterion to stop the iteration we require that the changes $\widehat{\Delta\beta}^{(i)}$ to the estimation should be less than 1% with respect to their uncertainty, thus $\widehat{\Delta\beta}^{(i)}/\mathrm{trace}(\boldsymbol{\Sigma}_{bb}) < 10^{-2}$.

Iterative Optimization using the Gauß-Helmert model

Objective: Compute an estimate for an unknown vector β and an observation vector y using the Gauss-Helmert Model, see section C.2.

1 $W = \dim(w);\ H = \dim(h);\ U = \dim(\widehat{\beta})$

2 $\widehat{y}^{(0)} = y$ // given observations

3 $\widehat{\beta}^{(0)}$ // initial value of unknowns

4 for i from 0 to MaxIter1 do

5   // compute Jacobians of constraint

6    $A = \dfrac{\partial w(y,\beta)}{\partial \beta}\Big|_{(\widehat{y}^{(i)},\widehat{\beta}^{(i)})}$

7    $B = \dfrac{\partial w(y,\beta)}{\partial y}\Big|_{(\widehat{y}^{(i)},\widehat{\beta}^{(i)})}$

8    $H = \dfrac{\partial h(\beta)}{\partial \beta}\Big|_{(\widehat{\beta}^{(i)})}$

9    $c_w = w_j(\widehat{y}^{(i)},\widehat{\beta}^{(i)}) + B(y - \widehat{y}^{(i)})$

10   $c_h = h(\widehat{\beta}^{(i)})$

11   $\Sigma_{ww}^{-1} = (B\,\Sigma_{yy}\,B^{\mathsf{T}})^{-1}$

12   $\begin{pmatrix} \widehat{\Delta\beta}^{(i)} \\ \mu \end{pmatrix} = \begin{pmatrix} A^{\mathsf{T}}\Sigma_{ww}^{-1}A & H^{\mathsf{T}} \\ H & 0 \end{pmatrix}^{-1} \begin{pmatrix} -A^{\mathsf{T}}\Sigma_{ww}^{-1}c_w \\ -c_h \end{pmatrix}$

13   $\widehat{\beta}^{(i+1)} = \widehat{\beta}^{(i)} + \widehat{\Delta\beta}^{(i)}$

14   $\lambda = \Sigma_{ww}^{-1}(c_w + A\widehat{\Delta\beta}^{(i)})$

15   $\widehat{e} = \Sigma_{yy}B^{\mathsf{T}}\lambda$

16   $\widehat{y}^{(i+1)} = y - \widehat{e}$

17   $\widehat{\widehat{y}}^{(0)} = \widehat{y}^{(i+1)}$

18   for l from 0 to MaxIter2 do

19     $G = \dfrac{\partial g(y)}{\partial y}\Big|_{(\widehat{\widehat{y}}^{(l)})}$

20     $\widehat{\widehat{y}}^{(l+1)} = \widehat{y}^{(i+1)} - \Sigma_{yy}G^{\mathsf{T}}(G\Sigma_{yy}G^{\mathsf{T}})^{-1}g(\widehat{\widehat{y}}^{(l)})$

21   done

22   $\Omega^2 = \left(\widehat{e}^{\mathsf{T}}\Sigma_{yy}^{-1}\widehat{e}\ \underline{\text{or}}\ \widehat{e}^{\mathsf{T}}B^{\mathsf{T}}\lambda\ \underline{\text{or}}\ \widehat{e}^{\mathsf{T}}B^{\mathsf{T}}\Sigma_{ww}^{-1}B\widehat{e}\right)$

23   $\widehat{\sigma}^2 = \Omega^2/(W + H - U)$

24   last if $\widehat{\Delta\beta}^{(i)}/\text{trace}(\Sigma_{bb}) < 10^{-2}$

25 done

26 return $(\widehat{\beta}^{(i)}, \widehat{\widehat{y}}^{(l)}, \widehat{\sigma}^2)$

Algorithm C.2: Iterative Solution for the optimization problem using the Gauß-Helmert model, see text for details.

C.2.2 General Gauss Helmert Model with Block Structure

We now want to introduce a variant of the above solution, which is given as follows: assume an unknown object $\tilde{\boldsymbol{\beta}}$ with n blocks of observations, where each block contains m_j observation sub-vector $\boldsymbol{y}_{(j,k)}$ of the same type, where $j = 1, \cdots, n$, $k = 1, \cdots, m_j$. The observation model function $\boldsymbol{w}_j(\boldsymbol{y}_{j,k}, \boldsymbol{\beta})$ are given for each observation sub-vector, where the function remains identical within each block j. The observation model function is bilinear of the form

$$\boldsymbol{w}_j(\boldsymbol{y}_{j,k}, \boldsymbol{\beta}) = \mathsf{A}_j(\boldsymbol{y}_{j,k})\boldsymbol{\beta} = \mathsf{B}_j(\boldsymbol{\beta})\boldsymbol{y}_{j,k}$$

The covariance matrices for each observation sub-vector $\boldsymbol{\Sigma}_{yy,(j,k)}$, that are uncorrelated among each other.
With the above assumptions, the computational cost of the iterative estimation can be greatly reduced due to the special structures of the Jacobians A, B and the covariance matrix $\boldsymbol{\Sigma}_{yy}$.

$$\begin{aligned} \mathsf{A} &= \operatorname{diag}\left(\mathsf{A}_1(\boldsymbol{y}_{1,1}), \cdots, \mathsf{A}_1(\boldsymbol{y}_{1,m_1}), \cdots \mathsf{A}_n(\boldsymbol{y}_{n,1}), \cdots, \mathsf{A}_1(\boldsymbol{y}_{n,m_n})\right) \\ \mathsf{B} &= \operatorname{diag}\left(\mathsf{B}_1(\boldsymbol{\beta}), \cdots, \mathsf{A}_1(\boldsymbol{\beta}), \cdots \mathsf{B}_n(\boldsymbol{\beta}), \cdots, \mathsf{B}_1(\boldsymbol{\beta})\right) \\ \boldsymbol{\Sigma}_{yy} &= \operatorname{diag}\left(\boldsymbol{\Sigma}_{yy,(1,1)}, \cdots, \boldsymbol{\Sigma}_{yy,(1,m_1)}, \cdots \boldsymbol{\Sigma}_{yy,(n,m_n)}, \cdots, \boldsymbol{\Sigma}_{yy,(n,m_n)}\right) \end{aligned} \tag{C.20}$$

Then the upper left sub-matrix of the normal equation (C.16) can be written as the sum of components:

$$\mathsf{A}^{\mathsf{T}}(\mathsf{B}\mathsf{Q}_{yy}\mathsf{B}^{\mathsf{T}})^{-1}\mathsf{A} = \sum_{(j,k)} \mathsf{A}^{\mathsf{T}}_{(j,k)} \left(\mathsf{B}_{(j,k)}\ \boldsymbol{\Sigma}_{yy,(j,k)}\ \mathsf{B}^{\mathsf{T}}_{(j,k)}\right)^{-1} \mathsf{A}_{(j,k)}$$

Additionally, one can efficiently compute the fitted observations $\widehat{\mathbf{y}}$ and obtains a significant speed-up, especially when the redundancy is large. The modified algorithm is listed in table C.3.

Iterative Block-Optimization using the Gauß-Helmert model

Objective: Compute an estimate for an unknown vector $\boldsymbol{\beta}$ and an observation vector $\boldsymbol{y}$ using the Gauss-Helmert Model, see section C.2. This algorithm takes advantage of the existence of blocks of independent observations, see text for details.

$W = \sum_j \dim(\boldsymbol{w}_j)$; $H = \dim(\boldsymbol{h})$; $U = \dim(\widehat{\boldsymbol{\beta}})$; $G = \sum_j \dim(\boldsymbol{w}_j)$

$\sum_{(j,k)} \widehat{\boldsymbol{y}}_{j,k}^{(0)} = \boldsymbol{y}_{(j,k)}$ // given observations

$\widehat{\boldsymbol{\beta}}^{(0)}$ // initial value of unknowns

for i from 0 to MaxIter1 do

    // compute Jacobians of constraint

    $A_{(j,k)} = \left.\frac{\partial \mathbf{w}_j(\boldsymbol{y},\boldsymbol{\beta})}{\partial \boldsymbol{\beta}}\right|_{\widehat{\boldsymbol{y}}^{(i)}_{(j,k)},\widehat{\boldsymbol{\beta}}^{(i)}}$

    $B_{(j,k)} = \left.\frac{\partial \mathbf{w}_j(\boldsymbol{y},\boldsymbol{\beta})}{\partial \boldsymbol{y}}\right|_{\widehat{\boldsymbol{y}}^{(i)}_{(j,k)},\widehat{\boldsymbol{\beta}}^{(i)}}$

    $H_{(j,k)} = \left.\frac{\partial \boldsymbol{h}(\boldsymbol{\beta})}{\partial \boldsymbol{\beta}}\right|_{(\widehat{\boldsymbol{\beta}}^{(i)})}$

    $\boldsymbol{c}_{w,(j,k)} = \boldsymbol{w}_j(\widehat{\boldsymbol{y}}^{(i)(j,k)}, \widehat{\boldsymbol{\beta}}^{(i)}) + B(\boldsymbol{y} - \widehat{\boldsymbol{y}}^{(i)}_{(j,k)})$

    $\boldsymbol{c}_h = \boldsymbol{h}(\widehat{\boldsymbol{\beta}}^{(i)})$

    $\boldsymbol{\Sigma}^{-1}_{ww,(j,k)} = (B_{(j,k)}\, \boldsymbol{\Sigma}_{yy,(j,k)}\, B^{\mathsf{T}}_{(j,k)})^{-1}$

$$\begin{pmatrix} \widehat{\Delta\boldsymbol{\beta}}^{(i)} \\ \boldsymbol{\mu} \end{pmatrix} = \begin{pmatrix} \sum_{(j,k)} A^{\mathsf{T}}_{(j,k)} \boldsymbol{\Sigma}^{-1}_{ww} A_{(j,k)} & H^{\mathsf{T}} \\ H & 0 \end{pmatrix}^{-1} \begin{pmatrix} \sum_{(j,k)} -A^{\mathsf{T}}_{(j,k)} \boldsymbol{\Sigma}^{-1}_{ww,(j,k)} \boldsymbol{c}_{w,(j,k)} \\ -\boldsymbol{c}_h \end{pmatrix}$$

    $\widehat{\boldsymbol{\beta}}^{(i+1)} = \widehat{\boldsymbol{\beta}}^{(i)} + \widehat{\Delta\boldsymbol{\beta}}^{(i)}$

    $\boldsymbol{\lambda}_{(j,k)} = \boldsymbol{\Sigma}^{-1}_{ww,(j,k)} (\boldsymbol{c}_{w,(j,k)} + A_{(j,k)} \widehat{\Delta\boldsymbol{\beta}}^{(i)})$

    $\widehat{\boldsymbol{e}} = \boldsymbol{\Sigma}_{yy,(j,k)}\, B^{\mathsf{T}}_{(j,k)} \boldsymbol{\lambda}$

    $\widehat{\boldsymbol{y}}^{(i+1)}_{(j,k)} = \boldsymbol{y}_{(j,k)} - \widehat{\boldsymbol{e}}_{(j,k)}$

    $\widehat{\widehat{\boldsymbol{y}}}^{(0)}_{(j,k)} = \widehat{\boldsymbol{y}}^{(i+1)}_{(j,k)}$

    for l from 0 to MaxIter do

        $G_{(j,k)} = \left.\frac{\partial \mathbf{g}_j(\boldsymbol{y})}{\partial \boldsymbol{y}}\right|_{(\widehat{\widehat{\boldsymbol{y}}}^{(l)}_{(j,k)})}$

        $\widehat{\widehat{\boldsymbol{y}}}^{(l+1)}_{(j,k)} = \widehat{\boldsymbol{y}}^{(i+1)}_{(j,k)} - \boldsymbol{\Sigma}_{\mathrm{yy}} \mathsf{G}^{\mathsf{T}}_{(j,k)} (\mathsf{G}_{(j,k)} \boldsymbol{\Sigma}_{\mathrm{y}_{(j,k)}\mathrm{y}_{(j,k)}} \mathsf{G}_{(j,k)^{\mathsf{T}}})^{-1} \mathbf{g}_j(\widehat{\widehat{\boldsymbol{y}}}^{(l)}_{(j,k)})$

    done

    $\Omega^2 = \sum_{(j,k)} \widehat{\boldsymbol{e}}^{\mathsf{T}}_{(j,k)} B^{\mathsf{T}}_{(j,k)} \boldsymbol{\lambda}_{(j,k)}$

    $\widehat{\sigma}^2 = \Omega^2/(W + H - U)$

    last if $\widehat{\Delta\boldsymbol{\beta}}^{(i)}/\mathrm{trace}(\boldsymbol{\Sigma}_{bb}) < 10^{-2}$

done

return $(\widehat{\boldsymbol{\beta}}^{(i)}, \widehat{\boldsymbol{y}}^{(i)}_{(j,k)}, \widehat{\sigma}^2)$

Algorithm C.3: Iterative Solution for the optimization problem using the Gauß-Helmert model. In contrast to the algorithm in table C.2, we assume blocks of independent observations, see text for details.

References

ABDEL-AZIZ, Y.I.; KARARA, H.M. (1971): Direct linear transformation from comparator coordinates into object space coordinates in close-range photogrammetry. In: *Proceedings of the Symposium on Close-Range Photogrammetry*, pages 1–18, Falls Church, VA, U.S.A., 1971. American Society of Photogrammetry.

ABRAHAM, S.; FÖRSTNER, W. (1997): Zur automatischen Modellwahl bei der Kalibrierung von CCD-Kameras. In: *19. DAGM-Symposium Mustererkennung*, pages 147–155. Springer-Verlag, 1997.

ALOIMONOS, Y. (August 1990): Perspective approximations. *Image and Vision Computing*, 8(3):177–192, August 1990.

BAARDA, W. (1968): *A Testing Procedure for Use in Geodetic Networks*, Volume 5 of series 1. Netherlands Geodetic Commission, 1968.

BAILLARD, C.; ZISSERMAN, A. (June 1999): Automatic reconstruction of piecewise planar models from multiple views. In: *Proc. IEEE Conference on Computer Vision and Pattern Recognition*, pages 559–565, June 1999.

BAILLARD, C.; SCHMID, C.; ZISSERMAN, A.; FITZGIBBON, A. (September 1999): Automatic line matching and 3D reconstruction of buildings from multiple views. In: *ISPRS Conference on Automatic Extraction of GIS Objects from Digital Imagery, IAPRS Vol.32, Part 3-2W5*, pages 69–80, September 1999.

BALTSAVIAS, E.P.; GRÜN, A.; VANGOOL, L. (Eds.) (June 2001): *Automatic Extraction of Man-Made Objects from Aerial and Satellite Images III.* A.A.Balkema Publishers, Ascona, June 2001. Proceedings of an International Workshop.

BARTOLI, A.; STURM, P. (2001): The 3D Line Motion Matrix and Alignment of Line Reconstructions. In: *Conference for Computer Vision and Pattern Recognition (CVPR)*, Volume 1, pages 287–292. IEEE, 2001.

BIEDERMAN, I. (1987): Recognition by Components: A Theory of Human Image Understanding. *Psychological Review*, 94(2):115–147, 1987.

BIGNONE, F.; HENRICSSON, O.; FUA, P.; STRICKER, M. (1996): Automatic Extraction of Generic House Roofs from High Resolution Aerial Imagery. In: *Computer Vision '96*, 1996.

BLAKE, A.; BASCLE, B.; ISARD, M.; MACCORMICK, J. (1998): Statistical models of visual shape and motion. In Lasenby et al. LASENBY *et al.* 1998, pages 1283–1302.

BLAKE, A.; LEUNG, T.; REHG, J.; TOYAMA, K. (Eds.) (December 2001): *Workshop on Models versus Exemplars in Computer Vision*, Kauai, Hawaii, December 2001. in conjunction with IEEE CVPR'2001.

BRAND, L. (1966): *Vector and Tensor Analysis.* John Wiley & Sons, Inc., 1966. Tenth printing.

BRAUN, C.; KOLBE, T.H.; LANG, F.; SCHICKLER, W.; STEINHAGE, V.; CREMERS, A.B.; FÖRSTNER, W.; PLÜMER, L. (1995): Models for Photogrammetric Building Reconstruction. *Computer & Graphics*, 19(1):109–118, 1995.

BRAUN, C. (1994): *Interpretation von Einzelbildern zur Gebäudeerfassung.* PhD thesis, Institut für Photogrammetrie, Universität Bonn, 1994.

BRONSTEIN, I. N.; HACKBUSCH, W.; SCHWARZ, H. R.; ZEIDLER, E. (1996): *Teubner-Taschenbuch der Mathematik.* Teubner, 1996.

BRÜGELMANN, R.; FÖRSTNER, W. (1992): Noise Estimation for Color Edge Extraction. In: FÖRSTNER, W.; RUWIEDEL, S. (Eds.), *Robust Computer Vision*, pages 90–107. Wichmann, Karlsruhe, 1992.

BRUNN, A.; WEIDNER, U. (1998): Hierarchical Bayesian nets for building extraction using dense digital surface models. *JPRS*, 53(5):296–307, 1998.

BUXTON, B.; CIPOLLA, R. (Eds.) (April 1996): *Computer Vision - ECCV '96*, number 1064 in LNCS, Cambridge, UK, April 1996. Springer Verlag.

CARLSSON, S. (1994): The Double Algebra: An effective Tool for Computing Invariants in Computer Vision. In: MUNDY, J.; A., ZISSERMAN; FORSYTH, D. (Eds.), *Applications of Invariance in Computer Vision*, number 825 in LNCS. Springer, 1994.

CLARKE, J.C. (1998): Modelling uncertainty: A primer. Technical Report 2161/98, University of Oxford, Dept. Engineering Science, 1998.

CLOWES, M. B. (1971): On Seeing Things. *Artificial Intelligence*, 2(1):79–116, 1971.

COLLINS, R. (September 1993): *Model Acquisition Using Stochastic Projective Geometry.* PhD thesis, Department of Computer Science, University of Massachusetts, September 1993. Also published as UMass Computer Science Technical Report TR95-70.

COURTNEY, P.; THACKER, N.; CLARK, A.F. (1997): Algorithmic modelling for performance evaluation. *Machine Vision and Applications*, 9(5/06):219–228, 1997.

COVER, T.M. THOMAS, J.A. (1991): *Elements of Information Theory.* John Wiley & Sons, 1991.

COXETER, H.S.M. (1989): *Introduction to Geometry.* John Wiley & Sons, 2. Edition, 1989.

CREVIER, D.; LEPAGE, R. (August 1997): Knowledge-Based Image Understanding Systems: A Survey. *Computer Vision and Image understanding*, 67(2):161–185, August 1997.

CRIMINISI, A. (August 2001): *Accurate Visual Metrology from Single and Multiple Uncalibrated Images.* Springer-Verlag London Ltd., August 2001.

CYBERCITY (2002): `http://www.cybercity.tv`. Homepage, 2002.

DAS, G.B. (July 1949): A Mathematical Approach to Problems in Photogrammetry. *Empire Survey Review*, X(73), July 1949.

DAVID, B.; HERREWEGEN, M. VAN DEN; SALGÈ, F. (1996): Conceptual Models for Geometry and Quality of Geographic Information. In: BURROUGH, PETER A.; FRANK, ANDREW U. (Eds.), *Geographic Objects with Indeterminate Boundaries*, Volume 2 of series GISDATA, Kapitel 13, pages 193–206. Taylor & Francis, 1996.

DEBEVEC, P.; TAYLOR, C.J.; MALIK, J. (August 1996): Modeling and Rendering Architecture from Photographs. In: *Proc. SIGGRAPH '96*, pages 11–21, August 1996.

DUDA, R.; HART, P. (1973): *Pattern Classification and Scene Analysis.* Wiley, New York, 1973.

DURRANT-WHYTE, H.F. (1989): Uncertain Geometry. In Kapur and Mundy KAPUR AND MUNDY 1989A, pages 447–482.

ENCARNACAO, J.; STRASSER, W.; KLEIN, R. (1997): *Graphische Datenverarbeitung*, Volume 2. Oldenbourg, München, 4. Edition, 1997.

FAUGERAS, O.; LUONG, QUANG-TUAN (2001): *The Geometry of Multiple Images*. MIT Press, 2001. with contributions from T. Papadopoulo.

FAUGERAS, O.; PAPADOPOULO, T. (1998): Grassmann-Cayley Algebra for Modeling Systems of Cameras and the Algebraic Equations of the Manifold of Trifocal Tensors. In: *Trans. of the ROYAL SOCIETY A, 365*, pages 1123–1152, 1998.

FAUGERAS, O.D.; LUONG, Q.; MAYBANK, S. (1992): Camera self-calibration: Theory and Experiments. In: *Proc. European Conference on Computer Vision*, Volume 588 of series LNCS, pages 321–334. Springer Verlag, 1992.

FAUGERAS, O. D. (May 1992): What can be seen in three dimensions with an uncalibrated stereo rig? In: *2nd European Conference on Computer Vision*, pages 563–578. ECCV, May 1992.

FAUGERAS, O. (1993): *Three-Dimensional Computer Vision, A Geometric Viewpoint*. MIT Press, 1993.

FINSTERWALDER, S. (1899): *Die geometrischen Grundlagen der Photogrammetrie*, pages 1–41. Teubner Verlag, Leipzig, Berlin, 1899. Jahresbericht Deutsche Mathem. Vereinigung, VI, 2.

FISHER, NICOLAS; LEWIS, TOBY; EMBLETON, BRIAN J.J. (1987): *Statistical Analysis of Sphercial Data*. Cambridge University Press, New York, 1987.

FÖRSTNER, W.; BRUNN, A.; HEUEL, S. (September 2000): Statistically Testing Uncertain Geometric Relations. In: SOMMER, G.; KRÜGER, N.; PERWASS, CH. (Eds.), *Mustererkennung 2000*, pages 17–26. DAGM, Springer, September 2000.

FÖRSTNER, W. (1987): Reliability Analysis of Parameter Estimation in Linear Models with Applications to Mensuration Problems in Computer Vision. *Computer Vision, Graphics & Image Processing*, 40:273–310, 1987.

FÖRSTNER, W. (1992): Uncertain Spatial Relationships and their Use for Object Location in Digital Images. In: FÖRSTNER, W.; HARALICK, R. M.; RADIG, B. (Eds.), *Robust Computer Vision - Tutorial Notes*. Institut für Photogrammetrie, Universität Bonn, 1992.

FÖRSTNER, W. (1994): A Framework for Low Level Feature Extraction. In: EKLUNDH, J. O. (Ed.), *Computer Vision - ECCV 94, Vol. II*, Volume 802 of series LNCS, pages 383–394. Springer, 1994.

FÖRSTNER, W. (1995): The Role of Robustness in Computer Vision. In: PINZ, A.; BURGER, W. (Eds.), *Vision Milestones 95*, Kapitel 6. Österr. Gesellsch. f. Künstliche Intelligenz, 1995.

FÖRSTNER, W. (1996): 10 Pros and Cons Against Performance Characterization of Vision Algorithms. In: CHRISTENSEN H. I., FÖRSTNER W., MADSEN C. B. (Ed.), *Workshop "Performance Characterisics of Vision Algorithms"*, 1996.

FÖRSTNER, W. (2000): Choosing Constraints in Image Triplets. In: VERNON, DAVID (Ed.), *Computer Vision - ECCV 2000*, Volume 1843, II of series Lecture Notes in computer science, pages 669–684. Springer, 2000.

FÖRSTNER, W. (2001): On Estimating 2D Points and Lines from 2D Points and Lines. In: *Festschrift anläßlich des 60. Geburtstages von Prof. Dr.-Ing. Bernhard Wrobel*, pages 69 – 87. Technische Universität Darmstadt, 2001.

FÖRSTNER, W. (2001): Vorlesungsskript Photogrammetrie I. University of Bonn, 2001.

FÖRSTNER, W. (2002): Precision of the Inverse. Technical report, Note, Institute for Photogrammetry, University of Bonn, 2002. Internal.

FUA, P.; BRECHBÜHLER, C. (April 1996): Imposing hard constraints on soft snakes. In Buxton and Cipolla BUXTON AND CIPOLLA 1996, pages 495–506.

FUCHS, C. (1998): *Extraktion polymorpher Bildstrukturen und ihre topologische und geometrische Gruppierung.* DGK, Bayer. Akademie der Wissenschaften, Reihe C, Heft 502, 1998.

GELERNTER, H. (1963): Realization of a geometry theorem proving machine. In: FEIGENBAUM, E.A.; FELDMAN, J. (Eds.), *Computers and Thought*, pages 134–152. McGraw-Hill, New York, 1963.

GRÜN, A.; WANG, X. (August 1999): CyberCity Modeler, a tool for interactive 3-D city model generation. In: *Proceedings of Germany Photogrammetry Week*, Stuttgart, Germany, August 1999.

GRÜN, A.; WANG, X. (June 2001): News from CyberCity-Modeler. In Baltsavias et al. BALTSAVIAS *et al.* 2001, pages 93–102. Proceedings of an International Workshop.

GRÜN, A.; KÜBLER, O.; AGOURIS, P. (Eds.) (1995): *Automatic Extraction of Man-Made Objects from Aerial and Space Images.* Birkhäuser, 1995.

GÜLCH, E.; MÜLLER, H.; LÄBE, T. (2000): Semi-automatische Verfahren in der photogrammetrischen Objekterfassung. *Photogrammetrie Fernerkundung Geoinformation, Wichmann Verlag*, 3, 2000.

HADDON, J.; FORSYTH, D. A. (July 2001): Noise in bilinear problems. In: *Proceedings of ICCV*, Volume II, pages 622–627, Vancouver, July 2001. IEEE Computer Society.

HARTLEY, R. I.; ZISSERMAN, A. (2000): *Multiple View Geometry in Computer Vision.* Cambridge University Press, 2000.

HARTLEY, R.; GUPTA, R.; CHANG, T. (jun 1992): Stereo from Uncalibrated Cameras. In: *Computer Society Conference on Computer Vision and Pattern Recognition.* IEEE, jun 1992.

HARTLEY, R. (1995): A Linear Method for Reconstruction from Lines and Points. In: *Proc. ICCV*, pages 882–887, 1995.

HARTLEY, R. I. (1995): In Defence of the 8-point algorithm. In: *ICCV 1995*, pages 1064–1070. IEEE CS Press, 1995.

HARTLEY, R.I. (1997): Lines and points in three views and the trifocal tensor. *International Journal of Computer Vision*, 22(2):125–140, 1997.

HARTLEY, R. I. (1998): Minimizing algebraic error. In Lasenby et al. LASENBY *et al.* 1998, pages 1175–1192.

HELMERT, F. R. (1872): *Die Ausgleichungsrechnung nach der Methode der Kleinsten Quadrate.* Teubner, Leipzig, 1872.

HEUEL, S.; FÖRSTNER, W. (1998): A Dual, Scalable and Hierarchical Representation for Perceptual Organization of Binary Images. In: *Workshop on Perceptual Organization in Computer Vision.* IEEE Computer Society, 1998.

HEUEL, S.; FÖRSTNER, W. (2001): Matching, Reconstructing and Grouping 3D Lines From Multiple Views Using Uncertain Projective Geometry. In: *Conference for Computer Vision and Pattern Recognition (CVPR).* IEEE, 2001.

HEUEL, S.; KOLBE, T.H. (July 2001): Building Reconstruction: The Dilemma of Generic Versus Specific Models. *KI -Künstliche Intelligenz*, 3:57–62, July 2001.

HEUEL, S.; NEVATIA, R. (1995): Including Interaction in an Automated Modelling System. In: *Symposium on Computer Vision '95*, pages 383–388, 1995.

HEUEL, S.; LANG, F.; FÖRSTNER, W. (2000): Topological and Geometrical Reasoning in 3D Grouping for Reconstructing Polyhedral Surfaces. In: *International Archives of Photogrammetry and Remote Sensing*, Volume XXXIII, B3, pages 397–404, Amsterdam, 2000. ISPRS.

HEUEL, S. (2001): Points, Lines and Planes and their Optimal Estimation. In Radig and Florczyk RADIG AND FLORCZYK 2001, pages 92–99.

HUBER, P. J. (1981): *Robust Statistics.* Wiley NY, 1981.

HUERTAS, A.; LIN, C.; NEVATIA, R. (1993): Detection of Buildings from Monocular Views of Aerial Scenes Using Perceptual Grouping and Shadows. In: *DARPA Workshop, Washington D.C.*, pages 253–260, 1993.

HUFFMAN, D. A. (1971): Impossible Objects as Nonsense Sentences. *Machine Intelligence*, 6:295–323, 1971.

INPHO (2002): `http://www.inpho.de`. Homepage, 2002.

JAYNES, C.; HANSON, A.; RISEMAN, E. (1997): Building Reconstruction from Optical and Range Images. In: FÖRSTNER, W.; PLÜMER, L. (Eds.), *Workshop on Semantic Modeling for the Acquisition of Topographic Information from Images and Maps, SMATI '97*, 1997. To appear.

KANATANI, K. (1996): *Statistical Optimization for Geometric Computation: Theory and Practice.* Elsevier Science, 1996.

KAPUR, D.; MUNDY, J.L. (Eds.) (1989): *Geometric Reasoning.* The MIT Press, Cambridge, Massachusets, USA, 1989.

KAPUR, D.; MUNDY, J.L. (1989): Geometric Reasoning and Artificial Intelligence: Introduction to the Special Volume. In *Geometric Reasoning* KAPUR AND MUNDY 1989A, pages 1–14.

KLEIN, H.; FÖRSTNER, W. (1984): Realization of automatic error detection in the block adjustment program PAT-M43 using robust estimators. In: *Proceedings of the 15th ISPRS Congress*, Volume XXV-3a, pages 234–245, Rio, Brazil, 1984. International Archives of Photogrammetry and Remote Sensing.

KLEIN, F. (1939): *Elementary Mathematics from an Advance Standpoint.* MacMillan, New York, 1939.

KLIR, G.J.; FOLGER, T.A. (1988): *Fuzzy Sets, Uncertainty, and Information.* Prentice-Hall International, Inc., 1988.

KOCH, K.R. (1999): *Parameter Estimation and Hypothesis Testing in Linear Models.* Springer, Berlin, 2. Edition, 1999.

KOLBE, THOMAS H. (1999): *Identifikation und Rekonstruktion von Gebäuden in Luftbildern mittels unscharfer Constraints.* PhD thesis, Institut für Umweltwissenschaften, Hochschule Vechta, 1999.

KRAUS, K. (1997): *Photogrammetry.* Dümmler Verlag Bonn, 1997. Vol.1: Fundamentals and Standard Processes. Vol.2: Advanced Methods and Applications.

LANG, F. (1999): *Geometrische und Semantische Rekonstruktion von Gebäuden durch Ableitung von 3D Gebäudeecken.* Reihe Informatik. Shaker Verlag, 1999. PhD Thesis.

LASENBY, J.; ZISSERMAN, A.; CIPOLLA, R.; LONGUET-HIGGINS, H.C. (Eds.) (1998): *New Geometric Techniques in Computer Vision*, Volume 356 (1740) of series A. Philosophical Transactions of the Royal Society of London, Mathematical, Philosophical and Engineering Sciences, 1998.

LEEDAN, Y.; MEER, P. (2000): Heteroscedastic Regression in Computer Vision: Problems with Bilinear Constraint. *International Journal of Computer Vision*, 37:127–150, 2000.

LIN, C.; HUERTAS, A.; NEVATIA, R. (1994): Detection of Buildings Using Perceptual Groupings and Shadows. In: *CVPR94*, pages 62–69, 1994.

LIU, X.; KANUNGO, T.; HARALICK, R. (February 1996): Statistical validation of computer vision software. In: *DARPA Image Understanding Workshop*, Volume II, pages 1533–40, Palm Springs, CA, February 1996.

LONGUET-HIGGINS, H. C. (1981): A computer program for reconstructing a scene from two projections. *Nature*, 293:133–135, 1981.

LOWE, D. (May 1991): Fitting Parameterized Three-Dimensional Models to Images. *IEEE Transactions on Pattern Analysis and Machine Intelligence*, 13(5):441–450, May 1991.

LUXEN, M.; FÖRSTNER, W. (2001): Optimal Camera Orientation from Observed Lines. In Radig and Florczyk RADIG AND FLORCZYK 2001, pages 84–91.

MATEI, B.; MEER, P. (June 2000): A General Method for Errors-in-Variables Problems in Computer Vision. In: *Computer Vision and Pattern Recognition Conference*, Volume II, pages 18–25. IEEE, June 2000.

MAYBANK, S.J. (1995): Probabilistic analysis of the application of the cross ratio to model based vision. *International Journal of Computer Vision*, 16:5–33, 1995.

MAYER, H. (1999): Automatic Object Extraction from Aerial Imagery - A Survey Focusing on Buildings. *Computer Vision and Image Understanding*, 74(2):138–149, 1999.

MIKHAIL, E. M.; ACKERMANN, F. (1976): *Observations and Least Squares*. University Press of America, 1976.

MOHR, R.; QUAN, L.; ET AL. (June 1992): Relative 3D Reconstruction Using Multiple Uncalibrated Images. Technical Report RT 84-IMAG-12 LIFIA, LIFIA, Institut IMAG, 46, Avenue Felix Viallet, 38031 Granoble Cedex, France, June 1992.

MOONS, T.; FRERE, D.; VANDEKERCKHOVE, J.; GOOL, L. VAN (1998): Automatic Modelling and 3D Reconstruction of Urban House Roofs from High Resolution Aerial Imagery. In: BURKHARDT, H.; NEUMANN, B. (Eds.), *Computer Vision - ECCV '98, Vol . I*, pages 410–425. Lecture Notes in Computer Science, 1406, Springer-Verlag, 1998.

MUMFORD, D. (2000): The Dawning Age of Stochasticity. In: ARNOLD, V.; ATIYAH, M. (Eds.), *Mathematics, Frontiers and Perspectives*, pages 197 – 218. American Math. Soc., Providence, R. I, 2000.

MUNDY, J. L.; ZISSERMAN, A. P. (1992): *Geometric Invariance in Computer Vision*. MIT Press, 1992.

MUNDY, J. (May 1992): The Application of Geometric Invariants. In: MUNDY, J.; ZISSERMAN, A. (Eds.), *Proc. of ESPRIT Basic Research Workshop " Invariants for Recognition ", ECCV 92, Santa Margherita Ligure*. ESPRIT & ECCV 92, MIT Press, May 1992.

NAVAB, N.; FAUGERAS, O.D.; VIEVILLE., T. (May 1993): The critical sets of lines for camera displacement estimation: A mixed euclidean-projective and constructive approach. In: *Proceedings of the 4th International Conference on Computer Vision,*, pages 712–723, Berlin, Germany, May 1993. IEEE Computer Society.

NORONHA, S.; NEVATIA, R. (May 2001): Detection and Modeling of Buildings from Multiple Aerial Images. *IEEE Trans. Pattern Analysis and Machine Intelligence*, 23(5):501–518, May 2001.

PAPOULIS, A.; PILLAI, S.U. (2002): *Probability, Random Variables and Stochastic Processes.* McGraw-Hill, 4. Edition, 2002.

PEARL, J. (1988): *Probablistic Reasonin in Intelligent Systems.* Morgan Kaufman Publishers, Inc., San Mateo, California, 1988.

PEIRCE, C. S. (1931): *Elements of Logic.* Harvard University Press, Cambridge, Mass.,, 1931. Collected Papers of Charles Sanders Peirce, Hartshorne, C. and Weiss, P. (eds.).

PHOTOMODELER (2002): `http://www.photomodeler.com/`. Homepage, 2002.

PIAGET, J. (1954): *The Construction of Reality in the Child.* New York: Basic Books,Inc., 1954.

POLLEFEYS, M.; KOCH, R.; VERGAUWEN, M.; GOOL, L. VAN (2000): Automated reconstruction of 3D scenes from sequences of images. *ISPRS Journal Of Photogrammetry And Remote Sensing*, 55(4):251–267, 2000.

PONCE, J.; ZISSERMAN, A.; HEBERT, M. (Eds.) (1996): *Object Representation in Computer Vision II*, number 1144 in LNCS. Springer-Verlag, 1996.

PORRILL, J. (December 1988): Optimal Combination and Constraints for Geometrical Sensor Data. *The International Journal of Robotics Research*, 7(6):66–77, December 1988.

PREGIBON, D. (1986): A DIY Guide to Statistical Strategy. In: *Artificial Intelligence and Statistics*, pages 389–400. Addison-Wesley, 1986.

PRESS, W.H.; TEUKOLSKY, S.A.; VETTERLING, W.T.; FLANNERY, B.P. (1992): *Numerical Recipes in C.* Cambridge University Press, second. Edition, 1992.

RADIG, BERND; FLORCZYK, STEFAN (Eds.) (2001): *Pattern Recognition, 23rd DAGM-Symposium, Munich, Germany, September 12-14, 2001, Proceedings*, Volume 2191 of series Lecture Notes in Computer Science. Springer, 2001.

REALVIZ (2002): `http://www.realviz.com/products/im/index.php`. Product page, 2002.

RICHTER-GEBERT, J.; WANG, D. (Eds.) (September 25-27 2000): *Proceedings of ADG 2000 - The third international workshop on automated deduction in geometry*, Zürich, Switzerland, September 25-27 2000.

ROBERTS, L.G. (1963): *Machine perception of three-dimensional solids.* PhD thesis, MIT Lincoln Laboratory, 1963. Technical Report 315.

ROBERTS, L. G. (1965): Machine Perception of Three Dimensional Solids. *Optical and Electro-Optical Information Processing*, pages 159–197, 1965.

ROTHWELL, CHARLIE; STERN, JULIEN P. (April 1996): Understanding the Shape Properties of Trihedral Polyhedra. In Buxton and Cipolla BUXTON AND CIPOLLA 1996, pages 175–185.

RUSSELL, S.J.; NORVIG, P. (1995): *Artificial Intelligence: A Modern Approach.* Prentice Hall, 1995.

SARKAR, S.; BOYER, K.L. (1994): *Computing Perceptual Organization in Computer Vision.* World Scientific, 1994.

SARKAR, S.; BOYER, K.L. (July 1995): Using Perceptual Inference Networks To Manage Vision Processes. *CVIU*, 62(1):27–46, July 1995.

SCHMID, C.; ZISSERMAN, A. (1997): Automatic line matching across views. *Proceedings Conference on Computer Vision and Pattern Recognition, Puerto Rico*, pages 666–671, 1997.

SEITZ, S. M.; ANANDAN, P. (1999): Implicit Representation and Scene Reconstruction from Probability Density Functions. In: *Proc. Computer Vision and Pattern Recognition Conf.*, pages 28–34. IEEE Computer Society, 1999.

Shashua, A.; Werman, M. (1995): Trilinearity of Three Perspective Views and its Associated Tensor. In: *Proc. ICCV*, pages 920–925, 1995.

Spetsakis, M.E.; Aloimonos, J.Y. (1990): A Unified Theory Of Structure From Motion. In: *Proceedings Image Understanding Workshop*, pages 271–283, 1990.

Stilla, U.; Soergel, U.; Thoennessen, U.; Michaelsen, E. (June 2001): Segmentation of LIDAR and INSAR elevation data for building reconstruction. In Baltsavias et al. Baltsavias *et al.* 2001, pages 297–308. Proceedings of an International Workshop.

Stolfi, J. (1991): *Oriented Projective Geometry: A Framework for Geometric Computations.* Academic Press, Inc., San Diego, 1991.

Sugihara, K. (1986): *Machine Interpretation of Line Drawings.* MIT-Press, 1986.

Sutherland, I.E. (1963): Sketchpad: A man-machine graphical communications system. Technical Report 296, MIT Lincoln Laboratories, 1963. Also published by Garland Publishing, New York, 1980.

Teleki, L. (1997): *Dreidimensionale qualitative Gebäuderekonstruktion.* PhD thesis, Deutsche Geodätische Kommission, München, Vol. C 489, 1997.

Thompson, E.H. (1968): The projective theory of relative orientation. *Photogrammetria*, 23:67–75, 1968.

Torr, P. H. S.; Murray, D. W. (1997): The Development and Comparison of Robust Methods for Estimating the Fundamental Matrix. *International Journal of Computer Vision*, 24(3):271–300, 1997.

Triggs, B.; McLauchlan, P.; Hartley, R.; Fitzgibbon, A. (2000): Bundle Adjustment A Modern Synthesis. In: Triggs, B.; Zisserman, A.; Szeliski, R. (Eds.), *Vision Algorithms: Theory & Practice*, number 1883 in LNCS. Springer-Verlag, 2000.

Triggs, B. (1995): Matching constraints and the joint image. In: *Proceedings of the Fifth International Conference on Computer Vision*, pages 338–343, Boston, MA, 1995. IEEE Computer Society Press.

Tsai, R. Y.; Huang, T. S. (Jan. 1984): Uniqueness and Estimation of Three-Dimensional Motion Parameters of Rigid Objects with Curved Surfaces. *IEEE T-PAMI*, PAMI-6(1):13–27, Jan. 1984.

van den Hout, C.M.A.; Stefanovic, P. (1976): Efficient analytical relative orientation. *Intern. Archives of Photogrammetry and Remote Sensing*, 21, 1976. Comm. III, Helsinki.

Weidner, U.; Förstner, W. (1995): Towards Automatic Building Extraction from High Resolution Digital Elevation Models. *ISPRS Journal*, 50(4):38–49, 1995.

Weidner, U. (1997): *Gebäudeerfassung aus Digitalen Oberflächenmodellen.* Deutsche Geodätische Kommission, Reihe C, No. 474, Munich, 1997.

Winter, S. (1996): *Unsichere topologische Beziehungen zwischen ungenauen Flächen.* PhD thesis, Deutsche Geodätische Kommission, München, Vol. C 465, 1996.

Witte, B.; Schmidt, H. (1995): *Vermessungskunde und Grundlagen der Statistik für das Bauwesen.* Wittwer, 3. Edition, 1995.

Wrobel, B.P. (2001): Minimum Solutions for Orientation. Volume 34 of series Springer Series in Information Sciences, Kapitel 2. Springer Verlag, Berlin Heidelberg, 2001.

Zadeh, L.A. (1979): A theory of approximate reasoning. In: Hyes, J.; Michie, D.; Mikulich, L. (Eds.), *Machine Intelligence 9*, pages 149–194, Halstead, New York, 1979.

ZEIDLER, E. (1996): *Teubner Taschenbuch der Mathematik*, Volume I. Teubner Verlag, 1996.

ZHANG, Z. (1994): Token tracking in a cluttered scene. *Image and Vision Computing*, 12(2):110–120, 1994.

ZHANG, Z. (1998): Determining the Epipolar Geometry and its Uncertainty: A Review. *International Journal of Computer Vision*, 27(2):161–195, 1998.

ZISSERMAN, A.; BEARDSLEY, P.; I., REID. (1995): Metric calibration of a stereo-rig. *IEEE Workshop on Representation of Visual Scenes*, pages 93–100, 1995. Boston.

Lecture Notes in Computer Science

For information about Vols. 1–2929

please contact your bookseller or Springer-Verlag

Vol. 3060: A.Y. Tawfik, S.D. Goodwin (Eds.), Advances in Artificial Intelligence. XIII, 582 pages. 2004. (Subseries LNAI).

Vol. 3053: J. Davies, D. Fensel, C. Bussler, R. Studer (Eds.), The Semantic Web: Research and Applications. XIII, 490 pages. 2004.

Vol. 3042: N. Mitrou, K. Kontovasilis, G.N. Rouskas, I. Iliadis, L. Merakos (Eds.), NETWORKING 2004, Networking Technologies, Services, and Protocols; Performance of Computer and Communication Networks; Mobile and Wireless Communications. XXXIII, 1519 pages. 2004.

Vol. 3034: J. Favela, E. Menasalvas, E. Chávez (Eds.), Advances in Web Intelligence. XIII, 227 pages. 2004. (Subseries LNAI).

Vol. 3033: M. Li, X.-H. Sun, Q. Deng, J. Ni (Eds.), Grid and Cooperative Computing. XXXVIII, 1076 pages. 2004.

Vol. 3032: M. Li, X.-H. Sun, Q. Deng, J. Ni (Eds.), Grid and Cooperative Computing. XXXVII, 1112 pages. 2004.

Vol. 3031: A. Butz, A. Krüger, P. Olivier (Eds.), Smart Graphics. X, 165 pages. 2004.

Vol. 3027: C. Cachin, J. Camenisch (Eds.), Advances in Cryptology - EUROCRYPT 2004. XI, 628 pages. 2004.

Vol. 3026: C. Ramamoorthy, R. Lee, K.W. Lee (Eds.), Software Engineering Research and Applications. XV, 377 pages. 2004.

Vol. 3025: G.A. Vouros, T. Panayiotopoulos (Eds.), Methods and Applications of Artificial Intelligence. XV, 546 pages. 2004. (Subseries LNAI).

Vol. 3024: T. Pajdla, J. Matas (Eds.), Computer Vision - ECCV 2004. XXVIII, 621 pages. 2004.

Vol. 3023: T. Pajdla, J. Matas (Eds.), Computer Vision - ECCV 2004. XXVIII, 611 pages. 2004.

Vol. 3022: T. Pajdla, J. Matas (Eds.), Computer Vision - ECCV 2004. XXVIII, 621 pages. 2004.

Vol. 3021: T. Pajdla, J. Matas (Eds.), Computer Vision - ECCV 2004. XXVIII, 633 pages. 2004.

Vol. 3019: R. Wyrzykowski, J. Dongarra, M. Paprzycki, J. Wasniewski (Eds.), Parallel Processing and Applied Mathematics. XIX, 1174 pages. 2004.

Vol. 3015: C. Barakat, I. Pratt (Eds.), Passive and Active Network Measurement. XI, 300 pages. 2004.

Vol. 3012: K. Kurumatani, S.-H. Chen, A. Ohuchi (Eds.), Multi-Agnets for Mass User Support. X, 217 pages. 2004. (Subseries LNAI).

Vol. 3011: J.-C. Régin, M. Rueher (Eds.), Integration of AI and OR Techniques in Constraint Programming for Combinatorial Optimization Problems. XI, 415 pages. 2004.

Vol. 3010: K.R. Apt, F. Fages, F. Rossi, P. Szeredi, J. Váncza (Eds.), Recent Advances in Constraints. VIII, 285 pages. 2004. (Subseries LNAI).

Vol. 3009: F. Bomarius, H. Iida (Eds.), Product Focused Software Process Improvement. XIV, 584 pages. 2004.

Vol. 3008: S. Heuel, Uncertain Projective Geometry. XVII, 205 pages. 2004.

Vol. 3007: J.X. Yu, X. Lin, H. Lu, Y. Zhang (Eds.), Advanced Web Technologies and Applications. XXII, 936 pages. 2004.

Vol. 3006: M. Matsui, R. Zuccherato (Eds.), Selected Areas in Cryptography. XI, 361 pages. 2004.

Vol. 3005: G.R. Raidl, S. Cagnoni, J. Branke, D.W. Corne, R. Drechsler, Y. Jin, C.G. Johnson, P. Machado, E. Marchiori, F. Rothlauf, G.D. Smith, G. Squillero (Eds.), Applications of Evolutionary Computing. XVII, 562 pages. 2004.

Vol. 3004: J. Gottlieb, G.R. Raidl (Eds.), Evolutionary Computation in Combinatorial Optimization. X, 241 pages. 2004.

Vol. 3003: M. Keijzer, U.-M. O'Reilly, S.M. Lucas, E. Costa, T. Soule (Eds.), Genetic Programming. XI, 410 pages. 2004.

Vol. 3002: D.L. Hicks (Ed.), Metainformatics. X, 213 pages. 2004.

Vol. 3001: A. Ferscha, F. Mattern (Eds.), Pervasive Computing. XVII, 358 pages. 2004.

Vol. 2999: E.A. Boiten, J. Derrick, G. Smith (Eds.), Integrated Formal Methods. XI, 541 pages. 2004.

Vol. 2998: Y. Kameyama, P.J. Stuckey (Eds.), Functional and Logic Programming. X, 307 pages. 2004.

Vol. 2997: S. McDonald, J. Tait (Eds.), Advances in Information Retrieval. XIII, 427 pages. 2004.

Vol. 2996: V. Diekert, M. Habib (Eds.), STACS 2004. XVI, 658 pages. 2004.

Vol. 2995: C. Jensen, S. Poslad, T. Dimitrakos (Eds.), Trust Management. XIII, 377 pages. 2004.

Vol. 2994: E. Rahm (Ed.), Data Integration in the Life Sciences. X, 221 pages. 2004. (Subseries LNBI).

Vol. 2993: R. Alur, G.J. Pappas (Eds.), Hybrid Systems: Computation and Control. XII, 674 pages. 2004.

Vol. 2992: E. Bertino, S. Christodoulakis, D. Plexousakis, V. Christophides, M. Koubarakis, K. Böhm, E. Ferrari (Eds.), Advances in Database Technology - EDBT 2004. XVIII, 877 pages. 2004.

Vol. 2991: R. Alt, A. Frommer, R.B. Kearfott, W. Luther (Eds.), Numerical Software with Result Verification. X, 315 pages. 2004.

Vol. 2989: S. Graf, L. Mounier (Eds.), Model Checking Software. X, 309 pages. 2004.

Vol. 2988: K. Jensen, A. Podelski (Eds.), Tools and Algorithms for the Construction and Analysis of Systems. XIV, 608 pages. 2004.

Vol. 2987: I. Walukiewicz (Ed.), Foundations of Software Science and Computation Structures. XIII, 529 pages. 2004.

Vol. 2986: D. Schmidt (Ed.), Programming Languages and Systems. XII, 417 pages. 2004.

Vol. 2985: E. Duesterwald (Ed.), Compiler Construction. X, 313 pages. 2004.

Vol. 2984: M. Wermelinger, T. Margaria-Steffen (Eds.), Fundamental Approaches to Software Engineering. XII, 389 pages. 2004.

Vol. 2983: S. Istrail, M.S. Waterman, A. Clark (Eds.), Computational Methods for SNPs and Haplotype Inference. IX, 153 pages. 2004. (Subseries LNBI).

Vol. 2982: N. Wakamiya, M. Solarski, J. Sterbenz (Eds.), Active Networks. XI, 308 pages. 2004.

Vol. 2981: C. Müller-Schloer, T. Ungerer, B. Bauer (Eds.), Organic and Pervasive Computing – ARCS 2004. XI, 339 pages. 2004.

Vol. 2980: A. Blackwell, K. Marriott, A. Shimojima (Eds.), Diagrammatic Representation and Inference. XV, 448 pages. 2004. (Subseries LNAI).

Vol. 2979: I. Stoica, Stateless Core: A Scalable Approach for Quality of Service in the Internet. XVI, 219 pages. 2004.

Vol. 2978: R. Groz, R.M. Hierons (Eds.), Testing of Communicating Systems. XII, 225 pages. 2004.

Vol. 2977: G. Di Marzo Serugendo, A. Karageorgos, O.F. Rana, F. Zambonelli (Eds.), Engineering Self-Organising Systems. X, 299 pages. 2004. (Subseries LNAI).

Vol. 2976: M. Farach-Colton (Ed.), LATIN 2004: Theoretical Informatics. XV, 626 pages. 2004.

Vol. 2973: Y. Lee, J. Li, K.-Y. Whang, D. Lee (Eds.), Database Systems for Advanced Applications. XXIV, 925 pages. 2004.

Vol. 2972: R. Monroy, G. Arroyo-Figueroa, L.E. Sucar, H. Sossa (Eds.), MICAI 2004: Advances in Artificial Intelligence. XVII, 923 pages. 2004. (Subseries LNAI).

Vol. 2971: J.I. Lim, D.H. Lee (Eds.), Information Security and Cryptology -ICISC 2003. XI, 458 pages. 2004.

Vol. 2970: F. Fernández Rivera, M. Bubak, A. Gómez Tato, R. Doallo (Eds.), Grid Computing. XI, 328 pages. 2004.

Vol. 2968: J. Chen, S. Hong (Eds.), Real-Time and Embedded Computing Systems and Applications. XIV, 620 pages. 2004.

Vol. 2967: S. Melnik, Generic Model Management. XX, 238 pages. 2004.

Vol. 2966: F.B. Sachse, Computational Cardiology. XVIII, 322 pages. 2004.

Vol. 2965: M.C. Calzarossa, E. Gelenbe, Performance Tools and Applications to Networked Systems. VIII, 385 pages. 2004.

Vol. 2964: T. Okamoto (Ed.), Topics in Cryptology – CT-RSA 2004. XI, 387 pages. 2004.

Vol. 2963: R. Sharp, Higher Level Hardware Synthesis. XVI, 195 pages. 2004.

Vol. 2962: S. Bistarelli, Semirings for Soft Constraint Solving and Programming. XII, 279 pages. 2004.

Vol. 2961: P. Eklund (Ed.), Concept Lattices. IX, 411 pages. 2004. (Subseries LNAI).

Vol. 2960: P.D. Mosses (Ed.), CASL Reference Manual. XVII, 528 pages. 2004.

Vol. 2959: R. Kazman, D. Port (Eds.), COTS-Based Software Systems. XIV, 219 pages. 2004.

Vol. 2958: L. Rauchwerger (Ed.), Languages and Compilers for Parallel Computing. XI, 556 pages. 2004.

Vol. 2957: P. Langendoerfer, M. Liu, I. Matta, V. Tsaoussidis (Eds.), Wired/Wireless Internet Communications. XI, 307 pages. 2004.

Vol. 2956: A. Dengel, M. Junker, A. Weisbecker (Eds.), Reading and Learning. XII, 355 pages. 2004.

Vol. 2954: F. Crestani, M. Dunlop, S. Mizzaro (Eds.), Mobile and Ubiquitous Information Access. X, 299 pages. 2004.

Vol. 2953: K. Konrad, Model Generation for Natural Language Interpretation and Analysis. XIII, 166 pages. 2004. (Subseries LNAI).

Vol. 2952: N. Guelfi, E. Astesiano, G. Reggio (Eds.), Scientific Engineering of Distributed Java Applications. X, 157 pages. 2004.

Vol. 2951: M. Naor (Ed.), Theory of Cryptography. XI, 523 pages. 2004.

Vol. 2949: R. De Nicola, G. Ferrari, G. Meredith (Eds.), Coordination Models and Languages. X, 323 pages. 2004.

Vol. 2948: G.L. Mullen, A. Poli, H. Stichtenoth (Eds.), Finite Fields and Applications. VIII, 263 pages. 2004.

Vol. 2947: F. Bao, R. Deng, J. Zhou (Eds.), Public Key Cryptography – PKC 2004. XI, 455 pages. 2004.

Vol. 2946: R. Focardi, R. Gorrieri (Eds.), Foundations of Security Analysis and Design II. VII, 267 pages. 2004.

Vol. 2943: J. Chen, J. Reif (Eds.), DNA Computing. X, 225 pages. 2004.

Vol. 2941: M. Wirsing, A. Knapp, S. Balsamo (Eds.), Radical Innovations of Software and Systems Engineering in the Future. X, 359 pages. 2004.

Vol. 2940: C. Lucena, A. Garcia, A. Romanovsky, J. Castro, P.S. Alencar (Eds.), Software Engineering for Multi-Agent Systems II. XII, 279 pages. 2004.

Vol. 2939: T. Kalker, I.J. Cox, Y.M. Ro (Eds.), Digital Watermarking. XII, 602 pages. 2004.

Vol. 2937: B. Steffen, G. Levi (Eds.), Verification, Model Checking, and Abstract Interpretation. XI, 325 pages. 2004.

Vol. 2936: P. Liardet, P. Collet, C. Fonlupt, E. Lutton, M. Schoenauer (Eds.), Artificial Evolution. XIV, 410 pages. 2004.

Vol. 2934: G. Lindemann, D. Moldt, M. Paolucci (Eds.), Regulated Agent-Based Social Systems. X, 301 pages. 2004. (Subseries LNAI).

Vol. 2930: F. Winkler (Ed.), Automated Deduction in Geometry. VII, 231 pages. 2004. (Subseries LNAI).